THE TELL-TALE
BRAIN

UNLOCKING THE MYSTERY
OF HUMAN NATURE

To my dearest acoosh!

May u succeeed in 'unlocking' all the mysteries of the human brain.... and go on to win the NOBEL Prize!

with tons & tons of luv,

appa

8/Jan/2011

Also by

V. S. RAMACHANDRAN

———

A Brief Tour of Human Consciousness

Phantoms in the Brain

THE TELL-TALE
BRAIN

UNLOCKING THE MYSTERY
OF HUMAN NATURE

V.S. RAMACHANDRAN

RANDOM HOUSE INDIA

Published by Random House India in 2010
1

Copyright © V.S. Ramachandran

Random House Publishers India Private Limited
MindMill Corporate Tower, 2nd Floor, Plot No 24A
Sector 16A, Noida 201301, UP

Random House Group Limited
20 Vauxhall Bridge Road
London SW1V 2SA
United Kingdom

978 81 8400 119 8

Printed and bound in India by Replika Press Pvt. Ltd.

For sale in the Indian subcontinent only

For my mother, V. S. Meenakshi, and

my father, V. M. Subramanian

For Jaya Krishnan, Mani, and Diane

And for my ancestral sage Bharadhwaja,

who brought medicine down from the gods to mortals

———

CONTENTS

———

PREFACE

———

There is not, within the wide range of philosophical inquiry, a sub-
ject more intensely interesting to all who thirst for knowledge, than the
precise nature of that important mental superiority which elevates the
human being above the brute . . .

—EDWARD BLYTH

FOR THE PAST QUARTER CENTURY I HAVE HAD THE MARVELOUS privilege of being able to work in the emerging field of cognitive neuroscience. This book is a distillation of a large chunk of my life's work, which has been to unravel—strand by elusive strand—the mysterious connections between brain, mind, and body. In the chapters ahead I recount my investigations of various aspects of our inner mental life that we are naturally curious about. How do we perceive the world? What is the so-called mind-body connection? What determines your sexual identity? What is consciousness? What goes wrong in autism? How can we account for all of those mysterious faculties that are so quintessentially human, such as art, language, metaphor, creativity, self-awareness, and even religious sensibilities? As a scientist I am driven by an intense curiosity to learn how the brain of an ape—an ape!—managed to evolve such a godlike array of mental abilities.

My approach to these questions has been to study patients with damage or genetic quirks in different parts of their brains that produce bizarre effects on their minds or behavior. Over the years I have worked with hundreds of patients afflicted (though some feel they are blessed) with a great diversity of unusual and curious neurological disorders. For example, people who "see" musical tones or "taste" the textures of everything they touch, or the patient who experiences himself leaving his body and viewing it from above near the ceiling. In this book I describe what

I have learned from these cases. Disorders like these are always baffling at first, but thanks to the magic of the scientific method we can render them comprehensible by doing the right experiments. In recounting each case I will take you through the same step-by-step reasoning—occasionally navigating the gaps with wild intuitive hunches—that I went through in my own mind as I puzzled over how to render it explicable. Often when a clinical mystery is solved, the explanation reveals something new about how the normal, healthy brain works, and yields unexpected insights into some of our most cherished mental faculties. I hope that you, the reader, will find these journeys as interesting as I did.

Readers who have assiduously followed my whole oeuvre over the years will recognize some of the case histories that I presented in my previous books, *Phantoms in the Brain* and *A Brief Tour of Human Consciousness*. These same readers will be pleased to see that I have new things to say about even my earlier findings and observations. Brain science has advanced at an astonishing pace over the past fifteen years, lending fresh perspectives on—well, just about everything. After decades of floundering in the shadow of the "hard" sciences, the age of neuroscience has truly dawned, and this rapid progress has directed and enriched my own work.

The past two hundred years saw breathtaking progress in many areas of science. In physics, just when the late nineteenth-century intelligentsia were declaring that physical theory was all but complete, Einstein showed us that space and time were infinitely stranger than anything formerly dreamed of in our philosophy, and Heisenberg pointed out that at the subatomic level even our most basic notions of cause and effect break down. As soon as we moved past our dismay, we were rewarded by the revelation of black holes, quantum entanglement, and a hundred other mysteries that will keep stoking our sense of wonder for centuries to come. Who would have thought the universe is made up of strings vibrating in tune with "God's music"? Similar lists can be made for discoveries in other fields. Cosmology gave us the expanding universe, dark matter, and jaw-dropping vistas of endless billions of galaxies. Chemistry explained the world using the periodic table of the elements and gave us plastics and a cornucopia of wonder drugs. Mathematics gave us computers—although many "pure" mathematicians would rather not see their discipline sullied by such practical uses. In biology, the anatomy and physiology of the body were worked out in exquisite detail, and the mechanisms that drive evolution finally started to become clear. Diseases that had literally plagued humankind since the dawn of history were at last understood for what they really were (as opposed to, say, acts of witchcraft or divine retribution). Revolutions occurred in surgery, pharmacology, and public health, and human life spans in the developed world doubled in the space of just four or five generations. The ultimate revolution was the deciphering of the genetic code in the 1950s, which marks the birth of modern biology.

By comparison, the sciences of the mind—psychiatry, neurology, psychology—languished for centuries. Indeed, until the last quarter of the twentieth century, rigorous theories of perception, emotion, cognition, and intelligence were nowhere to be found (one notable exception being color vision). For most of the twentieth century, all we had to offer in the way of explaining human behavior was two theoretical edifices—Freudianism and behaviorism—both of which would be dramatically eclipsed in the 1980s and 1990s, when neuroscience finally managed to advance beyond the Bronze Age. In historical terms that isn't a very long time. Compared with physics and chemistry, neuroscience is still a young upstart. But progress is progress, and what a period of progress it has been! From genes to cells to circuits to cognition, the depth and breadth of today's neuroscience—however far short of an eventual Grand Unified Theory it may be—is light-years beyond where it was when I started working in the field. In the last decade we have even seen neuroscience becoming self-confident enough to start offering ideas to disciplines that have traditionally been claimed by the humanities. So we now for instance have neuroeconomics, neuromarketing, neuroarchitecture, neuroarcheology, neurolaw, neuropolitics, neuroesthetics (see Chapters 4 and 8), and even neurotheology. Some of these are just neurohype, but on the whole they are making real and much-needed contributions to many fields.

As heady as our progress has been, we need to stay completely honest with ourselves and acknowledge that we have only discovered a tiny fraction of what there is to know about the human brain. But the modest amount that we have discovered makes for a story more exciting than any Sherlock Holmes novel. I feel certain that as progress continues through the coming decades, the conceptual twists and technological turns we are in for are going to be at least as mind bending, at least as intuition shaking, and as simultaneously humbling and exalting to the human spirit as the conceptual revolutions that upended classical physics a century ago. The adage that fact is stranger than fiction seems to be especially true for the workings of the brain. In this book I hope I can convey at least some of the wonder and awe that my colleagues and I have felt over the years as we have patiently peeled back the layers of the mind-brain mystery. Hopefully it will kindle your interest in what the pioneering neurosurgeon Wilder Penfield called "the organ of destiny" and Woody Allen, in a less reverential mood, referred to as man's "second favorite organ."

Overview

Although this book covers a wide spectrum of topics, you will notice a few important themes running through all of them. One is that humans are truly unique and special, not "just" another species of primate. I still find it a little bit surprising that this position needs as much defense as it does—and not just against the ravings of antievolutionists, but against no small number of my colleagues who seem comfortable stating that we are "just apes" in a casual, dismissive tone that seems to revel in our lowliness. I sometimes wonder: Is this perhaps the secular humanists' version of original sin?

Another common thread is a pervasive evolutionary perspective. It is impossible to understand how the brain works without also understanding how it evolved. As the great biologist Theodosius Dobzhansky said, "Nothing in biology makes sense except in the light of evolution." This stands in marked contrast to most other reverse-engineering problems. For example when the great English mathematician Alan Turing cracked the code of the Nazis' Enigma machine—a device used to encrypt secret messages—he didn't need to know anything about the research and development history of the device. He didn't need to know anything about the prototypes and earlier product models. All he needed was one working sample of the machine, a notepad, and his own brilliant brain. But in biological systems there is a deep unity between structure, function, and origin. You cannot make very much progress understanding any one of these unless you are also paying close attention to the other two.

You will see me arguing that many of our unique mental traits seem to have evolved through the novel deployment of brain structures that originally evolved for other reasons. This happens all the time in evolution. Feathers evolved from scales whose original role was insulation rather than flight. The wings of bats and pterodactyls are modifications of forelimbs originally designed for walking. Our lungs developed from the swim bladders of fish which evolved for buoyancy control. The opportunistic, "happenstantial" nature of evolution has been championed by many authors, most notably Stephen Jay Gould in his famous essays on natural history. I argue that the same principle applies with even greater force to the evolution of the human brain. Evolution found ways to radically repurpose many functions of the ape brain to create entirely new functions. Some of them—language comes to mind—are so powerful that I would go so far as to argue they have produced a species that transcends apehood to the same degree by which life transcends mundane chemistry and physics.

And so this book is my modest contribution to the grand attempt to crack the code of the human brain, with its myriad connections and modules that make it infinitely more enigmatic than any Enigma machine. The Introduction offers perspectives and history on the uniqueness of the human mind, and also provides a quick primer on the basic anatomy of the human brain. Drawing on my early experiments with the phantom limbs experienced by many amputees, Chapter 1 highlights the human brain's amazing capacity for change and reveals how a more expanded form of plasticity may have shaped the course of our evolutionary and cultural development. Chapter 2 explains how the brain processes incoming sensory information, visual information in particular. Even here, my focus is on human uniqueness: Although our brains employ the same basic sensory-processing mechanisms as those of other mammals, we have taken these mechanisms to a new level. Chapter 3 deals with an intriguing phenomenon called synesthesia, a strange blending of the senses that some people experience as a result of unusual brain wiring. Synesthesia opens a window into the genes and brain connectivity that make some people especially creative, and may hold clues about what makes us such a profoundly creative species to begin with.

The next triad of chapters investigates a type of nerve cell that I argue is especially crucial in making us human. Chapter 4 introduces these special cells, called mirror neurons, which lie at the heart of our ability to adopt each other's point of view and empathize with one another.

Human mirror neurons achieve a level of sophistication that far surpasses that of any lower primate, and appear to be the evolutionary key to our attainment of full-fledged culture. Chapter 5 explores how problems with the mirror-neuron system may underlie autism, a developmental disorder characterized by extreme mental aloneness and social detachment. Chapter 6 explores how mirror neurons may have also played a role in humanity's crowning achievement, language. (More technically, protolanguage, which is language minus syntax.)

Chapters 7 and 8 move on to our species' unique sensibilities about beauty. I suggest that there are laws of aesthetics that are universal, cutting across cultural and even species boundaries. On the other hand, Art with a capital A is probably unique to humans.

In the final chapter I take a stab at the most challenging problem of all, the nature of self-awareness, which is undoubtedly unique to humans. I don't pretend to have solved the problem, but I will share the intriguing insights that I have managed to glean over the years based on some truly remarkable syndromes that occupy the twilight zone between psychiatry and neurology, for example, people who leave their bodies temporarily, see God during seizures, or even deny that they exist. How can someone deny his own existence? Doesn't the denial itself imply existence? Can he ever escape from this Gödelian nightmare? Neuropsychiatry is full of such paradoxes, which cast their spell on me when I wandered the hospital corridors as medical student in my early twenties. I could see that these patients' troubles, deeply saddening as they were, were also rich troves of insight into the marvelously unique human ability to apprehend one's own existence.

Like my previous books, *The Tell-Tale Brain* is written in a conversational style for a general audience. I presume some degree of interest in science and curiosity about human nature, but I do not presume any sort of formal scientific background or even familiarity with my previous works. I hope this book proves instructive and inspiring to students of all levels and backgrounds, to colleagues in other disciplines, and to lay readers with no personal or professional stake in these topics. Thus in writing this book I faced the standard challenge of popularization, which is to tread the fine line between simplification and accuracy. Oversimplification can draw ire from hard-nosed colleagues and, worse, can make readers feel like they are being talked down to. On the other hand, too much detail can be off-putting to nonspecialists. The casual reader wants a thought-provoking guided tour of an unfamiliar subject—not a treatise, not a tome. I have done my best to strike the right balance.

Speaking of accuracy, let me be the first to point out that some of the ideas I present in this book are, shall we say, on the speculative side. Many of the chapters rest on solid foundations, such as my work on phantom limbs, visual perception, synesthesia, and the Capgras delusion. But I also tackle a few elusive and less well-charted topics, such as the origins of art and the nature of self-awareness. In such cases I have let educated guess-work and intuition steer my thinking wherever solid empirical data are spotty. This is nothing to be ashamed of: Every virgin area of scientific inquiry must first be explored in this way. It is a fundamental element of the scientific process that when data are scarce or sketchy and existing theories are anemic, scientists must brainstorm. We need to roll out our best hypotheses, hunches, and hare-brained, half-baked intuitions, and then rack our brains for ways to test them. You see this all the time in the history of science. For instance, one of the earliest models of the atom likened it to plum pudding, with electrons nested like plums in the thick "batter" of the atom. A few decades later physicists were thinking of atoms as miniature solar systems, with orderly electrons that orbit the nucleus like planets around a star. Each of these models was useful, and each got us a little bit closer to the final (or at least, the current) truth. So it goes. In my own field my colleagues and I are making our best effort to advance our understanding of some truly mysterious and hard-to-pin-down faculties. As the biologist Peter Medawar pointed out, "All good science emerges from an imaginative conception of what *might* be true." I realize, however, that in spite of this disclaimer I will probably annoy at least some of my colleagues. But as Lord Reith, the first director-general of the BBC, once pointed out, "There are some people whom it is one's duty to annoy."

Boyhood Seductions

"You know my methods, Watson," says Sherlock Holmes before explaining how he has found the vital clue. And so before we journey any further into the mysteries of the human brain, I feel that I should outline the methods behind my approach. It is above all a wide-ranging, multidisciplinary approach, driven by curiosity and a relentless question: What if? Although my current interest is neurology, my love affair with science dates back to my boyhood in Chennai, India. I was perpetually fascinated by natural phenomena, and my first passion was chemistry. I was enchanted by the idea that the whole universe is based on simple interactions between elements in a finite list. Later I found myself drawn to biology, with all its frustrating yet fascinating complexities. When I was twelve, I remember reading about axolotls, which are basically a species of salamander that has evolved to remain permanently in the aquatic larval stage. They manage to keep their gills (rather than trading them in for lungs, like salamanders or frogs) by shutting down metamorphosis and becoming sexually mature in the water. I was completely flabbergasted when I read that by simply giving these creatures the "metamorphosis hormone" (thyroid extract) you could make the axolotl revert back into the extinct, land-dwelling, gill-less adult ancestor that it had evolved from. You could go back in time, resurrecting a prehistoric animal that no longer exists anywhere on Earth. I also knew that for some mysterious reason adult salamanders don't regenerate amputated legs but the tadpoles do. My curiosity took me one step further, to the question of whether an axolotl—which is, after all, an "adult tadpole"—would retain its ability to regenerate a lost leg just as a modern frog tadpole does. And how many other axolotl-like beings exist on Earth, I wondered, that could be restored to their ancestral forms by simply giving them hormones? Could humans—who are after all apes that have evolved to retain many juvenile qualities—be made to revert to an ancestral form, perhaps something resembling *Homo erectus*, using the appropriate cocktail of hormones? My mind reeled out a stream of questions and speculations, and I was hooked on biology forever.

I found mysteries and possibilities everywhere. When I was eighteen, I read a footnote in some obscure medical tome that when a person with a sarcoma, a malignant cancer that affects soft tissues, develops high fever from an infection, the cancer sometimes goes into complete remission. Cancer shrinking as a result of fever? Why? What could explain it, and might it just possibly lead to a practical cancer therapy?[1] I was enthralled by the possibility of such odd, unexpected connections, and I learned an important lesson: Never take the obvious for granted. Once upon a time, it was so obvious that a four-pound rock would plummet earthward twice as fast as a two-pound rock that no one ever bothered to test it. That is, until Galileo Galilei came along and took ten minutes to perform an elegantly simple experiment that yielded a counterintuitive result and changed the course of history.

I had a boyhood infatuation with botany too. I remember wondering how I might get ahold of my own Venus flytrap, which Darwin had called "the most wonderful plant in the world." He had shown that it closes shut when you touch two hairs inside its trap in rapid succession. The double trigger makes it much more likely that it will be responding to the motions of insects as opposed to inanimate detritus falling or drifting in at random. Once it has clamped down on its prey, the plant stays shut and secretes digestive enzymes, but only if it has caught actual food. I was curious. What defines food? Will it stay shut for amino acids? Fatty acid? Which acids? Starch? Pure sugar? Saccharin? How sophisticated are the food detectors in its digestive system? Too bad, I never did manage to acquire one as a pet at that time.

My mother actively encouraged my early interest in science, bringing me zoological specimens from all over the world. I remember particularly well the time she gave me a tiny dried seahorse. My father also approved of my obsessions. He bought me a Carl Zeiss research microscope when I was still in my early teens. Few things could match the joy of looking at paramecia and volvox through a high-power objective lens. (Volvox, I learned, is the only biological creature on the planet that actually has a wheel.) Later, when I headed off to university, I told my father my heart was set on basic science. Nothing else stimulated my mind half as much. Wise man that he was, he persuaded me to study medicine. "You can become a second-rate doctor and still make a decent living," he said, "but you can't be second-rate scientist; it's an oxymoron." He pointed out that if I studied medicine I could play it safe, keeping both doors open and decide after graduation whether I was cut out for research or not.

All my arcane boyhood pursuits had what I consider to be a pleasantly antiquated, Victorian flavor. The Victorian era ended over a century ago (technically in 1901) and might seem remote from twenty-first-century neuroscience. But I feel compelled to mention my early romance with nineteenth-century science because it was a formative influence on my style of thinking and conducting research.

Simply put, this "style" emphasizes conceptually simple and easy-to-do experiments. As a student I read voraciously, not only about modern biology but also about the history of science. I remember reading about Michael Faraday, the lower-class, self-educated man who discovered the principle of electromagnetism. In the early 1800s he placed a bar magnet behind a sheet of paper and threw iron filings on the sheet. The filings instantly aligned themselves into arcing lines. He had rendered the magnetic field visible! This was about as direct a demonstration as possible that such fields are real and not just mathematical abstractions. Next Faraday moved a bar magnet to and fro through a coil of copper wire, and lo and behold, an electric current started running through the coil. He had demonstrated a link between two entirely separate areas of physics: magnetism and electricity. This paved the way not only for practical applications—such as hydroelectric power, electric motors, and electromagnets—but also for the deep theoretical insights of James Clerk Maxwell. With nothing more than bar magnets, paper, and copper wire, Faraday had ushered in a new era in physics.

I remember being struck by the simplicity and elegance of these experiments. Any schoolboy or -girl can repeat them. It was not unlike Galileo dropping his rocks, or Newton using two prisms to explore the nature of light. For better or worse, stories like these made me a technophobe early in life. I still find it hard to use an iPhone, but my technophobia has served me well in other respects. Some colleagues have warned me that this phobia might have been okay in the nineteenth century when biology and physics were in their infancy, but not in this era of "big science," in which major advances can only be made by large teams employing high-tech machines. I disagree. And even if it is partly true, "small science" is much more fun and can often turn up big discoveries. It still tickles me that my early experiments with phantom limbs (see Chapter 1) required nothing more than Q-tips, glasses of warm and cold water, and ordinary mirrors. Hippocrates, Sushruta, my ancestral sage Bharadwaja, or any other physicians between ancient times and the present could have performed these same basic experiments. Yet no one did.

Or consider Barry Marshall's research showing that ulcers are caused by bacteria—not acid or stress, as every doctor "knew." In a heroic experiment to convince skeptics of his theory, he actually swallowed a culture of the bacterium *Helicobacter pylori* and showed that his stomach lining became studded with painful ulcers, which he promptly cured by consuming antibiotics. He and others later went on to show that many other disorders, including stomach cancer and even heart attacks, might be triggered by microorganisms. In just a few weeks, using materials and methods that had been available for decades, Dr. Marshall had ushered in a whole new era of medicine. Ten years later he won a Nobel Prize.

My preference for low-tech methods has both strengths and drawbacks, of course. I enjoy it—partly because I'm lazy—but it isn't everyone's cup of tea. And this is a good thing. Science needs a variety of styles and approaches. Most individual researchers need to specialize, but the scientific enterprise as a whole is made more robust when scientists march to different drumbeats. Homogeneity breeds weakness: theoretical blind spots, stale paradigms, an echo-chamber mentality, and cults of personality. A diverse dramatis personae is a powerful tonic against these ailments. Science benefits from its inclusion of the abstraction-addled, absent-minded professors, the control-freak obsessives, the cantankerous bean-counting statistics junkies, the congenitally contrarian devil's advocates, the hard-nosed data-oriented literalists, and the starry-eyed romantics who embark on high-risk, high-payoff ventures, stumbling frequently along the way. If every scientist were like me, there would be no one to clear the brush or demand periodic reality checks. But if every scientist were a brush-clearing, never-stray-beyond-established-fact type, science would advance at a snail's pace and would have a hard time unpainting itself out of corners. Getting trapped in narrow cul-de-sac specializations and "clubs" whose membership is open only to those who congratulate and fund each other is an occupational hazard in modern science.

When I say I prefer Q-tips and mirrors to brain scanners and gene sequencers, I don't mean to give you the impression that I eschew technology entirely. (Just think of doing biology without a microscope!) I may be a technophobe, but I'm no Luddite. My point is that science should be question driven, not methodology driven. When your department has spent millions of dollars on a state-of-the-art liquid-helium-cooled brain-imaging machine, you come under pressure to use it all the time. As the old saying goes, "When the only tool you have is a hammer, everything starts to look like a nail." But I have nothing against high-tech brain scanners (nor against hammers). Indeed, there is so much brain imaging going on these days that some significant discoveries are bound to be made, if only by accident. One could justifiably argue that the modern toolbox of state-of-the-art gizmos has a vital and indispensable place in research. And indeed, my low-tech-leaning colleagues and I often do take advantage of brain imaging, but only to test specific hypotheses. Sometimes it works, sometimes it doesn't, but we are always grateful to have the high technology available—if we feel the need.

ACKNOWLEDGMENTS

———

ALTHOUGH IT IS LARGELY A PERSONAL ODYSSEY, THIS BOOK RELIES heavily on the work of many of my colleagues who have revolutionized the field in ways we could not have even imagined even just a few years ago. I cannot overstate the extent to which I have benefited from reading their books. I will mention just a few of them here: Joe LeDoux, Oliver Sacks, Francis Crick, Richard Dawkins, Stephen Jay Gould, Dan Dennett, Pat Churchland, Gerry Edelman, Eric Kandel, Nick Humphrey, Tony Damasio, Marvin Minsky, Stanislas Dehaene. If I have seen further, it is by standing on the shoulders of these giants. Some of these books resulted from the foresight of two enlightened agents—John Brockman and Katinka Matson—who have created a new scientific literacy in America and the world beyond. They have successfully reignited the magic and awe of science in the age of Twitter, Facebook, YouTube, sound-bite news, and reality TV—an age when the hard-won values of the Enlightenment are sadly in decline.

Angela von der Lippe, my editor, suggested major reorganization of chapters and provided valuable feedback throughout every stage of revision. Her suggestions improved the clarity of presentation enormously.

Special thanks to four people who have had a direct influence on my scientific career: Richard Gregory, Francis Crick, John D. Pettigrew, and Oliver Sacks.

I would also like to thank the many people who either goaded me on to pursue medicine and science as a career or influenced my thinking over the years. As I intimated earlier, I would not be where I am were it not for my mother and father. When my father was convincing me to go into medicine, I received similar advice from Drs. Rama Mani and M. K. Mani. I have never once regretted letting them talk me into it. As I often tell my students, medicine gives you a certain breadth of vision while at the same time imparting an intensely pragmatic attitude. If your theory is right, your patient gets better. If your theory is wrong—no matter how elegant or convincing it may be—she gets worse or dies. There is no better test of whether you are on the right track or not. And this no-nonsense attitude then spills over into your research as well.

I also owe an intellectual debt to my brother V. S. Ravi, whose vast knowledge of English and Telugu literature (especially Shakespeare and Thyagaraja) is unsurpassed. When I had just entered medical school (premed), he would often read me passages from Shakespeare and Omar Khayyam's *Rubaiyat,* which had a deep impact on my mental development. I remember hearing him quote Macbeth's famous "sound and fury" soliloquy and thinking, "Wow, that pretty much says it all." It impressed on me the importance of economy of expression, whether in literature or in science.

I thank Matthew Blakeslee, who did a superb job in helping edit the book. Over fifteen years ago, as my student, he also assisted me in constructing the very first crude but effective prototype of the "mirror box" which inspired the subsequent construction of elegant, ivory-inlaid mahogany ones at Oxford (and which are now available commercially, although I have no personal financial stake in them). Various drug companies and philanthropic organizations have distributed thousands of such boxes to war veterans from Iraq and amputees in Haiti.

I also owe a debt of gratitude to the many patients who cooperated with me over the years. Many of them were in depressing situations, obviously, but most of them were unselfishly willing to help advance basic science in whatever way they could. Without them this book could not have been written. Naturally, I care about protecting their privacy. In the interest of confidentiality, all names, dates, and places, and in some instances the circumstances surrounding the admission of the patient, have been disguised. The conversations with patients (such as those with language problems) are literal transcripts of videotapes, except in a few cases where I had to re-create our exchanges based on memory. In one case ("John," in Chapter 2, who developed embolic stroke originating from veins around an inflamed appendix) I have described appendicitis as it usually presents itself since notes on this particular case were unavailable. And the conversation with this patient is an edited summary of the conversation as recounted by the physician who originally saw him. In all cases the key symptoms and signs and history that are relevant to the neurological aspect of patients' problems are presented as accurately as possible. But other aspects have been changed—for example, a patient who is fifty rather than fifty-five may have had an embolism originating in the heart rather than leg—so that even a close friend or relative would be unable to recognize the patient from the description.

I turn now to thank friends and colleagues with whom I have had productive conversations over the years. I list them in alphabetical order: Krishnaswami Alladi, John Allman, Eric Altschuler, Stuart Anstis, Carrie Armel, Shai Azoulai, Horace Barlow, Mary Beebe, Roger Bingham, Colin Blakemore, Sandy Blakeslee, Geoff Boynton, Oliver Braddick, David Brang, Mike Calford, Fergus Campbell, Pat Cavanagh, Pat and Paul Churchland, Steve Cobb, Francis Crick, Tony and Hanna Damasio, Nikki de Saint Phalle, Anthony Deutsch, Diana Deutsch, Paul Drake, Gerry Edelman, Jeff Elman, Richard Friedberg, Sir Alan Gilchrist, Beatrice Golomb, Al Gore (the "real" president), Richard Gregory, Mushirul Hasan, Afrei Hesam, Bill Hirstein, Mikhenan ("Mikhey") Horvath, Ed Hubbard, David Hubel, Nick Humphrey, Mike Hyson, Sudarshan Iyengar, Mumtaz Jahan, Jon Kaas, Eric Kandel, Dorothy Kleffner, E. S. Krishnamoorthy, Ranjit Kumar, Leah Levi, Steve Link, Rama Mani, Paul McGeoch, Don McLeod, Sarada Menon, Mike Merzenich, Ranjit Nair, Ken Nakayama, Lindsay Oberman, Ingrid Olson, Malini Parthasarathy, Hal Pashler, David Peterzell, Jack Pettigrew, Jaime Pineda, Dan Plummer, Alladi Prabhakar, David Presti, N. Ram and N. Ravi (editors of *The Hindu*), Alladi Ramakrishnan, V. Madhusudhan Rao, Sushila Ravindranath, Beatrice Ring, Bill Rosar, Oliver Sacks, Terry Sejnowski, Chetan Shah, Naidu ("Spencer") Sitaram, John Smythies, Allan Snyder, Larry Squire, Krishnamoorthy Srinivas, A. V. Srinivasan, Krishnan Sriram, Subramaniam Sriram, Lance Stone, Somtow ("Cookie") Sucharitkul, K. V. Thiruvengadam, Chris Tyler, Claude Valenti, Ajit Varki, Ananda Veerasurya, Nairobi Venkataraman, Alladi Venkatesh, T. R. Vidyasagar, David Whitteridge, Ben Williams, Lisa Williams, Chris Wills, Piotr Winkielman, and John Wixted.

Thanks to Elizabeth Seckel and Petra Ostermuencher for their help.

I also thank Diane, Mani, and Jaya, who are an endless source of delight and inspiration. The *Nature* paper they published with me on flounder camouflage made a huge splash in the ichthyology world.

Julia Kindy Langley kindled my passion for the science of art.

Last but not least, I am grateful to the National Institutes of Health for funding much of the research reported in the book, and to private donors and patrons: Abe Pollin, Herb Lurie, Dick Geckler, and Charlie Robins.

INTRODUCTION

No Mere Ape

———

Now I am quite sure that if we had these three creatures fossilized or preserved in spirits for comparison and were quite unprejudiced judges, we should at once admit that there is very little greater interval as animals between the gorilla and the man than exists between the gorilla and the baboon.

—THOMAS HENRY HUXLEY,
*lecturing at the Royal
Institution, London*

"I know, my dear Watson, that you share my love of all that is bizarre and outside the conventions and humdrum routine of everyday life."

—SHERLOCK HOLMES

Is MAN AN APE OR AN ANGEL (AS BENJAMIN DISRAELI ASKED IN A famous debate about Darwin's theory of evolution)? Are we merely chimps with a software upgrade? Or are we in some true sense *special*, a species that transcends the mindless fluxions of chemistry and instinct? Many scientists, beginning with Darwin himself, have argued the former: that human mental abilities are merely elaborations of faculties that are ultimately of the same *kind* we see in other apes. This was a radical and controversial proposal in the nineteenth century—some people are still not over it—but ever since Darwin published his world-shattering treatise on the theory of evolution, the case for man's primate origins has been bolstered a thousandfold. Today it is impossible to seriously refute this point: We are anatomically, neurologically, genetically, physiologically apes. Anyone who has ever been struck by the uncanny near-humanness of the great apes at the zoo has felt the truth of this.

I find it odd how some people are so ardently drawn to either-or dichotomies. "Are apes self-aware *or* are they automata?" "Is life meaningful *or* is it meaningless?" "Are humans 'just' animals *or* are we exalted?" As a scientist I am perfectly comfortable with settling on categorical conclusions—when it makes sense. But with many of these supposedly urgent metaphysical dilemmas, I must admit I don't see the conflict. For instance, why can't we be a branch of the animal kingdom *and* a wholly unique and gloriously novel *phenomenon* in the universe?

I also find it odd how people so often slip words like "merely" and "nothing but" into statements about our origins. Humans are apes. So too we are mammals. We are vertebrates. We are pulpy, throbbing colonies of tens of trillions of cells. We are all of these things, but we are not "merely" these things. And we are, in addition to all these things, something unique, something unprecedented, something transcendent. We are something truly new under the sun, with uncharted and perhaps limitless potential. We are the first and only species whose fate has rested in its own hands, and *not* just in the hands of chemistry and instinct. On the great Darwinian stage we call Earth, I would argue there has not been an upheaval as big as us since the origin of life itself. When I think about what we are and what we may yet achieve, I can't see any place for snide little "merelies."

Any ape can reach for a banana, but only humans can reach for the stars. Apes live, contend, breed, and die in forests—end of story. Humans write, investigate, create, and quest. We splice genes, split atoms, launch rockets. We peer upward into the heart of the Big Bang and delve deeply into the digits of pi. Perhaps most remarkably of all, we gaze inward, piecing together the puzzle of our own unique and marvelous brain. It makes the mind reel. How can a three-pound mass of jelly that you can hold in your palm imagine angels, contemplate the meaning of infinity, and even question its own place in the cosmos? Especially awe inspiring is the fact that any single brain, including yours, is made up of atoms that were forged in the hearts of countless, far-flung stars billions of years ago. These particles drifted for eons and light-years until gravity and chance brought them together here, now. These atoms now form a con-glomerate—your brain—that can not only ponder the very stars that gave it birth but can also think about its own ability to think and wonder about its own ability to wonder. With the arrival of humans, it has been said, the universe has suddenly become conscious of itself. This, truly, is the greatest mystery of all.

It is difficult to talk about the brain without waxing lyrical. But how does one go about actually studying it? There are many methods, ranging from single-neuron studies to high-tech brain scanning to cross-species comparison. The methods I favor are unapologetically old-school. I generally see patients who have suffered brain lesions due to stroke, tumor, or head injury and as a result are experiencing disturbances in their perception and consciousness. I also sometimes meet people who do not appear brain damaged or impaired, yet report having wildly unusual perceptual or mental experiences. In either case, the procedure is the same: I interview them, observe their behavior, administer some simple tests, take a peek at their brains (when possible), and then come up with a hypothesis that bridges psychology and neurology—in other words, a hypothesis that connects strange behavior to what has gone wrong in the intricate wiring of the brain.[1] A decent percentage of the time I am successful. And so, patient by patient, case by case, I gain a stream of fresh insights into how the human mind and brain work—and how they are inextricably linked. On the coattails of such discoveries I often get evolutionary insights as well, which bring us that much closer to understanding what makes our species unique.

Consider the following examples:

- Whenever Susan looks at numbers, she sees each digit tinged with its own inherent hue. For example, 5 is red, 3 is blue. This condition, called synesthesia, is eight times more common in artists, poets, and novelists than in the general population, suggesting that it may be linked to creativity in some mysterious way. Could synesthesia be a neuropsychological fossil of sorts—a clue to understanding the evolutionary origins and nature of human creativity in general?

- Humphrey has a phantom arm following an amputation. Phantom limbs are a common experience for amputees, but we noticed something unusual in Humphrey. Imagine his amazement when he merely watches me stroke and tap a student volunteer's arm—and actually feels these tactile sensations in his phantom. When he watches the student fondle an ice cube, he feels the cold in his phantom fingers. When he watches her massage her own hand, he feels a "phantom massage" that relieves the painful cramp in his phantom hand! Where do his body, his phantom body, and a stranger's body meld in his mind? What or where is his real sense of self?

• A patient named Smith is undergoing neurosurgery at the University of Toronto. He is fully awake and conscious. His scalp has been perfused with a local anesthetic and his skull has been opened. The surgeon places an electrode in Smith's anterior cingulate, a region near the front of the brain where many of the neurons respond to pain. And sure enough, the doctor is able to find a neuron that becomes active whenever Smith's hand is poked with a needle. But the surgeon is astonished by what he sees next. The same neuron fires just as vigorously when Smith merely *watches* another patient being poked. It is as if the neuron (or the functional circuit of which it is a part) is empathizing with another person. A stranger's pain becomes Smith's pain, almost literally. Indian and Buddhist mystics assert that there is no essential difference between self and other, and that true enlightenment comes from the compassion that dissolves this barrier. I used to think this was just well-intentioned mumbo-jumbo, but here is a neuron that doesn't know the difference between self and other. Are our brains uniquely hardwired for empathy and compassion?

• When Jonathan is asked to imagine numbers he always sees each number in a particular spatial location in front of him. All numbers from *1* to *60* are laid out sequentially on a virtual number line that is elaborately twisted in three-dimensional space, even doubling back on itself. Jonathan even claims that this twisted line helps him perform arithmetic. (Interestingly, Einstein often claimed to see numbers spatially.) What do cases like Jonathan's tell us about our unique facility with numbers? Most of us have a vague tendency to image numbers from left to right, but why is Jonathan's warped and twisted? As we shall see, this a striking example of a neurological anomaly that makes no sense whatsoever except in evolutionary terms.

• A patient in San Francisco becomes progressively demented, yet starts creating paintings that are hauntingly beautiful. Has his brain damage somehow unleashed a hidden talent? A world away, in Australia, a typical undergraduate volunteer named John is participating in an unusual experiment. He sits down in a chair and is fitted with a helmet that delivers magnetic pulses to his brain. Some of his head muscles twitch involuntarily from the induced current. More amazingly, John starts producing lovely drawings—something he claims he couldn't do before. Where are these inner artists emerging from? Is it true that most of us "use only 10 percent of our brain"? Is there a Picasso, a Mozart, and a Srinivasa Ramanujan (a math prodigy) in all of us, waiting to be liberated? Has evolution suppressed our inner geniuses for a reason?

• Until his stroke, Dr. Jackson was a prominent physician in Chula Vista, California. Afterward he is left partially paralyzed on his right side, but fortunately only a small part of his cortex, the brain's seat of higher intelligence, has been damaged. His higher mental functions are largely intact: He can understand most of what is said to him and he can hold up a conversation reasonably well. In the course of probing his mind with various simple tasks and questions, the big surprise comes when we ask him to explain a proverb, "All that glitters is not gold."

"It means just because something is shiny and yellow doesn't mean it's gold, Doctor. It could be copper or some alloy."

"Yes," I say, "but is there a deeper meaning beyond that?"

"Yes," he replies, "it means you have to be very careful when you go to buy jewelry; they often rip you off. One could measure the metal's specific gravity, I suppose."

Dr. Jackson has a disorder that I call "metaphor blindness." Does it follow from this that the human brain has evolved a dedicated "metaphor center"?

• Jason is a patient at a rehabilitation center in San Diego. He has been in a semicomatose state called akinetic mutism for several months before he is seen by my colleague Dr. Subramaniam Sriram. Jason is bedridden, unable to walk, recognize, or interact with people—not even his parents—even though he is fully alert and often follows people around with his eyes. Yet if his father goes next door and phones him, Jason instantly becomes fully conscious, recognizes his dad, and converses with him. When his father returns to the room, Jason reverts at once to a zombie-like state. It is as if there are two Jasons trapped inside one body: the one connected to vision, who is alert but not conscious, and the one connected to hearing who is alert *and* conscious. What might these eerie comings and goings of conscious personhood reveal about how the brain generates self-awareness?

These may sound like phantasmagorical short stories by the likes of Edgar Allan Poe or Philip K. Dick. Yet they are all true, and these are only a few of the cases you will encounter in this book. An intensive study of these people can not only help us figure out why their bizarre symptoms occur, but also help us understand the functions of the normal brain—yours and mine. Maybe someday we will even answer the most difficult question of all: How does the human brain give rise to consciousness? What or who is this "I" within me that illuminates one tiny corner of the universe, while the rest of the cosmos rolls on indifferent to every human concern? A question that comes perilously close to theology.

WHEN PONDERING OUR uniqueness, it is natural to wonder how close other species before us might have come to achieving our cognitive state of grace. Anthropologists have found that the hominin family tree branched many times in the past several million years. At various times numerous protohuman and human-like ape species thrived and roamed the earth, but for some reason our line is the only one that "made it." What were the brains of those other hominins like? Did they perish because they didn't stumble on the right combination of neural adaptations? All we have to go on now is the mute testimony of their fossils and their scattered stone tools. Sadly, we may never learn much about how they behaved or what their minds were like.

We stand a much better chance of solving the mystery of the relatively recently extinct Neanderthals, a cousin-species of ours, who were almost certainly within a proverbial stone's throw of achieving full-blown humanhood. Though traditionally depicted as the archetypical brutish, slow-witted cave dweller, *Homo neanderthalensis* has been receiving a serious image makeover in recent years. Just like us they made art and jewelry, ate a rich and varied diet, and buried their dead. And evidence is mounting that their language was more complex than the stereotypical "cave man talk" gives them credit for. Nevertheless, around thirty thousand years ago they vanished from the earth. The reigning assumption has always been that the Neanderthals died and humans thrived on because humans were somehow superior: better language, better tools, better social organization, or something like that. But the matter is far from settled. Did we outcompete them? Did we murder them all? Did we—to borrow a phrase from the movie *Braveheart*—breed them out? Were we just plain lucky, and they unlucky? Could it as easily have been them instead of us who planted a flag on the moon? The Neanderthals' extinction is recent enough that we have been able to recover actual bones (not just fossils), and along with them some samples of Neanderthal DNA. As genetic studies continue, we will assuredly learn more about the fine line that divided us.

And then of course there were the hobbits.

Far away on a remote island near Java there lived, not so long ago, a race of diminutive creatures—or should I say, people—who were just three feet tall. They were very close to human and yet, to the astonishment of the world, turn out to have been a different species who coexisted alongside us almost up until historical times. On the Connecticut-sized island of Flores they eked out a living hunting twenty-foot dragonlizards, giant rats, and pigmy elephants. They manufactured miniature tools to wield with their tiny hands and apparently had enough planning skills and foresight to navigate the open seas. And yet incredibly, their brains were about one-third the size of a human's brain, smaller than that of a chimp.[2]

If I were to give you this story as a script for a science fiction movie, you would probably reject it as too farfetched. It sounds like something straight out of H. G. Welles or Jules Verne. Yet remarkably, it happens to be true. Their discoverers entered them into the scientific record as *Homo floresiensis*, but many people refer to them by their nickname, hobbits. The bones are only about fifteen thousand years old, which implies that these strange human cousins lived side by side with our ancestors, perhaps as friends, perhaps as foes—we do not know. Nor again do we know why they vanished, although given our species' dismal record as responsible stewards of nature, it's a decent bet that we drove them to extinction. But many islands in Indonesia are still unexplored, and it is not inconceivable that an isolated pocket of them has survived somewhere. (One theory holds that the CIA has spotted them already but the information is being withheld until it is ruled out that they are hoarding weapons of mass destruction like blowpipes.)

The hobbits challenge all our preconceived notions about our supposed privileged status as *Homo sapiens*. If the hobbits had had the resources of the Eurasian continent at their disposal, might they have invented agriculture, civilization, the wheel, writing? Were they self-conscious? Did they have a moral sense? Were they aware of their mortality? Did they sing and dance? Or are these mental functions (and ipso facto, are their corresponding neural circuits) found only in humans? We still know precious little about the hobbits, but their similarities to and differences from humans might help us further understand what makes us different from the great apes and monkeys, and whether there was a quantum leap in our evolution or a gradual change. Indeed, getting ahold of some samples of hobbit DNA would be a discovery of far greater scientific import than any DNA recovery scenario à la Jurassic Park.

This question of our special status, which will reappear many times in this book, has a long and contentious history. It was a major preoccupation of intellectuals in Victorian times. The protagonists were some of the giants of nineteenth-century science, including Thomas Huxley, Richard Owen, and Alfred Russel Wallace. Even though Darwin started it all, he himself shunned controversy. But Huxley, a large man with piercing dark eyes and bushy eyebrows, was renowned for his pugnacity and wit and had no such compunctions. Unlike Darwin, he was outspoken about the implications of evolutionary theory for humans, earning him the epithet "Darwin's bulldog."

Huxley's adversary, Owen, was convinced that humans were unique. The founding father of the science of comparative anatomy, Owen inspired the often-satirized stereotype of a paleontologist who tries to reconstruct an entire animal from a single bone. His brilliance was matched only by his arrogance. "He knows that he is superior to most men," wrote Huxley, "and does not conceal that he knows." Unlike Darwin, Owen was more impressed by the differences than by similarities between different animal groups. He was struck by the absence of living intermediate forms between species, of the kind you might expect to find if one species gradually evolved into another. No one saw elephants with one-foot trunks or giraffes with necks half as long their modern counterparts. (The okapi, which have such necks, were discovered much later.) Observations like these, together with his strong religious views, led him to regard Darwin's ideas as both implausible and heretical. He emphasized the huge gap between the mental abilities of apes and humans and pointed out (mistakenly) that the human brain had a unique anatomical structure called the "hippocampus minor," which he said was entirely absent in apes.

Huxley challenged this view; his own dissections failed to turn up the hippocampus minor. The two titans clashed over this for decades. The controversy occupied center stage in the Victorian press, creating the kind of media sensation that is reserved these days for the likes of Washington sex scandals. A parody of the hippocampus minor debate, published in Charles Kingsley's children's book *The Water-Babies*, captures the spirit of the times:

> [Huxley] held very strange theories about a good many things. He . . . declared that apes had hippopotamus majors [*sic*] in their brains just as men have. Which was a shocking thing to say; for, if it were so, what would become of the faith, hope, and charity of immortal millions? You may think that there are other more important differences between you and an ape, such as being able to speak, and make machines, and know right from wrong, and say your prayers, and other little matters of that kind; but that is a child's fancy, my dear. Nothing is to be depended on but the great hippopotamus test. If you have a hippopotamus major in your brain, you are no ape, though you had four hands, no feet, and were more apish than the apes of all aperies.

Joining the fray was Bishop Samuel Wilberforce, a staunch creationist who often relied on Owen's anatomical observations to challenge Darwin's theory. The battle raged on for twenty years until, tragically, Wilberforce was thrown off a horse and died instantly when his head hit the pavement. It is said that Huxley was sipping his cognac at the Athenaeum in London when the news reached him. He wryly quipped to the reporter, "At long last the Bishop's brain has come into contact with hard reality, and the result has been fatal."

Modern biology has amply demonstrated that Owen was wrong: There is no hippocampus minor, no sudden discontinuity between apes and us. The view that we are special is generally thought to be held only by creationist zealots and religious fundamentalists. Yet I am prepared to defend the somewhat radical view that on this particular issue Owen was right after all—although for reasons entirely different from those he had in mind. Owen was correct in asserting that the human brain—unlike, say, the human liver or heart—is indeed unique and distinct from that of the ape by a huge gap. But this view is entirely compatible with Huxley and Darwin's claim that our brain evolved piecemeal, sans divine intervention, over millions of years.

But if this is so, you may wonder, where does our uniqueness come from? As Shakespeare and Parmenides had already stated long before Darwin, nothing can come of nothing.

It is a common fallacy to assume that gradual, small changes can only engender gradual, incremental results. But this is linear thinking, which seems to be our default mode for thinking about the world. This may be due to the simple fact that most of the phenomena that are perceptible to humans, at everyday human scales of time and magnitude and within the limited scope of our naked senses, tend to follow linear trends. Two stones feel twice as heavy as one stone. It takes three times as much food to feed three times as many people. And so on. But outside of the sphere of practical human concerns, nature is full of nonlinear phenomena. Highly complex processes can emerge from deceptively simple rules or parts, and small changes in one underlying factor of a complex system can engender radical, qualitative shifts in other factors that depend on it.

Think of this very simple example: Imagine you have block of ice in front of you and you are gradually warming it up: 20 degrees Fahrenheit . . . 21 degrees . . . 22 degrees . . . Most of the time, heating the ice up by one more degree doesn't have any interesting effect: all you have that you didn't have a minute ago is a slightly warmer block of ice. But then you come to 32 degrees Fahrenheit. As soon as you reach this critical temperature, you see an abrupt, dramatic change. The crystalline structure of the ice decoheres, and suddenly the water molecules start slipping and flowing around each other freely. Your frozen water has turned into liquid water, thanks to that one critical degree of heat energy. At that key point, incremental changes stopped having incremental effects, and precipitated a sudden qualitative change called a phase transition.

Nature is full of phase transitions. Frozen water to liquid water is one. Liquid water to gaseous water (steam) is another. But they are not confined to chemistry examples. They can occur in social systems, for example, where millions of individual decisions or attitudes can interact to rapidly shift the entire system into a new balance. Phase transitions are afoot during speculative bubbles, stock market crashes, and spontaneous traffic jams. On a more positive note, they were on display in the breakup of the Soviet Bloc and the exponential rise of the Internet.

I would even suggest that phase transitions may apply to human origins. Over the millions of years that led up to *Homo sapiens*, natural selection continued to tinker with the brains of our ancestors in the normal evolutionary fashion—which is to say, gradual and piecemeal: a dime-sized expansion of the cortex here, a 5 percent thickening of the fiber tract connecting two structures there, and so on for countless generations. With each new generation, the results of these slight neural improvements were apes who were slightly better at various things: slightly defter at wielding sticks and stones; slightly cleverer at social scheming, wheeling and dealing; slightly more foresightful about the behaviors of game or the portents of weather and season; slightly better at remembering the distant past and seeing connections to the present.

Then sometime about a hundred and fifty thousand years ago there was an explosive development of certain key brain structures and functions whose fortuitous combinations resulted in the mental abilities that make us special in the sense that I am arguing for. We went through a *mental* phase transition. All the same old parts were there, but they started working together in new ways that were far more than the sum of their parts. This transition brought us things like full-fledged human language, artistic and religious sensibilities, and consciousness and self-awareness. Within the space of perhaps thirty thousand years we began to build our own shelters, stitch hides and furs into garments, create shell jewelry and rock paintings, and carve flutes out of bones. We were more or less finished with genetic evolution, but had embarked on a much (much!) faster-paced form of evolution that acted not on genes but on culture.

And just what structural brain improvements were the keys to all of this? I will be happy to explain. But before I do that, I should give you a survey of brain anatomy so you can best appreciate the answer.

A Brief Tour of Your Brain

The human brain is made up of about 100 billion nerve cells, or neurons (Figure Int.1). Neurons "talk" to each other through threadlike fibers that alternately resemble dense, twiggy thickets (dendrites) and long, sinuous transmission cables (axons). Each neuron makes from one thousand to ten thousand contacts with other neurons. These points of contact, called synapses, are where information gets shared between neurons. Each synapse can be excitatory or inhibitory, and at any given moment can be on or off. With all these permutations the number of possible brain states is staggeringly vast; in fact, it easily exceeds the number of elementary particles in the known universe.

Given this bewildering complexity, it's hardly surprising that medical students find neuroanatomy tough going. There are almost a hundred structures to reckon with, most of them with arcane-sounding names. The fimbria. The fornix. The indusium griseum. The locus coeruleus. The nucleus motoris dissipatus formationis of Riley. The medulla oblongata. I must say, I love the way these Latin names roll off the tongue. Meh-*dull*-a oblong-*gah*-ta! My favorite is the substantia innominata, which literally means "substance without a name." And the smallest muscle in the body, which is used to abduct the little toe, is the abductor ossis metatarsi digiti quinti minimi. I think it sounds like a poem. (With the first wave of the Harry Potter generation now coming up through medical school, perhaps soon we'll finally start hearing these terms pronounced with more of the relish they deserve.)

Fortunately, underlying all this lyrical complexity there is a basic plan of organization that's easy to understand. Neurons are connected into networks that can process information. The brain's many dozens of structures are ultimately all purpose-built networks of neurons, and often have elegant internal organization. Each of these structures performs some set of discrete (though not always easy to decipher) cognitive or physiological functions. Each structure makes patterned connections with other brain structures, thus forming circuits. Circuits pass

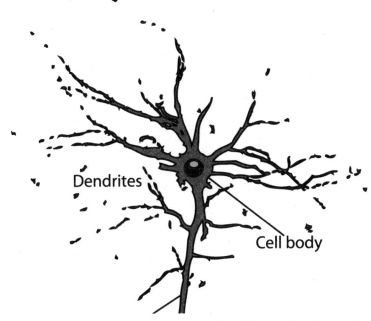

Dendrites

Cell body

FIGURE INT.1   Drawing of a neuron showing the cell body, dendrites, and axon. The axon transmits information (in the form of nerve impulses) to the next neuron (or set of neurons) in the chain. The axon is quite long, and only part of it is shown here. The dendrites receive information from the axons of other neurons. The flow of information is thus always unidirectional.

information back and forth and in repeating loops, and allow brain structures to work together to create sophisticated perceptions, thoughts, and behaviors.

The information processing that occurs both within and between brain structures can get quite complicated—this is, after all, the information-processing engine that generates the human mind—but there is plenty that can be understood and appreciated by nonspecialists. We will revisit many of these areas in greater depth in the chapters ahead, but a basic acquaintance now with each region will help you to appreciate how these specialized areas work together to determine mind, personality, and behavior.

The human brain looks like a walnut made of two mirror-image halves (Figure Int.2). These shell-like halves are the cerebral cortex. The cortex is split down the middle into two hemispheres: one on the left, one

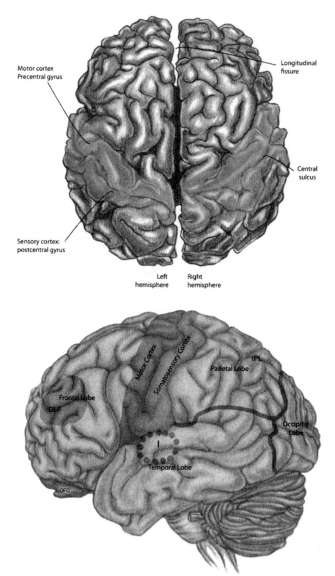

FIGURE INT.2   The human brain viewed from the top and from the left side.
The top view shows the two mirror-symmetric cerebral hemispheres, each of which
controls the movements of—and receives signals from—the opposite side of the
body (though there are some exceptions to this rule). Abbreviations: DLF, dorsolateral prefrontal cortex; OFC, orbitofrontal cortex; IPL, inferior parietal lobule;
I, insula, which is tucked away deep beneath the Sylvian fissure below the frontal
lobe. The ventromedial prefrontal cortex (VMF, not labeled) is tucked away in the
inner lower part of the frontal lobe, and the OFC is part of it.

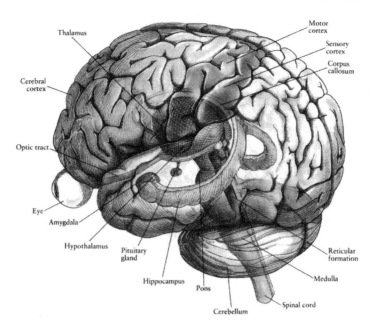

Thalamus

Motor cortex

Sensory cortex

Corpus callosum

Cerebral cortex

Optic tract

Eye

Amygdala

Hypothalamus   Pituitary gland

Hippocampus   Pons

Cerebellum

Reticular formation

Medulla

Spinal cord

FIGURE INT.3   A schematic drawing of the human brain showing internal structures such as the amygdala, hippocampus, basal ganglia, and hypothalamus.

on the right. In humans the cortex has grown so large that it has been forced to become convoluted (folded), giving it its famous cauliflower-like appearance. (In contrast, the cortex of most other mammals is smooth and flat for the most part, with few if any folds in the surface.) The cortex is essentially the seat of higher thought, the tabula (far from) rasa where all of our highest mental functions are carried out. Not surprisingly, it is especially well developed in two groups of mammals: dolphins and primates. We'll return to the cortex later in the chapter. For now let's look at the other parts of the brain.

Running up and down the core of the spinal column is a thick bundle of nerve fibers—the spinal cord—that conducts a steady stream of messages between brain and body. These messages include things like touch and pain flowing up from the skin, and motor commands *rat-a-tat*-tatting down to the muscles. At its uppermost extent the spinal cord pokes up out of its bony sheath of vertebrae, enters the skull, and grows thick and bulbous (Figure Int.3). This thickening is called the brainstem, and it is divided into three lobes: medulla, pons, and midbrain.

The medulla and nuclei (neural clusters) on the floor of the pons control important vital functions like breathing, blood pressure, and body temperature. A hemorrhage from even a tiny artery supplying this region can spell instant death. (Paradoxically, the higher areas of the brain can sustain comparatively massive damage and leave the patient alive and even fit. For example, a large tumor in the frontal lobe might produce barely detectable neurological symptoms.)

Sitting on the roof of the pons is the cerebellum (Latin for "little brain"), which controls the fine coordination of movements and is also involved in balance, gait, and posture. When your motor cortex (a higher brain region that issues voluntary movement commands) sends a signal to the muscles via the spinal cord, a copy of that signal—sort of like a cc email—gets sent to the cerebellum. The cerebellum also receives sensory feedback from muscle and joint receptors throughout the body. Thus the cerebellum is able to detect any mismatches that may occur between the intended action and the actual action, and in response can insert appropriate corrections into the outgoing motor signal. This sort of real-time, feedback-driven mechanism is called a servo-control loop. Damage to the cerebellum causes the loop to go into oscillation. For example, a patient may attempt to touch her nose, feel her hand overshooting, and attempt to compensate with an opposing motion, which causes her hand to overshoot even more wildly in the opposite direction. This is called an intention tremor.

Surrounding the top portion of the brainstem are the thalamus and the basal ganglia. The thalamus receives its major inputs from the sense organs and relays them to the sensory cortex for more sophisticated processing. Why we need a relay station is far from clear. The basal ganglia are a strangely shaped cluster of structures that are concerned with the control of automatic movements associated with complex volitional actions—for example, adjusting your shoulder when throwing a dart, or coordinating the force and tension in dozens of muscles throughout your body while you walk. Damage to cells in the basal ganglia results in disorders like Parkinson's disease, in which the patient's torso is stiff, his face is an expressionless mask, and he walks with a characteristic shuffling gait. (Our neurology professor in medical school used to diagnose Parkinson's by just listening to the patient's footsteps next door; if we couldn't do the same, he would fail us. Those were the days before high-tech medicine and magnetic resonance imaging, or MRI.) In contrast, excessive amounts of the brain chemical dopamine in the basal ganglia can lead to disorders known a choreas, which are characterized by uncontrollable movements that bear a superficial resemblance to dancing.

Finally we come to the cerebral cortex. Each cerebral hemisphere is subdivided into four lobes (see Figure Int.2): occipital, temporal, parietal, and frontal. These lobes have distinct domains of functioning, although in practice there is a great deal of interaction between them.

Broadly speaking, the occipital lobes are mainly concerned with visual processing. In fact, they are subdivided into as many as thirty distinct processing regions, each partially specialized for a different aspect of vision such as color, motion, and form.

The temporal lobes are specialized for higher perceptual functions, such as recognizing faces and other objects and linking them to appropriate emotions. They do this latter job in close cooperation with a structure called the amygdala ("almond"), which lies in the front ties (anterior poles) of the temporal lobes. Also tucked away beneath each temporal lobe is the hippocampus ("seahorse"), which lays down new memory traces. In addition to all this, the upper part of the left temporal lobe contains a patch of cortex known as Wernicke's area. In humans this area has ballooned to seven times the size of the same area in chimpanzees; it is one of the few brain areas that can be safely declared unique to our species. Its job is nothing less than the comprehension of meaning and the semantic aspects of language—functions that are prime differentiators between human beings and mere apes.

The parietal lobes are primarily involved in processing touch, muscle, and joint information from the body and combining it with vision, hearing, and balance to give you a rich "multimedia" understanding of your corporeal self and the world around it. Damage to the right parietal lobe commonly results in a phenomenon called hemispatial neglect: The patient loses awareness of the left half of visual space. Even more remarkable is somatoparaphrenia, the patient's vehement denial of ownership of her own left arm and insistence that it belongs to someone else. The parietal lobes have expanded greatly in human evolution, but no part of them has grown more than the inferior parietal lobules (IPL; see Figure Int.2). So great was this expansion that at some point in our past a large portion of it split into two new processing regions called the angular gyrus and the supramarginal gyrus. These uniquely human areas house some truly quintessential human abilities.

The right parietal lobe is involved in creating a mental model of the spatial layout of the outside world: your immediate environs, plus all the locations (but not identity) of objects, hazards, and people within it, along with your physical relationship to each of these things. Thus you can grab things, dodge missiles, and avoid obstacles. The right parietal, especially the right *superior* lobule (just above the IPL), is also responsible for constructing your body image—the vivid mental awareness you have of your body's configuration and movement in space. Note that even though it is called an "image," the body image is not a purely visual construct; it is also partly touch and muscle based. After all, a blind person has a body image too, and an extremely good one at that. In fact, if you zap the right angular gyrus with an electrode, you will have an out-of-body experience.

Now let's consider the left parietal lobe. The left angular gyrus is involved in important functions unique to humans such as arithmetic, abstraction, and aspects of language such as word finding and metaphor. The left supramarginal gyrus, on the other hand, conjures up a vivid image of intended skilled actions—for example, sewing with a needle, hammering a nail, or waving goodbye—and executes them. Consequently, lesions in the left angular gyrus eliminate abstract skills like reading, writing, and arithmetic, while injury to the left supramarginal gyrus hinders you from orchestrating skilled movements. When I ask you to salute, you conjure up a visual image of the salute and, in a sense, use the image to guide your arm movements. But if your left supramarginal gyrus is damaged, you will simply stare at your hand perplexed or flail it around. Even though it isn't paralyzed or weak and you clearly understand the command, you won't be able to make your hand respond to your intention.

The frontal lobes also perform several distinct and vital functions. Part of this region the motor cortex—the vertical strip of cortex running just in front of the big furrow in the middle of the brain (Figure Int.2)—is involved in issuing simple motor commands. Other parts are involved in planning actions and keeping goals in mind long enough to follow through on them. There is another small part of the frontal lobe that is required for holding things in memory long enough to know what to attend to. This faculty is called working memory or short-term memory.

So far so good. But when you move to the more anterior part of the frontal lobes you enter the most inscrutable terra incognita of the brain: the prefrontal cortex (parts of which are identified in Figure Int.2). Oddly enough, a person can sustain massive damage to this area and come out of it showing no obvious signs of any neurological or cognitive deficits. The patient may seem perfectly normal if you casually interact with her for a few minutes. Yet if you talk to her relatives, they will tell you that her personality has changed beyond recognition. "She isn't in there anymore. I don't even recognize this new person" is the sort of heart-wrenching statement you frequently hear from bewildered spouses and lifelong friends. And if you continue to interact with the patient for a few hours or days, you too will see that there is something profoundly deranged.

If the left prefrontal lobe is damaged, the patient may withdraw from the social world and show a marked reluctance to do anything at all. This is euphemistically called pseudodepression—"pseudo" because none of the standard criteria for identifying depression, such as feelings of bleakness and chronic negative thought patterns, are revealed by psychological or neurological probing. Conversely, if the right prefrontal lobe is damaged, a patient will seem euphoric even though, once again he really won't be. Cases of prefrontal damage are especially distressing to relatives. Such a patient seems to lose all interest in his own future and he shows no moral compunctions of any kind. He may laugh at a funeral or urinate in public. The great paradox is that he seems normal in most respects: his language, his memory, and even his IQ are unaffected. Yet he has lost many of the most quintessential attributes that define human nature: ambition, empathy, foresight, a complex personality, a sense of morality, and a sense of dignity as a human being. (Interestingly, a lack of empathy, moral standards, and self-restraint are also frequently seen in sociopaths, and the neurologist Antonio Damasio has pointed out they may have some clinically undetected frontal dysfunction.) For these reasons the prefrontal cortex has long been regarded as the "seat of humanity." As for the question of *how* such a relatively small patch of the brain manages to orchestrate such a sophisticated and elusive suite of functions, we are still very much at a loss.

Is it possible to isolate a given part of the brain, as Owen attempted, that makes our species unique? Not quite. There is no region or structure that appears to have been grafted into the brain *de novo* by an intelligent designer; at the anatomical level, every part of our brain has a direct analog in the brains of the great apes. However, recent research has identified a handful of brain regions that have been so radically elaborated that at the *functional* (or cognitive) level they actually can be considered novel and unique. I mentioned three of these areas above: Wernicke's area in the left temporal lobe, the prefrontal cortex, and the IPL in each parietal lobe. Indeed, the offshoots of the IPL—namely, the supramarginal and angular gyri, are anatomically nonexistent in apes. (Owen would have loved to have known about these.) The extraordinarily rapid development of these areas in humans suggests that *something* crucial must have been going on there, and clinical observations confirm this.

Within some of these regions, there is a special class of nerve cells called mirror neurons. These neurons fire not only when you perform an action, but also when you watch someone else perform the same action. This sounds so simple that its huge implications are easy to miss. What these cells do is effectively allow you to empathize with the other person and "read" her intentions—figure out what she is really up to. You do this by running a simulation of her actions using your own body image.

When you watch someone else reach for a glass of water, for example, your mirror neurons automatically simulate the same action in your (usually subconscious) imagination. Your mirror neurons will often go a step further and have you perform the action they *anticipate* the other person is about to take—say, to lift the water to her lips and take a drink. Thus you automatically form an assumption about her intentions and motivations—in this case, that she is thirsty and is taking steps to quench that thirst. Now, you could be wrong in this assumption—she might intend to use the water to douse a fire or to fling in the face of a boorish suitor—but usually your mirror neurons are reasonably accurate guessers of others' intentions. As such, they are the closest thing to telepathy that nature was able to endow us with.

These abilities (and the underlying mirror-neuron circuitry) are also seen in apes, but only in humans do they seem to have developed to the point of being able to model aspects of others' *minds* rather than merely their actions. Inevitably this would have required the development of additional connections to allow a more sophisticated deployment of such circuits in complex social situations. Deciphering the nature of these connections—rather than just saying, "It's done by mirror neurons"—is one of the major goals of current brain research.

It is difficult to overstate the importance of understanding mirror neurons and their function. They may well be central to social learning, imitation, and the cultural transmission of skills and attitudes—perhaps even of the pressed-together sound clusters we call "words." By hyper-developing the mirror-neuron system, evolution in effect turned culture into the new genome. Armed with culture, humans could adapt to hostile new environments and figure out how to exploit formerly inaccessible or poisonous food sources in just one or two generations—instead of the hundreds or thousands of generations such adaptations would have taken to accomplish through genetic evolution.

Thus culture became a significant new source of evolutionary pressure, which helped select for brains that had even better mirror-neuron systems and the imitative learning associated with them. The result was one of the many self-amplifying snowball effects that culminated in *Homo sapiens*, the ape that looked into its own mind and saw the whole cosmos reflected inside.

CHAPTER 1

Phantom Limbs and Plastic Brains

———

I love fools' experiments. I am always making them.

—CHARLES DARWIN

As a medical student I examined a patient named Mikhey during my neurology rotation. Routine clinical testing required me to poke her neck with a sharp needle. It should have been mildly painful, but with each poke she laughed out loud, saying it was ticklish. This, I realized, was the ultimate paradox: laughter in the face of pain, a microcosm of the human condition itself. I was never able to investigate Mikhey's case as I would have liked.

Soon after this episode, I decided to study human vision and perception, a decision largely influenced by Richard Gregory's excellent book *Eye and Brain*. I spent several years doing research on neurophysiology and visual perception, first at the University of Cambridge's Trinity College, and then in collaboration with Jack Pettigrew at Caltech.

But I never forgot the patients like Mikhey whom I had encountered during my neurology rotation as a medical student. In neurology, it seemed, there were so many questions left unresolved. Why did Mikhey laugh when poked? Why does the big toe go up when you stroke the outer border of the foot of a stroke patient? Why do patients with temporal lobe seizures believe they experience God and exhibit hypergraphia (incessant, uncontrollable writing)? Why do otherwise intelligent, perfectly lucid patients with damage to the right parietal lobe deny that their left arm belongs to them? Why does an emaciated anorexic with perfectly normal eyesight look in a mirror and claim she looks obese? And so, after years of specializing in vision, I returned to my first love:

neurology. I surveyed the many unanswered questions of the field and decided to focus on a specific problem: phantom limbs. Little did I know that my research would yield unprecedented evidence of the amazing plasticity and adaptability of the human brain.

It had been known for over a century that when a patient loses an arm to amputation, she may continue to feel vividly the presence of that arm—as though the arm's ghost were still lingering, haunting its former stump. There had been various attempts to explain this baffling phenomenon, ranging from flaky Freudian scenarios involving wish fulfillment to invocations of an immaterial soul. Not being satisfied with any of these explanations, I decided to tackle it from a neuroscience perspective.

I remember a patient named Victor on whom I conducted nearly a month of frenzied experiments. He came to see me because his left arm had been amputated below the elbow about three weeks prior to his visit. I first verified that there was nothing wrong with him neurologically: His brain was intact, his mind was normal. Based on a hunch I blindfolded him and started touching various parts of his body with a Q-tip, asking him to report what he felt, and where. His answers were all normal and correct until I started touching the left side of his face. Then something very odd happened.

He said, "Doctor, I feel that on my phantom hand. You're touching my thumb."

I used my knee hammer to stroke the lower part of his jaw. "How about now?" I asked.

"I feel a sharp object moving across the pinky to the palm," he said.

By repeating this procedure I discovered that there was an entire map of the missing hand on his face. The map was surprisingly precise and consistent, with fingers clearly delineated (Figure 1.1). On one occasion I pressed a damp Q-tip against his cheek and sent a bead of water trickling down his face like a tear. He felt the water move down his cheek in the normal fashion, but claimed he could also feel the droplet trickling down the length of his phantom arm. Using his right index finger, he even traced the meandering path of the trickle through the empty air in front of his stump. Out of curiosity I asked him to elevate his stump and point the phantom upward toward the ceiling. To his astonishment he felt the next drop of water flowing *up* along the phantom, defying the law of gravity.

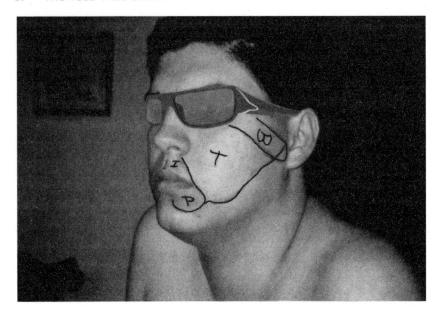

FIGURE 1.1   A patient with a phantom left arm. Touching different parts of his face evoked sensations in different parts of the phantom: P, pinky; T, thumb; B, ball of thumb; I, index finger.

Victor said he had never discovered this virtual hand on his face before, but as soon as he knew about it he found a way to put it to good use: Whenever his phantom palm itches—a frequent occurrence that used to drive him crazy—he says he can now relieve it by scratching the corresponding location on his face.

Why does all this happen? The answer, I realized, lies in the brain's anatomy. The entire skin surface of the left side of the body is mapped onto a strip of cortex called the postcentral gyrus (see Figure Int.2 in the Introduction) running down the right side of the brain. This map is often illustrated with a cartoon of a man draped on the brain surface (Figure 1.2). Even though the map is accurate for the most part, some portions of it are scrambled with respect to the body's actual layout. Notice how the map of the face is located next to the map of the hand instead of being near the neck where it "should" be. This provided the clue I was looking for.

Think of what happens when an arm is amputated. There is no longer an arm, but there is still a *map* of the arm in the brain. The job of this map, its raison d'être, is to represent its arm. The arm may be gone

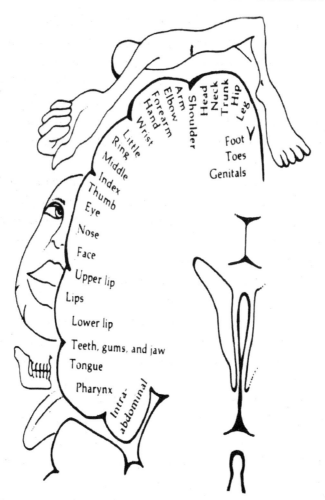

FIGURE 1.2   The Penfield map of the skin surface on the postcentral gyrus (see Figure Int.2). The drawing shows a coronal section (roughly, a cross section) going through the middle of the brain at the level of the postcentral gyrus. The artist's whimsical depiction of a person draped on the brain surface shows the exaggerated representations of certain body parts (face and hand) and the fact that the hand map is above the face map.

but the brain map, having nothing better to do, soldiers on. It keeps representing the arm, second by second, day after day. This map persistence explains the basic phantom limb phenomenon—why the felt presence of the limb persists long after the flesh-and-blood limb has been severed.

Now, how to explain the bizarre tendency to attribute touch sensations arising from the face to the phantom hand? The orphaned brain map continues to represent the missing arm and hand in absentia, but it is not receiving any actual touch inputs. It is listening to a dead channel, so to speak, and is hungry for sensory signals. There are two possible explanations for what happens next. The first is that the sensory input flowing from the facial skin to the face map in the brain begins to actively invade the vacated territory corresponding the missing hand. The nerve fibers from the facial skin that normally project to the face cortex sprout thousands of neural tendrils that creep over into the arm map and establish strong, new synapses. As a result of this cross-wiring, touch signals applied to the face not only activate the face map, as they normally do, but also activate the hand map in the cortex, which shouts "hand!" to higher brain areas. The net result is that the patient feels that his phantom hand is being touched every time his face is touched.

A second possibility is that even prior to amputation, the sensory input from the face not only gets sent to the face area but partially encroaches into the hand region, almost as if they are reserve troops ready to be called into action. But these abnormal connections are ordinarily silent; perhaps they are continuously inhibited or damped down by the normal baseline activity from the hand itself. Amputation would then unmask these ordinarily silent synapses so that touching the face activates cells in the hand area of the brain. That in turn causes the patient to experience the sensations as arising from the missing hand.

Independent of which of these two theories—sprouting or unmasking—is correct, there is an important take-home message. Generations of medical students were told that the brain's trillions of neural connections are laid down in the fetus and during early infancy and that adult brains lose their ability to form new connections. This lack of plasticity—this lack of ability to be reshaped or molded—was often used as an excuse to tell patients why they could expect to recover very little function after a stroke or traumatic brain injury. Our observations flatly contradicted this dogma by showing, for the first time, that even the basic sensory maps in the adult human brain can change over distances of several centimeters. We were then able to use brain-imaging techniques to show directly that our theory was correct: Victor's brain maps had indeed changed as predicted (Figure 1.3).

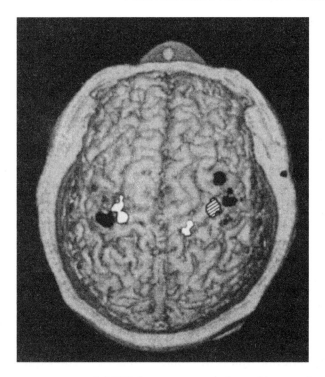

FIGURE 1.3   A MEG (magnetoencephalograph) map of
the body surface in a right-arm amputee. Hatched area,
hand; black areas, face; white areas, upper arm. Notice that
the region corresponding to the right hand (hatched area)
is missing from the left hemisphere, but this region gets
activated by touching the face or upper arm.

   Soon after we published, evidence confirming and extending these
findings started to come in from many groups. Two Italian researchers, Giovanni Berlucchi and Salvatore Aglioti, found that after amputation of a finger there was a "map" of a single finger draped neatly
across the face as expected. In another patient the trigeminal nerve (the
sensory nerve supplying the face) was severed and soon a map of the
face appeared on the palm: the exact converse of what we had seen.
Finally, after amputation of the foot of another patient, sensations from
the penis were felt in the phantom foot. (Indeed, the patient claimed
that his orgasm spread into his foot and was therefore "much bigger than it used to be.") This occurs because of another of these odd

discontinuities in the brain's map of the body: The map of the genitals is right next to the map of the foot.

MY SECOND EXPERIMENT on phantom limbs was even simpler. In a nutshell, I created a simple setup using ordinary mirrors to mobilize paralyzed phantom limbs and reduce phantom pain. To understand how this works, I first need to explain why some patients are able to "move" their phantoms but others are not.

Many patients with phantoms have a vivid sense of being able to move their missing limbs. They say things like "It's waving goodbye" or "It's reaching out to answer the phone." Of course, they know perfectly well that their hands aren't really doing these things—they aren't delusional, just armless—but subjectively they have a realistic sensation that they *are* moving the phantom. Where do these feelings come from?

I conjectured that they were coming from the motor command centers in the front of the brain. You might recall from the Introduction how the cerebellum fine-tunes our actions through a servo-loop process. What I didn't mention is that the parietal lobes also participate in this servo-loop process through essentially the same mechanism. Again briefly: Motor output signals to the muscles are (in effect) cc'ed to the parietal lobes, where they are compared to sensory feedback signals from the muscles, skin, joints, and eyes. If the parietal lobes detect any mismatches between the intended movements and the hand's actual movements, they make corrective adjustments to the next round of motor signals. You use this servo-guided system all the time. This is what allows you, for instance, to maneuver a heavy juice pitcher into a vacant spot on the breakfast table without spilling or knocking over the surrounding tableware. Now imagine what happens if the arm is amputated. The motor command centers in the front of the brain don't "know" the arm is gone—they are on autopilot—so they continue to send motor command signals to the missing arm. By the same token, they continue to cc these signals to the parietal lobes. These signals flow into the orphaned, input-hungry hand region of your body-image center in the parietal lobe. These cc'ed signals from motor commands are misinterpreted by the brain as actual movements of the phantom.

Now you may wonder why, if this is true, you don't experience the same sort of vivid phantom movement when you *imagine* moving your hand while deliberately holding it still. Here is the explanation I proposed several years ago, which has been since confirmed by brain-imaging studies. When your arm is intact, the sensory feedback from the skin, muscles, and joint sensors in your arm, as well as the visual feedback from your eyes, are all testifying in unison that your arm is not in fact moving. Even though your motor cortex is sending "move" signals to your parietal lobe, the countervailing testimony of the sensory feedback acts as a powerful veto. As a result, you don't experience the imagined movement as though it were real. If the arm is gone, however, your muscles, skin, joints, and eyes cannot provide this potent reality check. Without the feedback veto, the strongest signal entering your parietal lobe is the motor command to the hand. As a result, you experience actual movement sensations.

Moving phantom limbs is bizarre enough, but it gets even stranger. Many patients with phantom limbs report the exact opposite: Their phantoms are paralyzed. "It's frozen, Doctor." "It's in a block of cement." For some of these patients the phantom is twisted into an awkward, extremely painful position. "If only I could move it," a patient once told me, "it might help alleviate the pain."

When I first saw this, I was baffled. It made no sense. They had lost their limbs, but the sensory-motor connections in their brains were presumably the same as they had been before their amputations. Puzzled, I started examining some of these patients' charts and quickly found the clue I was looking for. Prior to amputation, many of these patients had had real paralysis of their arm caused by a peripheral nerve injury: the nerve that used to innervate the arm had been ripped out of the spinal cord, like a phone cord being yanked out of its wall jack, by some violent accident. So the arm had lain intact but paralyzed for many months prior to amputation. I started to wonder if perhaps this period of real paralysis could lead to a state of learned paralysis, which I conjectured could come about in the following way.

During the preamputation period, every time the motor cortex sent a movement command to the arm, the sensory cortex in the parietal lobe would receive negative feedback from the muscles, skin, joints, and eyes. The entire feedback loop had gone dead. Now, it is well established that experience modifies the brain by strengthening or weakening the synapses that link neurons together. This modification process is known as learning. When patterns are constantly reinforced—when the brain sees that event B invariably follows event A, for instance—the synapses between the neurons that represent A and the neurons that represent B are strengthened. On the other hand, if A and B stop having any apparent relationship to each other, the neurons that represent A and B will shut down their mutual connections to reflect this new reality.

So here we have a situation where the motor cortex was continually sending out movement commands to the arm, which the parietal lobe continually saw as having absolutely zero muscular or sensory effect. The synapses that used to support the strong correlations between motor commands and the sensory feedback they should generate were shown to be liars. Every new, impotent motor signal reinforced this trend, so the synapses grew weaker and weaker and eventually became moribund. In other words, the paralysis was learned by the brain, stamped into the circuitry where the patient's body image was constructed. Later, when the arm was amputated, the learned paralysis got carried over into the phantom so the phantom felt paralyzed.

How could one test such an outlandish theory? I hit on the idea of constructing a mirror box (Figure 1.4). I placed an upright mirror in the center of a cardboard box whose top and front had been removed. If you stood in front of the box, held your hands on either side of the mirror and looked down at them from an angle, you would see the reflection of one hand precisely superimposed on the *felt* location of your other hand. In other words, you would get the vivid but false impression that you were looking at both of your hands; in fact, you would only be looking at one actual hand and one reflection of a hand.

If you have two normal, intact hands, it can be entertaining to play around with this illusion in the mirror box. For example, you can move your hands synchronously and symmetrically for a few moments—pretending to conduct an orchestra works well—and then suddenly move them in different ways. Even though you know it's an illusion, a jolt of mild surprise invariably shoots through your mind when you do this. The surprise comes from the sudden mismatch between two streams of feedback: The skin-and-muscle feedback you get from the hand behind the mirror says one thing, but the visual feedback you get from the reflected hand—which your parietal lobe had become convinced is the hidden hand itself—reports some other movement.

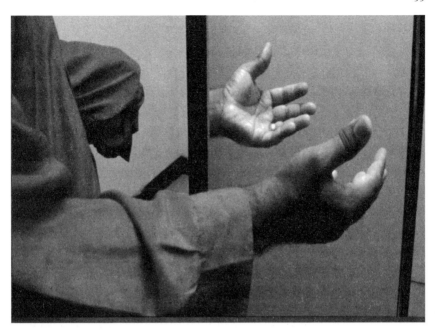

FIGURE 1.4   The mirror arrangement for animating the phantom limb. The patient "puts" his paralyzed and painful phantom left arm behind the mirror and his intact right hand in front of the mirror. If he then views the mirror reflection of the right hand by looking into the right side of the mirror, he gets the illusion that the phantom has been resurrected. Moving the real hand causes the phantom to appear to move, and it then feels like it is moving—sometimes for the first time in years. In many patients this exercise relieves the phantom cramp and associated pain. In clinical trials, mirror visual feedback has also been shown to be more effective than conventional treatments for chronic regional pain syndrome and paralysis resulting from stroke.

Now let's look at what this mirror-box setup does for a person with a paralyzed phantom limb. The first patient we tried this on, Jimmie, had an intact right arm, phantom left arm. His phantom jutted like a mannequin's resin-cast forearm out of his stump. Far worse, it was also subject to painful cramping that his doctors could do nothing about. I showed him the mirror box and explained to him this might seem like a slightly off-the-wall thing we were about to try, with no guarantee that it would have any effect, but he was cheerfully willing to give it a try. He held out his paralyzed phantom on the left side of the mirror, looked into the right side of the box and carefully positioned his right hand so that its image was congruent with (superimposed on) the felt position of the phantom. This immediately gave him the startling visual impression that the phantom had been resurrected. I then asked him to perform mirror-symmetric movements of both arms and hands while he continued looking into the mirror. He cried out, "It's like it's plugged back in!" Now he not only had a vivid impression that the phantom was obeying his commands, but to his amazement, it began to relieve his painful phantom spasms for the first time in years. It was as though the mirror visual feedback (MVF) had allowed his brain to "unlearn" the learned paralysis.

Even more remarkably, when one of our patients, Ron, took the mirror box home and played around with it for three weeks in his spare time, his phantom limb vanished completely, along with the pain. All of us were shocked. A simple mirror box had exorcised a phantom. How? No one has proven the mechanism yet, but here is how I suspect it works. When faced with such a welter of conflicting sensory inputs—no joint or muscle feedback, impotent copies of motor-command signals, and now discrepant visual feedback thrown in via the mirror box—the brain just gives up and says, in effect, "To hell with it; there is no arm." The brain resorts to denial. I often tell my medical colleagues that this is the first case in the history of medicine of a successful amputation of a phantom limb. When I first observed this disappearance of the phantom using MVF, I myself didn't quite believe it. The notion that you could amputate a phantom with a mirror seemed outlandish, but it has now been replicated by other groups of researchers, especially Herta Flor, a neuroscientist at the University of Heidelberg. The reduction of phantom pain has also been confirmed by Jack Tsao's group at the at the Walter Reed Army Medical Center in Maryland. They conducted a placebo-controlled clinical study on 24 patients (including 16 placebo controls). The phantom pain vanished after just three weeks in the 8 patients using the mirror, whereas none of the control patients (who used Plexiglas and visual imagery instead of mirrors) showed any improvement. Moreover, when the control patients were switched over to the mirror, they showed the same substantial pain reduction as the original experimental group.

More important, MVF is now being used for accelerating recovery from paralysis following stroke. My postdoctoral colleague Eric Altschuler and I first reported this in *The Lancet* in 1998, but our sample size was small—just 9 patients. A German group led by Christian Dohle has recently tried the technique on 50 stroke patients in a triple-blind controlled study, and shown that a majority of them regained both sensory and motor functions. Given that one in six people will suffer from a stroke, this is an important discovery.

More clinical applications for MVF continue to emerge. One pertains to a curious pain disorder with an equally curious name—complex regional pain syndrome–Type II (CRPS-II)—which is simply a verbal smoke screen for "Sounds awful! I have no idea what it is." Whatever you call it, this affliction is actually quite common: It manifests in about 10 percent of stroke victims. The better-known variant of the disorder occurs after a minor injury such as an ordinarily innocuous hairline fracture in one of the metacarpals (hand bones). There is initially pain, of course, as one would expect to accompany a broken hand. Ordinarily the pain gradually goes away as the bone heals. But in an unfortunate subset of patients this doesn't happen. They end up with chronic, excruciating pain that is unrelenting and persists indefinitely long after the original wound has healed. There is no cure—or at least, that's what I had been taught in medical school.

It occurred to me that an evolutionary approach to this problem might be useful. We usually think of pain as a single thing, but from a functional point of view there are at least two kinds of pain. There is acute pain—as when you accidentally put your hand on a hot stove, yelp, and yank your hand away—and then there is chronic pain: pain that persists or recurs over long or indefinite periods, such as might accompany a bone fracture in the hand. Although the two feel the same (painful), they have different biological functions and different evolutionary origins. Acute pain causes you to instantly remove your hand from the stove to prevent further tissue damage. Chronic pain motivates you to keep your fractured hand immobilized to prevent reinjury while it heals.

I began to wonder: If learned paralysis could explain immobilized phantoms, perhaps CRPS-II is a form of "learned pain." Imagine a patient with a fractured hand. Imagine how, during his long convalescence, pain shoots through his hand every time he moves it. His brain is seeing a constant "if A then B" pattern of events, where A is movement and B is pain. Thus the synapses between the various neurons that represent these two events are strengthened daily—for months on end. Eventually the very attempt to move the hand elicits excruciating pain. This pain may even spread to the arm, causing it to freeze up. In some such cases, the arm not only develops paralysis but actually becomes swollen and inflamed, and in the case of Sudek's atrophy the bone may even start atrophying. All of this can be seen as a strange manifestation of mind-body interactions gone horribly awry.

At the "Decade of the Brain" symposium that I organized at the University of California, San Diego, in October 1996, I suggested that the mirror box might help alleviate learned pain in the same way that it affects phantom pain. The patient could try moving her limbs in synchrony while looking in the mirror, creating the illusion that the afflicted arm is moving freely, with no pain being evoked. Watching this repeatedly may lead to an "unlearning" of learned pain. A few years later the mirror box was tested by two research groups and found to be effective in treating CRPS-II in a majority of patients. Both studies were conducted double-blind using placebo controls. To be honest I was quite surprised. Since that time, two other double-blind randomized studies have confirmed the striking effectiveness of the procedure. (There is a variant of CRPS-II seen in 15 percent of stroke victims, and the mirror is effective in them as well.)

I'll mention one last observation on phantom limbs that is even more remarkable than the cases mentioned so far. I used the conventional mirror box but added a novel twist. I had the patient, Chuck, looking at the reflection of his intact limb so as to optically resurrect the phantom as before. But this time, instead of asking him to move his arm, I asked him to hold it steady while I put a minifying (image-shrinking) concave lens between his line of sight and the mirror reflection. From Chuck's point of view, his phantom now appeared to be about one-half or one-third its "real" size.

Chuck looked surprised and said, "It's amazing, Doctor. My phantom not only looks small but feels small as well. And guess what—the pain has shrunk too! Down to about one-fourth the intensity it was before."

This raises the intriguing question of whether even real pain in a real arm evoked with a pinprick would also be diminished by optically shrinking the pin and the arm. In several of the experiments I just described, we saw just how potent a factor vision (or its lack) can be in influencing phantom pain and motor paralysis. If this sort of optically mediated anesthesia could be shown to work on an intact hand, it would be another astonishing example of mind-body interaction.

———

IT IS FAIR to say that these discoveries—together with the pioneering animal studies of Mike Merzenich and John Kaas and some ingenious clinical work by Leonardo Cohen and Paul Bach y Rita—ushered in a whole new era in neurology, and in neurorehabilitation especially. They led to a radical shift in the way we think about the brain. The old view, which prevailed through the 1980s, was that the brain consists of many specialized modules that are hardwired from birth to perform specific jobs. (The box-and-arrow diagrams of brain connectivity in anatomy textbooks have fostered this highly misleading picture in the minds of generations of medical students. Even today, some textbooks continue to represent this "pre-Copernican" view.)

But starting in the 1990s, this static view of the brain was steadily supplanted by a much more dynamic picture. The brain's so-called modules don't do their jobs in isolation; there is a great deal of back-and-forth interaction between them, far more than previously suspected. Changes in the operation of one module—say, from damage, or from maturation, or from learning and life experience—can lead to significant changes in the operations of many other modules to which it is connected. To a surprising extent, one module can even take over the functions of another. Far from being wired up according to rigid, prenatal genetic blueprints, the brain's wiring is highly malleable—and not just in infants and young children, but throughout every adult lifetime. As we have seen, even the basic "touch" map in the brain can be modified over relatively large distances, and a phantom can be "amputated" with a mirror. We can now say with confidence that the brain is an extraordinarily plastic biological system that is in a state of dynamic equilibrium with the external world. Even its basic connections are being constantly updated in response to changing sensory demands. And if you take mirror neurons into account, then we can infer that your brain is also in synch with other brains—analogous to a global Internet of Facebook pals constantly modifying and enriching each other.

As remarkable as this paradigm shift was, and leaving aside its vast clinical importance, you may be wondering at this point what these tales of phantom limbs and plastic brains have to do with human uniqueness. Is lifelong plasticity a distinctly human trait? In fact, it is not. Don't lower primates get phantom limbs? Yes, they do. Don't their cortical limb and face representations remap following amputation? Definitely. So what does plasticity tell us about our uniqueness?

The answer is that lifelong plasticity (not just genes) is one of the central players in the evolution of human uniqueness. Through natural selection our brains evolved the ability to exploit learning and culture to drive our mental phase transitions. We might as well call ourselves *Homo plasticus*. While other animal brains exhibit plasticity, we are the only species to use it as a central player in brain refinement and evolution. One of the major ways we managed to leverage neuroplasticity to such stratospheric heights is known as neoteny—our almost absurdly prolonged infancy and youth, which leaves us both hyperplastic and hyperdependent on older generations for well over a decade. Human childhood helps lay the groundwork of the adult mind, but plasticity remains a major force throughout life. Without neoteny and plasticity, we would still be naked savanna apes—without fire, without tools, without writing, lore, beliefs, or dreams. We really would be "nothing but" apes, instead of aspiring angels.

INCIDENTALLY, EVEN THOUGH I was never able to directly study Mikhey—the patient I met as a medical student who laughed when she should have yelped in pain—I never stopped pondering her case. Mikhey's laughter raises an interesting question: Why does anybody laugh at anything? Laughter—and its cognitive companion, humor—is a universal trait present in all cultures. Some apes are known to "laugh" when tickled, but I doubt if they would laugh upon seeing a portly ape slip on a banana peel and fall on his arse. Jane Goodall certainly has never reported anything about chimpanzees performing pantomime skits for each other à la the Three Stooges or the Keystone Kops. Why and how humor evolved in us is a mystery, but Mikhey's predicament gave me a clue.

Any joke or humorous incident has the following form. You narrate a story step-by-step, leading your listener along a garden path of expectation, and then you introduce an unexpected twist, a punch line, the comprehension of which requires a complete reinterpretation of the preceding events. But that's not enough: No scientist whose theoretical edifice is demolished by a single ugly fact entailing a complete overhaul is likely to find it amusing. (Believe me, I've tried!) Deflation of expectation is necessary but not sufficient. The extra key ingredient is that the new interpretation must be inconsequential. Let me illustrate. The dean of the medical school starts walking along a path, but before reaching his destination he slips on a banana peel and falls. If his skull is fractured and blood starts gushing out, you rush to his aid and call the ambulance. You don't laugh. But if he gets up unhurt, wiping the banana off his expensive trousers, you break out into a fit of laughter. It's called slapstick. The key difference is that in the first case, there is a true alarm requiring urgent attention. In the second case it's a *false* alarm, and by laughing you inform your kin in the vicinity not to waste their resources rushing to his aid. It is nature's "all's okay" signal. What is left unexplained is the slight *schadenfreude* aspect to the whole thing.

How does this explain Mikhey's laughter? I didn't know this at that time, but many years later I saw another patient named Dorothy with a similar "laughter from pain" syndrome. A CT (computed tomography) scan revealed that one of the pain pathways in her brain was damaged. Even though we think of pain as a single sensation, there are in fact several layers to it. The sensation of pain is initially processed in a small structure called the insula ("island"), which is folded deep beneath the temporal lobe on each side of the brain (see Figure Int.2, in the Introduction). From the insula the pain information is then relayed to the anterior cingulate in the frontal lobes. It is *here* you feel the actual unpleasantness—the agony and the awfulness of the pain—along with an expectation of danger. If this pathway is cut, as it was in Dorothy and presumably in Mikhey, the insula continues to provide the basic sensation of pain but it doesn't lead to the expected awfulness and agony: The anterior cingulate doesn't get the message. It says, in effect, "all's okay." So here we have the two key ingredients for laughter: A palpable and imminent indication that alarm is warranted (from the insula) followed by a "no big whoop" follow-up (from the silence of the anterior cingulate). So the patient laughs uncontrollably.

And the same holds for tickling. The huge adult approaches the child menacingly. She is clearly outmatched, prey, completely at the mercy of a hulking Grendel. Some instinctive part of her—her inner primate, primed to flee from the terrors of eagles and jaguars and pythons (oh my!)—cannot help but interpret the situation this way. But then the monster turns out be gentle. It deflates her expectation of danger. What might have been fangs and claws digging fatally into her ribs turn out to be nothing but firmly undulating fingers. And the child laughs. It may well be that tickling evolved as a early playful rehearsal for adult humor.

The false-alarm theory explains slapstick, and it is easy to see how it might have been evolutionarily coopted (exapted, to use the technical term) for *cognitive* slapstick—jokes, in other words. Cognitive slapstick may similarly serve to deflate falsely evoked expectations of danger which might otherwise result in resources being wasted on imaginary dangers. Indeed, one could go so far as to say that humor helps as an effective antidote against a useless struggle against the ultimate danger: the ever-present fear of death in self-conscious beings like us.

Lastly, consider that universal greeting gesture in humans: the smile. When an ape is approached by another ape, the default assumption is that it is being approached by a potentially dangerous stranger, so it signals its readiness to fight by protruding its canines in a grimace. This evolved further and became ritualized into a mock threat expression, an aggressive gesture warning the intruder of potential retaliation. But if the approaching ape is recognized as a friend, the threat expression (baring canines) is aborted halfway, and this halfway grimace (partly hiding the canines) becomes an expression of appeasement and friendliness. Once again a potential threat (attack) is abruptly aborted—the key ingredients for laughter. No wonder a smile has the same subjective feeling as laughter. It incorporates the same logic and may piggyback on the same neural circuits. How very odd that when your lover smiles at you, she is in fact half-baring her canines, reminding you of her bestial origins.

And so it is that we can begin with a bizarre mystery that could have come straight from Edgar Allan Poe, apply Sherlock Holmes's methods, diagnose and explain Mikhey's symptoms, and, as a bonus, illuminate the possible evolution and biological function of a much treasured but deeply enigmatic aspect of the human mind.

CHAPTER 2

Seeing and Knowing

———

"You see but you do not observe."

—SHERLOCK HOLMES

THIS CHAPTER IS ABOUT VISION. OF COURSE, EYES AND VISION ARE
not unique to humans—not by a long shot. In fact, the ability to see is so
useful that eyes have evolved many separate times in the history of life.
The eyes of the octopus are eerily similar to our own, despite the fact that
our last common ancestor was a blind aquatic slug- or snail-like creature
that lived well over half a billion years ago.[1] Eyes are not unique to us,
but vision does not occur in the eye. It occurs in the brain. And there is
no other creature on earth that sees objects quite the way we do. Some
animals have much higher visual acuity than we do. You sometimes hear
factoids like the fact that an eagle could read tiny newsprint from fifty
feet away. But of course, eagles can't read.

This book is about what makes humans special, and a recurring
theme is that our unique mental traits must have evolved from preexisting brain structures. We begin our journey with visual perception, partly
because more is known about its intricacies than about any other brain
function and partly because the development of visual areas accelerated
greatly in primate evolution, culminating in humans. Carnivores and
herbivores probably have fewer than a dozen visual areas and no color
vision. The same holds for our own ancestors, tiny nocturnal insectivores
scurrying up tree branches, little realizing that their descendents would
one day inherit—and possibly annihilate!—the earth. But humans have
as many as thirty visual areas instead of a mere dozen. What are they
doing, given that a sheep can get away with far fewer?

When our shrewlike ancestors became diurnal, evolving into pro-simians and monkeys, they began to develop extrasophisticated visuomotor capacities for precisely grasping and manipulating branches, twigs, and leaves. Furthermore, the shift in diet from tiny nocturnal insects to red, yellow, and blue fruits, as well as to leaves whose nutritional value was color coded in various shades of green, brown, and yellow, propelled the emergence of a sophisticated system for color vision. This rewarding aspect of color perception may have subsequently been exploited by female primates to advertise their monthly sexual receptivity and ovulation with estrus—a conspicuous colorful swelling of the rumps to resemble ripe fruits. (This feature has been lost in human females, who have evolved to be continuously receptive sexually throughout the month—something I have yet to observe personally.) In a further twist, as our ape ancestors evolved toward adopting a full-time upright bipedal posture, the allure of swollen pink rumps may have been transferred to plump lips. One is tempted to suggest—tongue in cheek—that our predilection for oral sex may also be an evolutionary throwback to our ancestors' days as frugivores (fruit eaters). It is an ironic thought that our enjoyment of a Monet or a Van Gogh or of Romeo's savoring Juliet's kiss may ultimately trace back to an ancient attraction to ripe fruits and rumps. (This is what makes evolutionary psychology so much fun: You can come up with an outlandishly satirical theory and get away with it.)

In addition to the extreme agility of our fingers, the human thumb developed a unique saddle joint allowing it to oppose the forefinger. This feature, which enables the so-called precision grip, may seem trivial, but it is useful for picking small fruits, nuts, and insects. It also turns out to be quite useful for threading needles, hafting hand axes, counting, or conveying Buddha's peace gesture. The requirement for fine independent finger movements, opposable thumbs, and exquisitely precise eye-hand coordination—the evolution of which was set in motion early in the primate line—may have been the final source of selection pressure that led us to develop our plethora of sophisticated visual and visuomotor areas in the brain. Without all these areas, it is arguable whether you could blow a kiss, write, count, throw a dart, smoke a joint, or—if you are a monarch—wield a scepter.

This link between action and perception has become especially clear in the last decade with the discovery of a new class of neurons in the frontal lobes called canonical neurons. These neurons are similar in some respects to the mirror neurons I introduced in the last chapter. Like mirror neurons, each canonical neuron fires during the performance of a specific action such as reaching for a vertical twig or an apple. But the same neuron will also fire at the mere *sight* of a twig or an apple. In other words, it is as though the abstract property of *graspability* were being encoded as an intrinsic aspect of the object's visual shape. The distinction between perception and action exists in our ordinary language, but it is one that the brain evidently doesn't always respect.

While the line between visual perception and prehensile action became increasingly blurred in primate evolution, so too did the line between visual perception and visual imagination in human evolution. A monkey, a dolphin, or a dog probably enjoys some rudimentary form of visual imagery, but only humans can create symbolic visual tokens and juggle them around in the mind's eye to try out novel juxtapositions. An ape can probably conjure up a mental picture of a banana or the alpha male of his troop, but only a human can mentally juggle visual symbols to create novel combinations, such as babies sprouting wings (angels) or beings that are half-horse, half-human (centaurs). Such imagery and "off-line" symbol juggling may, in turn, be a requirement for another unique human trait, language, which we take up in Chapter 6.

IN 1988 A sixty-year-old man was taken to the emergency room of a hospital in Middlesex, England. John had been a fighter pilot World War II. Until that fateful day, when he suddenly developed severe abdominal pain and vomiting, he had been in perfect health. The house officer, Dr. David McFee, elicited a history of the illness. The pain had begun near the navel and then migrated to the lower right side of his abdomen. This sounded to Dr. McFee like a textbook case of appendicitis: an inflammation of a tiny vestigial appendage protruding from the colon on the right side of the body. In the fetus the appendix first starts growing directly under the navel, but as the intestines lengthen and become convoluted the appendix gets pushed into the lower right quadrant of the abdomen. But the brain remembers its original location, so that is where it experiences the initial pain—under the belly button. Soon the inflammation spreads to the abdominal wall overlying it. That's when the pain migrates to the right.

Next Dr. McFee elicited a classic sign called rebound tenderness. With three fingers he very slowly compressed the lower right abdominal wall and noted that this caused no pain. But when he suddenly withdrew his hand to release the pressure, there was a short delay followed by sudden pain. This delay results from the inertial lag of the inflamed appendix as it rebounds to hit the abdominal wall.

Finally, Dr. McFee applied pressure in John's lower left quadrant, causing him to feel a sharp twinge of pain in the lower right, the true location of the appendix. The pain is caused by the pressure displacing the gas from the left to the right side of the colon, which causes the appendix to inflate slightly. This tell-tale sign, together with John's high fever and vomiting, clinched the diagnosis. Dr. McFee scheduled the appendectomy right away: The swollen, inflamed appendix could rupture anytime and spill its contents into the abdominal cavity, producing life-threatening peritonitis. The surgery went smoothly, and John was moved to the recovery room to rest and recuperate.

Alas, John's real troubles had only just begun.[2] What should have been a routine recovery became a waking nightmare when a small clot from a vein in his leg was released into his blood and clogged up one of his cerebral arteries, causing a stroke. The first sign of this was when his wife walked into the room. Imagine John's astonishment—and hers—when he could no longer recognize her face. The only way he knew who he was talking to was because he could still recognize her voice. Nor could he recognize anyone else's face—not even his own face in a mirror.

"I know it's me," he said. "It winks when I wink and it moves when I do. It's obviously a mirror. But it doesn't look like me."

John emphasized repeatedly that there was nothing wrong with his eyesight.

"My vision is fine, Doctor. Things are out of focus in my mind, not in my eye."

Even more remarkably, he couldn't recognize familiar objects.

When shown a carrot, he said, "It's a long thing with a tuft at the end—a paint brush?"

He was using fragments of the object to intellectually deduce what it was instead of recognizing it instantly as a whole like most of us do.

When shown a picture of a goat, he described it as "an animal of some kind. Maybe a dog." Often John could perceive the generic class the object belonged to—he could tells animals from plants, for example—but could not say what specific exemplar of that class it was. These symptoms were not caused by any limitation of intellect or verbal sophistication. Here is John's description of a carrot, which I'm sure you will agree is much more detailed than what most of us could produce:

> A carrot is a root vegetable cultivated and eaten as human consumption worldwide. Grown from seed as an annual crop, the carrot produces long thin leaves growing from a root head. This is deep growing and large in comparison with the leaf growth, sometimes gaining a length of twelve inches under a leaf top of similar height when grown in good soil. Carrots may be eaten raw or cooked and can be harvested during any size or state of growth. The general shape of a carrot is an elongated cone, and its color ranges between red and yellow.

John could no longer identify objects, but he could still deal with them in terms of their spatial extent, their dimensions, and their movement. He was able to walk around the hospital without bumping into obstacles. He could even drive short distances with some help—a truly amazing feat, given all the traffic he had to negotiate. He could locate and gauge the approximate speed of a moving vehicle, although he couldn't tell if it was a Jaguar, a Volvo, or even a truck. These distinctions prove to be irrelevant to actually driving.

When he reached home, he saw an engraving of St. Paul's Cathedral that had been hanging on the wall for decades. He said he knew someone had given it to him but had forgotten what it depicted. He could produce an astonishingly accurate drawing, copying its every detail—including printing flaws! But even after he had done so, he still couldn't say what it was. John could see perfectly clearly; he just didn't know what he was seeing—which is why the flaws weren't "flaws" for him.

John had been an avid gardener prior to his stroke. He walked out of his house and much to his wife's surprise picked up a pair of shears and proceeded to trim the hedge effortlessly. However, when he tried to tidy up the garden, he often plucked the flowers from the ground because he couldn't tell them from the weeds. Trimming the hedge, on the other hand, required only that John see where the unevenness was. No identification of objects was required. The distinction between seeing and knowing is illustrated well by John's predicament.

Although an inability to know what he was looking at was John's main problem, he had other subtler difficulties as well. For instance he had tunnel vision, often losing the proverbial forest for the trees. He could reach out and grab a cup of coffee when it was on an uncluttered table by itself, but got hopelessly muddled when confronted with a buffet service. Imagine his surprise when he discovered he had poured mayonnaise rather than cream into his coffee.

Our perception of the world ordinarily seems so effortless that we tend to take it for granted. You look, you see, you understand—it seems as natural and inevitable as water flowing downhill. Its only when something goes wrong, as in patients like John, that we realize how extraordinarily sophisticated it really is. Even though our picture of the world seems coherent and unified, it actually emerges from the activity those thirty (or more) different visual areas in the cortex, each of which mediates multiple subtle functions. Many of these areas are ones we share with other mammals but some of them "split" off at some point to become newly specialized modules in higher primates. Exactly how many of our visual areas are unique to humans isn't clear. But a great deal more is known about them than about other higher brain regions such as the frontal lobes, which are involved in such things as morality, compassion, and ambition. A thorough understanding of how the visual system really works may therefore provide insights into the more general strategies the brain uses to handle information, including the ones that are unique to us.

A FEW YEARS ago I was at an after-dinner speech given by David Attenborough at the university aquarium in La Jolla, California, near where I work. Sitting next to me was a distinguished-looking man with a walrus moustache. After his fourth glass of wine he told me that he worked for the creation science institute in San Diego. I was very tempted to tell him that creation science is an oxymoron, but before I could do so he interrupted me to ask where I worked and what I was currently interested in.

"Autism and synesthesia these days. But I also study vision."

"Vision? What's there to study?"

"Well, what do you think goes on in your head when you look at something—that chair for example?"

"There is an optical image of the chair in my eye—on my retina. The image is transmitted along a nerve to the visual area of the brain and you see it. Of course, the image in the eye is upside down, so it has to be made upright again in the brain before you see it."

His answer embodies a logical fallacy called the homunculus fallacy. If the image on the retina is transmitted to the brain and "projected" on some internal mental screen, then you would need some sort of "little man"—a homunculus—inside your head looking at the image and interpreting or understanding it for you. But how would the homunculus be able to understand the images flashing by on his screen? There would have to be another, even smaller chap looking at the image in *his* head—and so on. It is a situation of infinite regress of eyes, images, and little people, without really solving the problem of perception.

In order to understand perception, you need to first get rid of the notion that the image at the back of your eye simply gets "relayed" back to your brain to be displayed on a screen. Instead, you must understand that as soon as the rays of light are converted into neural impulses at the back of your eye, it no longer makes any sense to think of the visual information as being an image. We must think, instead, of symbolic descriptions that *represent* the scenes and objects that had been in the image. Say I wanted someone to know what the chair across the room from me looks like. I could take him there and point it out to him so he could see it for himself, but that isn't a symbolic description. I could show him a photograph or a drawing of the chair, but that is still not symbolic because it bears a physical resemblance. But if I hand the person a written note describing the chair, we have crossed over into the realm of symbolic description: The squiggles of ink on the paper bear no physical resemblance to the chair; they merely symbolize it.

Analogously, the brain creates symbolic descriptions. It does not re-create the original image, but represents the various features and aspects of the image in totally new terms—not with squiggles of ink, of course, but in its own alphabet of nerve impulses. These symbolic encodings are created partly in your retina itself but mostly in your brain. Once there, they are parceled and transformed and combined in the extensive network of visual brain areas that eventually let you recognize objects. Of course, the vast majority of this processing goes on behind the scenes without entering your conscious awareness, which is why it feels effortless and obvious, as it did to my dinner companion.

I've been glibly dismissing the homunculus fallacy by pointing out the logical problem of infinite regress. But is there any direct evidence that it is in fact a fallacy?

First, what you see can't just be the image on the retina because the retinal image can remain constant but your perception can change radically. If perception simply involves transmitting and displaying an image on an inner mental screen, how can this be true? Second, the converse is also true: The retinal image can change, yet your perception of the object remains stable. Third, despite appearances, perception takes time and happens in stages.

The first reason is the most easy to appreciate. It's the basis of many visual illusions. A famous example is the Necker cube, discovered accidentally by the Swiss crystallographer Louis Albert Necker (Figure 2.1). He was gazing at a cuboid crystal through a microscope one day, and imagine his amazement when the crystal suddenly seemed to flip! Without visibly moving, it switched its orientation right in front of his very eyes. Was the crystal itself changing? To find out he drew a wireframe cube on a scrap of paper and noticed that the drawing did the same thing. Conclusion: His perception was changing, not the crystal.

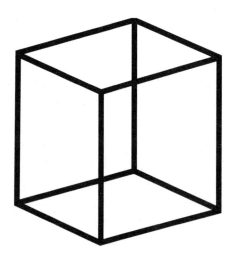

FIGURE 2.1 Skeleton outline drawing of a cube: You can see it in either of two different ways, as if it were above you or below you.

FIGURE 2.2   This picture has not been Photoshopped! It was taken with an ordinary camera from the special viewing point that makes the Ames room work. The fun part of this illusion comes when you have two people walk to opposite ends of the room: It looks for all the world as if they are standing just a few feet apart from each other and one of them has grown giant, with his head brushing the ceiling, while the other has shrunk to the size of a fairy.

You can try this on yourself. It is fun even if you have tried it dozens of times in the past. You will see that the drawing suddenly flips on you, and it's partly—but only partly—under voluntary control. The fact that your perception of an unchanging image can change and flip radically is proof that perception must involve more than simply displaying an image in the brain. Even the simplest act of perception involves judgment and interpretation. Perception is an actively formed opinion of the world rather than a passive reaction to sensory input from it.

Another striking example is the famous Ames room illusion (Figure 2.2). Imagine taking a regular room like the one you are in now and stretching out one corner so the ceiling is much taller in that corner than elsewhere. Now make a small hole in any of the walls and look inside the room. From nearly any viewing perspective you see a bizarrely deformed trapezoidal room. But there is one special vantage point from which, astonishingly, the room looks completely normal! The walls, floor, and ceiling all seem to be arranged at proper right angles to each other, and the windows and floor tiles seem to be of uniform size. The usual explanation for this illusion is that from this particular vantage point the image cast on your retina by the distorted room is identical to that which would be produced by a normal room—it's just geometric optics. But surely this begs the question. How does your visual system know what a normal room should look like from exactly this particular vantage point?

To turn the problem on its head, let's assume you are looking through a peephole into a normal room. There is in fact an infinity of distorted trapezoidal Ames rooms that could produce exactly the same image, yet you stably perceive a normal room. Your perception doesn't oscillate wildly between a million possibilities; it homes in instantly on the correct interpretation. The only way it can do this is by bringing in certain built-in knowledge or hidden assumptions about the world—such as walls being parallel, floor tiles being squares, and so on—to eliminate the infinity of false rooms.

The study of perception, then, is the study of these assumptions and the manner in which they are enshrined in the neural hardware of your brain. A life-size Ames room is hard to construct, but over the years psychologists have created hundreds of visual illusions that have been cunningly devised to help us explore the assumptions that drive perception. Illusions are fun to look at since they seem to violate common sense. But they have the same effect on a perceptual psychologist as the smell of burning rubber does on an engineer—an irresistible urge to discover the cause (to quote what biologist Peter Medawar said in a different context).

Take the simplest of illusions, foreshadowed by Isaac Newton and established clearly by Thomas Young (who, coincidentally, also deciphered the Egyptian hieroglyphics). If you project a red and a green circle of light to overlap on a white screen, the circle you see actually looks yellow. If you have three projectors—one shining red, another green, and another blue—with proper adjustment of each projector's brightness you can produce any color of the rainbow—indeed, hundreds of different hues just by mixing them in the right ratio. You can even produce white. This illusion is so astonishing that people have difficulty believing it when they first see it. It's also telling you something fundamental about vision. It illustrates the fact that even though you can distinguish thousands of colors, you have only three classes of color-sensitive cells in the eye: one for red light, one for green, and one for blue. Each of these responds optimally to just one wavelength but will continue to respond, though less well, to other wavelengths. Thus any observed color will excite the red, green, and blue receptors in different ratios, and higher brain mechanisms interpret each ratio as a different color. Yellow light, for example, falls halfway in the spectrum between red and green, so it activates red and green receptors equally and the brain has learned, or evolved to interpret, this as the color we call yellow. Using just colored lights to figure out the laws of color vision was one of the great triumphs of visual science. And it paved the way for color printing (economically using just three dyes) and color TV.

My favorite example of how we can use illusions to discover the hidden assumptions underlying perception is shape-from-shading (Figure 2.3). Although artists have long used shading to enhance the impression of depth in their pictures, it's only recently that scientists have begun to investigate it carefully. For example, in 1987 I created several

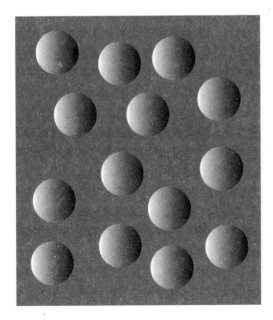

FIGURE 2.3   Eggs or cavities? You can flip between the two depending on which direction you decide the light is shining from, right or left. They always all flip together.

computerized displays like the one shown in Figure 2.3—arrays of randomly scattered disks in a field of gray. Each disk contains a smooth gradient from white at one end to black on the other, and the background is the exact "middle gray" between black and white. These experiments were inspired, in part, by the observations of the Victorian physicist David Brewster. If you inspect the disks in Figure 2.3, they will initially look like a set of eggs lit from the right side. With some effort you can also see them as cavities lit from the left side. But you cannot simultaneously see some as eggs and some as cavities even if you try hard. Why? One possibility is that the brain picks the simplest interpretation by default, seeing all of the disks the same way. It occurred to me that another possibility is that your visual system assumes that there is only a single light source illuminating the entire scene or large chunks of it. This isn't strictly true of an artificially lit environment with many lightbulbs, but it is largely true of the natural world, given that our planetary system has only one sun. If you ever catch hold of an alien, be sure to show her this display to find out if her solar system had a single sun like ours. A creature from a binary star system might be immune to the illusion.

So which explanation is correct—a preference for the simpler interpretation, or an assumption of a single light source? To find out I did the obvious experiment of creating the mixed display shown in Figure 2.4 in which the top and bottom rows have different directions of shading. You will notice that in this display, if you get yourself to see the top row as eggs, then the bottom row is always seen as cavities, and vice versa, and

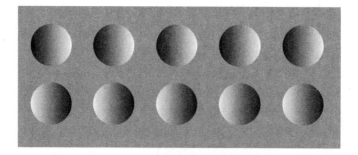

FIGURE 2.4   Two rows of shaded disks. When the top row is seen as eggs, the bottom row looks like cavities, and vice versa. It is impossible to see them all the same way. Illustrates the "single light source" assumption built into perceptual processing.

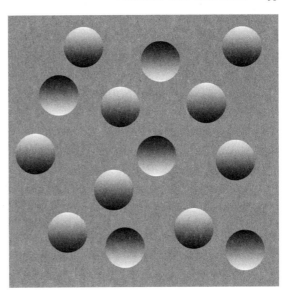

FIGURE 2.5   Sunny side up. Half the disks (light on top) are seen as eggs and half as cavities. This illusion shows that the visual system automatically assumes that light shines from above. View the page upside down, and the eggs and cavities will switch.

it is impossible to see them all simultaneously as eggs or simultaneously as cavities. This proves it's not simplicity but the assumption of a single light source.

It gets better. In Figure 2.5 the shaded disks have been shaded vertically rather than horizontally. You will notice that the ones that are light on top are nearly always seen as eggs bulging toward you, whereas the ones that are dark on top are seen as cavities. We may conclude that, in addition to the single-light-source assumption revealed in Figure 2.4, there is another even stronger assumption at work, which is that the light is shining from above. Again, makes sense given the position of the sun in the natural world. Of course, this isn't always true; the sun is sometimes on the horizon. But its true statistically—and it's certainly never below you. If you rotate the picture so it's upside down, you will find that all the bumps and cavities switch. On the other hand, if you rotate it exactly 90 degrees, you will find that the shaded disks are now ambiguous as in Figure 2.4, since you don't have a built-in bias for assuming light comes from the left or the right.

Now I'd like you to try another experiment. Go back to Figure 2.4, but this time, instead of rotating the page, hold it upright and tilt your body and head to the right, so your right ear almost touches your right shoulder and your head is parallel to the ground. What happens? The ambiguity disappears. The top row always looks like bumps and the bottom row as cavities. This is because the top row is now light on the top with reference to your head and retina, even though it's still light on the right in reference to the world. Another way of saying this is that the overhead lighting assumption is head centered, not world centered or body-axis centered. It's as if your brain assumes that the sun is stuck to the top of your head and remains stuck to it when you tilt your head 90 degrees! Why such a silly assumption? Because statistically speaking, your head is upright most of the time. Your ape ancestors rarely walked around looking at the world with their heads tilted. Your visual system therefore takes a shortcut; it makes the simplifying assumption that the sun is stuck to your head. The goal of vision is not to get things perfectly right all the time, but to do get it right often enough and quickly enough to survive as long as possible to leave behind as many babies as you can. As far as evolution is concerned, that's all that matters. Of course, this shortcut makes you vulnerable to certain incorrect judgments, as when you tilt your head, but this happens so rarely in real life that your brain can get away with being lazy like this. The explanation of this visual illusion illustrates how you can begin with a relatively simple set of displays, ask questions of the kind that your grandmother might ask, and gain real insights, in a matter of minutes, into how we perceive the world.

Illusions are an example of the black-box approach to the brain. The metaphor of the black box comes to us from engineering. An engineering student might be given a sealed box with electrical terminals and lightbulbs studding the surface. Running electricity through certain terminals causes certain bulbs to light up, but not in a straightforward or one-to-one relationship. The assignment is for the student to try different combinations of electrical inputs, noting which lightbulbs are activated in each case, and from this trial-and-error process deduce the wiring diagram of the circuit inside the box without opening it.

In perceptual psychology we are often faced with the same basic problem. To narrow down the range of hypotheses about how the brain processes certain kinds of visual information, we simply try varying the sensory inputs and noting what people see or believe they see. Such experiments enable us discover the laws of visual function, in much the same way Gregor Mendel was able to discover the laws heredity by cross-breeding plants with various traits, even though he had no way to know anything about the molecular and genetic mechanisms that made them true. In the case of vision, I think the best example is one we've already considered, in which Thomas Young predicted the existence of three kinds of color receptors in the eye based on playing around with colored lights.

When studying perception and discovering the underlying laws, sooner or later one wants to know how these laws actually arise from the activity of neurons. The only way to find out is by opening the black box—that is, by directly experimenting on the brain. Traditionally there are three ways to approach this: neurology (studying patients with brain lesions), neurophysiology (monitoring the activity of neural circuits or even of single cells), and brain imaging. Specialists in each of these areas are mutually contemptuous and have tended to see their own methodology as the most important window on brain functioning, but in recent decades there has been a growing realization that a combined attack on the problem is needed. Even philosophers have now joined the fray. Some of them, like Pat Churchland and Daniel Dennett, have a broad vision, which can be a valuable antidote to the narrow cul-de-sacs of specialization that the majority of neuroscientists find themselves trapped in.

IN PRIMATES, INCLUDING humans, a large chunk of the brain—comprising the occipital lobes and parts of the temporal and parietal lobes—is devoted to vision. Each of the thirty or so visual areas within this chunk contains either a complete or partial map of the visual world. Anyone who thinks vision is simple should look at one of David Van Essen's anatomical diagrams depicting the structure of the visual pathways in monkeys (Figure 2.6), bearing in mind that they are likely to be even more complex in humans.

Notice especially that there are at least as many fibers (actually many more!) coming back from each stage of processing to an earlier stage as there are fibers going forward from each area into the next area higher up in the hierarchy. The classical notion of vision as a stage-by-stage sequential analysis of the image, with increasing sophistication as you go along, is demolished by the existence of so much feedback. What these back projections are doing is anybody's guess, but my hunch is that at each stage in processing, whenever the brain achieves a partial solution

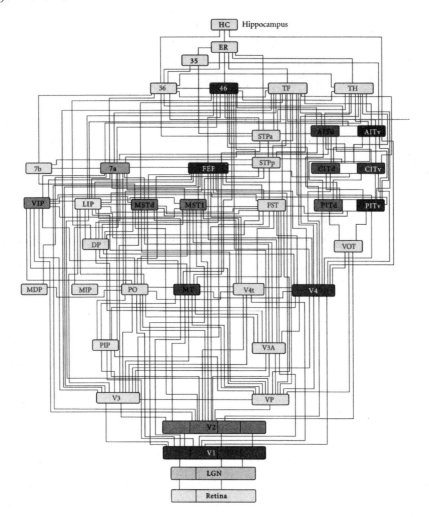

FIGURE 2.6   David Van Essen's diagram depicting the extraordinary complexity of the connections between the visual areas in primates, with multiple feedback loops at every stage in the hierarchy. The "black box" has been opened, and it turns out to contain . . . a whole labyrinth of smaller black boxes! Oh well, no deity ever promised us it would be easy to figure ourselves out.

to a perceptual "problem"—such as determining an object's identity, location, or movement—this partial solution is immediately fed back to earlier stages. Repeated cycles of such an iterative process help eliminate dead ends and false solutions when you look at "noisy" visual images such as camouflaged objects (like the scene "hidden" in Figure 2.7).[3] In other words, these back projections allow you to play a sort of "twenty

FIGURE 2.7   What do you see? It looks like random splatterings of black ink at first, but when you look long enough you can see the hidden scene.

questions" game with the image, enabling you to rapidly home in on the correct answer. It's as if each of us is hallucinating all the time and what we call perception involves merely selecting the one hallucination that best matches the current input. This is an overstatement, of course, but it has a large grain of truth. (And, as we shall see later, may help explain aspects of our appreciation of art.)

The exact manner in which object recognition is achieved is still quite mysterious. How do the neurons firing away when you look at an object recognize it as a face rather than, say, a chair? What are the defining attributes of a chair? In modern designer furniture shops a big blob of plastic with a dimple in the middle is recognized as a chair. It would appear that what is critical is its function—something that permits sitting—rather than whether it has four legs or a back rest. Somehow the nervous system translates the act of sitting as synonymous with the perception of chair. If it is a face, how do you recognize the person instantly even though you have encountered millions of faces over a lifetime and stored away the corresponding representations in your memory banks?

Certain features or signatures of an object can serve as a shortcut

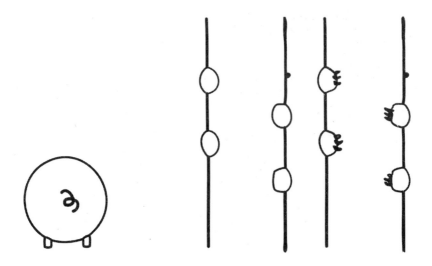

FIGURE 2.8   (a) A pig rump.   (b) A bear.

to recognizing it. In Figure 2.8a, for example, there is a circle with a squiggle in the middle but you see a pig's rump. Similarly, in Figure 2.8b you have four blobs on either side of a pair of straight vertical lines, but as soon as I add some features such as claws, you might see it as a bear climbing a tree. These images suggest that certain very simple features can serve as diagnostic labels for more complex objects, but they don't answer the even more basic question of how the features themselves are extracted and recognized. How is a squiggle recognized as a squiggle? And surely the squiggle in Figure 2.8a can only be a tail given the over-all context of being inside a circle. No rump is seen if the squiggle falls outside the circle. This raises the central problem in object recognition; namely, how does the visual system determine relationships between fea-tures to identify the object? We still have precious little understanding.

The problem is even more acute for faces. Figure 2.9a is a cartoon face. The mere presence of horizontal and vertical dashes can substitute for nose, eyes, and mouth, but only if the relationship between them is correct. The face in Figure 2.9b has the same exact features as the one in Figure 2.9a, but they're scrambled. No face is seen—unless you happen to be Picasso. Their correct arrangement is crucial.

But surely there is more to it. As Steven Kosslyn of Harvard Univer-sity has pointed out, the relationship between features (such as nose, eyes,

FIGURE 2.9   (a) A cartoon face.   (b) A scrambled face.

mouth in the right relative positions) tells you only that it's a face and not, say, a pig or a donkey; it doesn't tell you whose face it is. For recognizing individual faces you have to switch to measuring the relative sizes and distances between features. It's as if your brain has a created a generic template of the human face by averaging together the thousands of faces it has encountered. Then, when you encounter a novel face, you compare the new face with the template—that is, your neurons mathematically subtract the average face from the new one. The pattern of deviation from the average face becomes your specific template for the new face. For example, compared to the average face Richard Nixon's face would have a bulbous nose and shaggy eyebrows. In fact, you can deliberately exaggerate these deviations and produce a caricature—a face that can be said to look more like Nixon than the original. Again, we will see later how this has relevance to some types of art.

We have to bear in mind, though, that words such as "exaggeration," "template," and "relationships" can lull us into a false sense of having explained much more than we really have. They conceal depths of ignorance. We don't know how neurons in the brain perform any of these operations. Nonetheless, the scheme I have outlined might provide a useful place to start future research on these questions. For example, over twenty years ago neuroscientists discovered neurons in the temporal lobes of monkeys that respond to faces; each set of neurons firing when the monkey looks at a specific familiar face, such as Joe the alpha male or Lana the pride of his harem. In an essay on art that I published in

1998, I predicted that such neurons might, paradoxically, fire even more vigorously in response to an exaggerated caricature of the face in question than to the original. Intriguingly, this prediction has now been confirmed in an elegant series of experiments performed at Harvard. Such experiments are important because they will help us translate purely theoretical speculations on vision and art into more precise, testable models of visual function.

Object recognition is a difficult problem, and I have offered some speculations on what the steps involved are. The word "recognition," however, doesn't tell us anything much unless we can explain how the object or face in question evokes meaning—based on the memory associations of the face. The question of how neurons encode meaning and evoke all the semantic associations of an object is the holy grail of neuroscience, whether you are studying memory, perception, art, or consciousness.

AGAIN, WE DON'T really know why we higher primates have such a large number of distinct visual areas, but it seems that they are all specialized for different aspects of vision, such as color vision, seeing movement, seeing shapes, recognizing faces, and so on. The computational strategies for each of these might be sufficiently different that evolution developed the neural hardware separately.

A good example of this is the middle temporal (MT) area, a small patch of cortical tissue found in each hemisphere, that appears to be mainly concerned with seeing movement. In the late 1970s a woman in Zurich, whom I'll call Ingrid, suffered a stroke that damaged the MT areas on both sides of her brain but left the rest of her brain intact. Ingrid's vision was normal in most respects: She could read newspapers and recognize objects and people. But she had great difficulty seeing movement. When she looked at a moving car, it appeared like a long succession of static snapshots, as if seen under a strobe. She could read the number plate and tell you what color it was, but there was no impression of motion. She was terrified of crossing the street because she didn't know how fast the cars were approaching. When she poured water into a glass, the stream of water looked like a static icicle. She didn't know when to stop pouring because she couldn't see the rate at which the water level was rising, so it always overflowed. Even talking to people was like "talking on a phone," she said, because she couldn't see the lips moving. Life became a strange ordeal for her. So it would seem that the MT areas are concerned mainly with seeing motion but not with other aspects of vision. There are four other bits of evidence supporting this view.

First, you can record from single nerve cells in a monkey's MT areas. The cells signal the direction of moving objects but don't seem that interested in color or shape. Second, you can use microelectrodes to stimulate tiny clusters of cells in a monkey's MT area. This causes the cells to fire, and the monkey starts hallucinating motion when the current is applied. We know this because the monkey starts moving his eyes around tracking imaginary moving objects in its visual field. Third, in human volunteers, you can watch MT activity with functional brain imaging such as fMRI (functional MRI). In fMRI, magnetic fields in the brain produced by changes in blood flow are measured while the subject is doing or looking at something. In this case, the MT areas lights up while you are looking at moving objects, but not when you are shown static pictures, colors, or printed words. And fourth, you can use a device called a transcranial magnetic stimulator to briefly stun the neurons of volunteers' MT areas—in effect creating a temporary brain lesion. Lo and behold, the subjects become briefly motion blind like Ingrid while the rest of their visual abilities remain, to all appearances, intact. All this might seem like overkill to prove the single point that MT is the motion area of the brain, but in science it never hurts to have converging lines of evidence that prove the same thing.

Likewise, there is an area called V4 in the temporal lobe that appears to be specialized for processing color. When this area is damaged on both sides of the brain, the entire world becomes drained of color and looks like a black-and-white motion picture. But the patient's other visual functions seem to remain perfectly intact: She can still perceive motion, recognize faces, read, and so on. And just as with the MT areas, you can get converging lines of evidence through single-neuron studies, functional imaging, and direct electrical stimulation to show that V4 is the brain's "color center."

Unfortunately, unlike MT and V4, most of the rest of the thirty or so visual areas of the primate brain do not reveal their functions so cleanly when they are lesioned, imaged, or zapped. This may be because they are not as narrowly specialized, or their functions are more easily compensated for by other regions (like water flowing around an obstacle), or perhaps our definition of what constitutes a single function is murky ("ill posed," as computer scientists say). But in any case, beneath all the bewildering anatomical complexity there is a simple organizational pattern that is very helpful in the study of vision. This pattern is a division of the flow of visual information along (semi)separate, parallel pathways (Figure 2.10).

Let's first consider the two pathways by which visual information enters the cortex. The so-called old pathway starts in the retinas, relays through an ancient midbrain structure called the superior colliculus, and then projects—via the pulvinar—to the parietal lobes (see Figure 2.10). This pathway is concerned with spatial aspects of vision: where, but not what, an object is. The old pathway enables us to orient toward objects and track them with our eyes and heads. If you damage this pathway in a hamster, the animal develops a curious tunnel vision, seeing and recognizing only what is directly in front of its nose.

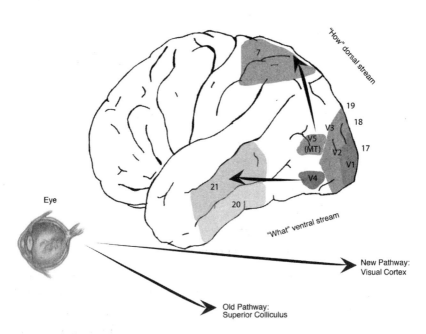

FIGURE 2.10   The visual information from the retina gets to the brain via two pathways. One (called the old pathway) relays through the superior colliculus, arriving eventually in the parietal lobe. The other (called the new pathway) goes via the lateral geniculate nucleus (LGN) to the visual cortex and then splits once again into the "how" and "what" streams.

The new pathway, which is highly developed in humans and in primates generally, allows sophisticated analysis and recognition of complex visual scenes and objects. This pathway projects from the retina to V1, the first and largest of our cortical visual maps, and from there splits into two subpathways, or streams: pathway 1, or what is often called the "how" stream, and pathway 2 the "what" stream. You can think of the "how" stream (sometimes called the "where" stream) as being concerned with the relationships *among* visual objects in space, while the "what" stream is concerned with the relationships of features *within* visual objects themselves. Thus the "how" stream's function overlaps to some extent with that of the old pathway, but it mediates much more sophisticated aspects of spatial vision—determining the overall spatial layout of the visual scene rather than just the location of an object. The "how" stream projects to the parietal lobe and has strong links to the motor system. When you dodge an object hurled at you, when you navigate around a room avoiding bumping into things, when you step gingerly over a tree branch or a pit, or when you reach out to grab an object or fend off a blow, you are relying on the "how" stream. Most of these computations are unconscious and highly automated, like a robot or a zombie copilot that follows your instructions without need of much guidance or monitoring.

Before we consider the "what" stream, let me first mention the fascinating visual phenomenon of blindsight. It was discovered in Oxford in the late 1970s by Larry Weizkrantz. A patient named Gy had suffered substantial damage to his left visual cortex—the origin point for both the "how" and the "what" streams. As a result he became completely blind in his right visual field—or so it seemed at first. In the course of testing Gy's intact vision, Weizkrantz told him to reach out and try to touch a tiny spot of light that he told Gy was to his right. Gy protested that he couldn't see it and there would be no point, but Weizkrantz asked him to try anyway. To his amazement, Gy correctly touched the spot. Gy insisted that he had been guessing, and was surprised when he was told that he had pointed correctly. But repeated trials proved that it had not been a lucky stab in the dark; Gy's finger homed in on target after target, even though he had no conscious visual experience of where they were or what they looked like. Weizkrantz dubbed the syndrome blindsight to emphasize its paradoxical nature. Short of ESP, how can we explain this? How can a person locate something he cannot see? The answer lies in the anatomical division between the old and new pathways in the brain. Gy's new pathway, running through V1, was damaged, but his old pathway was perfectly intact. Information about the spot's location traveled up smoothly to his parietal lobes, which in turn directed the hand to move to the correct location.

This explanation of blindsight is elegant and widely accepted, but it raises an even more intriguing question: Doesn't this imply that only the new pathway has visual consciousness? When the new pathway is blocked, as in Gy's case, visual awareness winks out. The old pathway, on the other hand, is apparently performing equally complex computations to guide the hand, but without a wisp of consciousness creeping in. This is one reason why I likened this pathway to a robot or a zombie. Why should this be so? After all, they are just two parallel pathways made up of identical-looking neurons, so why is only one of them linked to conscious awareness?

Why indeed. While I have raised it here as a teaser, the question of conscious awareness is a big one that we will leave for the final chapter.

Now let's have look at pathway 2, the "what" stream. This stream is concerned mainly with recognizing what an object is and what it means to you. This pathway projects from V1 to the fusiform gyrus (see Figure 3.6), and from there to other parts of the temporal lobes. Note that the fusiform area itself mainly performs a dry classification of objects: It discriminates Ps from Qs, hawks from handsaws, and Joe from Jane, but it does not assign significance to any of them. Its role is analogous to that of a shell collector (conchologist) or a butterfly collector (lepidopterist), who classifies and labels hundreds of specimens into discrete nonoverlapping conceptual bins without necessarily knowing (or caring) anything else about them. (This is approximately true but not completely; some aspects of meaning are probably fed back from higher centers to the fusiform.)

But as pathway 2 proceeds past the fusiform to other parts of the temporal lobes, it evokes not only the name of a thing but a penumbra of associated memories and facts about it—broadly speaking the semantics, or meaning, of an object. You not only recognize Joe's face as being "Joe," but you remember all sorts of things about him: He is married to Jane, has a warped sense of humor, is allergic to cats, and is on your bowling team. This semantic retrieval process involves widespread activation of the temporal lobes, but it seems to center on a handful of "bottlenecks" that include Wernicke's language area and the inferior parietal lobule (IPL), which is involved in quintessentially human abilities as such as naming, reading, writing, and arithmetic. Once meaning is extracted in these bottleneck regions, the messages are relayed to the amygdala, which lies embedded in the front tip of the temporal lobes, to evoke feelings about what (or whom) you are seeing.

In addition to pathways 1 and 2[4] there seems to be an alternate, somewhat more reflexive pathway for emotional response to objects that I call pathway 3. If the first two were the "how" and "what" streams, this one could be thought of as the "so what" stream. In this pathway, biologically salient stimuli such as eyes, food, facial expressions, and animate motion (such as someone's gait and gesturing) pass from the fusiform gyrus through an area in the temporal lobe called the superior temporal sulcus (STS) and then straight to the amygdala.[5] In other words, pathway 3 bypasses high-level object perception—and the whole rich penumbra of associations evoked through pathway 2—and shunts quickly to the amygdala, the gateway to the emotional core of the brain, the limbic system. This shortcut probably evolved to promote fast reaction to high-value situations, whether innate or learned.

The amygdala works in conjunction with past stored memories and other structures in the limbic system to gauge the emotional significance of whatever you are looking at: Is it friend, foe, mate? Food, water, danger? Or is it just something mundane? If it's insignificant—just a log, a piece of lint, the trees rustling in the wind—you feel nothing toward it and most likely will ignore it. But if it's important, you instantly feel something. If it is an intense feeling, the signals from the amygdala also cascade into your hypothalamus (see Figure Int.3), which not only orchestrates the release of hormones but also activates the autonomic nervous system to prepare you to take appropriate action, whether it's feeding, fighting, fleeing, or wooing. (Medical students use the mnemonic of the "four Fs" to remember these.) These autonomic responses include all the physiological signs of strong emotion such as increased heart rate, rapid shallow breathing, and sweating. The human amygdala is also connected with the frontal lobes, which add subtle flavors to this "four F" cocktail of primal emotions, so that you have not just anger, lust, and fear, but also arrogance, pride, caution, admiration, magnanimity, and the like.

———

LET US NOW return to John, our stroke patient from earlier in the chapter. Can we explain at least some of his symptoms based on the broad-brushstrokes layout of the visual system I have just painted? John was definitely not blind. Remember, he could almost perfectly copy an engraving of St. Paul's Cathedral even though he did not recognize what he was drawing. The earlier stages of visual processing were intact, so John's brain could extract lines and shapes and even discern relationships between them. But the crucial next link in the "what" stream—the fusiform gyrus—from which visual information could trigger recognition, memory, and feelings—had been cut off. This disorder is called agnosia, a term coined by Sigmund Freud meaning that the patient sees but doesn't know. (It would have been interesting to see if John had the right emotional response to a lion even while being unable to distinguish it consciously from a goat, but the researchers didn't try that. It would have implied a selective sparing of pathway 3.)

John could still "see" objects, could reach out and grab them, and walk around the room dodging obstacles because his "how" stream was largely intact. Indeed, anyone watching him walk around wouldn't even suspect that his perception had been profoundly deranged. Remember, when he returned home from the hospital, he could trim hedges with shears or pull out a plant from the soil. And yet he could not tell weeds from flowers, or for that matter recognize faces or cars or tell salad dressing from cream. Thus symptoms that would otherwise seem bizarre and incomprehensible begin to make sense in terms of the anatomical scheme with it's the multiple visual pathways that I've just outlined.

This is not to say that his spatial sense was completely intact. Recall that he could grab an isolated coffee cup easily enough but was befuddled by a cluttered buffet table. This suggests that he was also experiencing some disruption of a process vision researchers call segmentation: knowing which fragments of a visual scene belong together to constitute a single object. Segmentation is a critical prelude to object recognition in the "what" stream. For instance, if you see the head and hindquarters of a cow protruding from opposite sides of a tree trunk, you automatically perceive the entire animal—your mind's eye fills it in without question. We really have no idea how neurons in the early stages of visual processing accomplish this linking so effortlessly. Aspects of this process of segmentation were probably also damaged in John.

Additionally, John's lack of color vision suggests that there was damage to his color area, V4, which not surprisingly lies in the same brain region—the fusiform gyrus—as the face recognition area. John's main symptoms can be partially explained in terms of damage to specific aspects of visual function, but some of them cannot be. One of his most intriguing symptoms became manifest when he was asked to draw flowers from memory. Figure 2.11 shows the drawings he produced, which

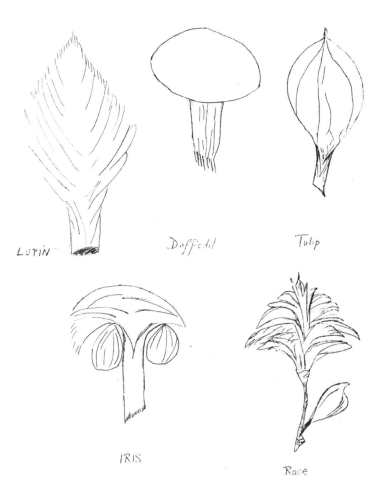

FIGURE 2.11   "Martian flowers." When asked to draw specific flowers, John instead produced generic flowers, conjured up, without realizing it, in his imagination.

he confidently labeled rose, tulip, and iris. Notice that the flowers are drawn well but they don't look like any real flowers that we know! It's as though he had a generic concept of a flower and, lacking access to memories of real flowers, produces what might be called Martian flowers that really don't exist.

A few years after John returned home, his wife died and he moved to a sheltered home for the rest of his life. (He died about three years before this book was printed.) While he was there, he managed to take care of himself by staying in a small room where everything was organized to facilitate his recognition. Unfortunately, as his physician Glyn Humphreys pointed out to me, he would still get terribly lost going outside—even getting lost in the garden once. Yet despite these handicaps he displayed considerable fortitude and courage, keeping up his spirits until the very end.

JOHN'S SYMPTOMS ARE strange enough but, not long ago, I encountered a patient named David who had an even more bizarre symptom. His problem was not with recognizing objects or faces but with responding to them emotionally—the very last step in the chain of events that we call perception. I described him in my previous book, *Phantoms in the Brain*. David was a student in one of my classes before he was involved in a car crash that left him comatose for two weeks. After he woke up from the coma, he made a remarkable recovery within a few months. He could think clearly, was alert and attentive, and could understand what was said to him. He could also speak, write, and read fluently even though his speech was slightly slurred. Unlike John he had no problem recognizing objects and people. Yet he had one profound delusion. Whenever he saw his mother, he would say, "Doctor, this woman looks exactly like my mother but she isn't—she's an imposter pretending to be my mother."

He had a similar delusion about his father but not about anyone else. David had what we now call the Capgras syndrome (or delusion), named after the physician who first described it. David was the first patient I had ever seen with this disorder, and I was transformed from skeptic to believer. Over the years I had learned to be wary of odd syndromes. A majority of them are real but sometimes you read about a syndrome that represents little more than a neurologist's or psychiatrist's vanity—an attempted shortcut to fame by having a disease named after him or being credited with its discovery.

But seeing David convinced me that the Capgras syndrome is bona fide. What could be causing such a bizarre delusion? One interpretation that can still be found in older psychiatry textbooks is a Freudian one. The explanation would run like this: Maybe David, like all men, had a strong sexual attraction to his mother when he was a baby—the so-called Oedipus complex. Fortunately, when he grew up his cortex became more dominant over his primitive emotional structures and began repressing or inhibiting these forbidden sexual impulses toward mom. But maybe the blow to David's head damaged his cortex, thereby removing the inhibition and allowing his dormant sexual urges to emerge into consciousness. Suddenly and inexplicably, David found himself being sexually turned on by his mother. Perhaps the only way he could "rationalize" this away was to assume she wasn't really his mother. Hence the delusion.

This explanation is ingenious but it never made much sense to me. For example, soon after I had seen David, I encountered another patient, Steve, who had the same delusion about his pet poodle! "This dog looks just like Fifi," he would say "but it really isn't. It just looks like Fifi." Now how can the Freudian theory account for this? You would have to posit latent bestial tendencies lurking in the subconscious minds of all men, or something equally absurd.

The correct explanation, it turns out, is anatomical. (Ironically Freud himself famously said, "Anatomy is destiny.") As noted previously, visual information is initially sent to the fusiform gyrus, where objects, including faces, are first discriminated. The output from the fusiform is relayed via pathway 3 to the amygdala, which performs an emotional surveillance of the object or face and generates the appropriate emotional response. What about David, though? It occurred to me that the car accident might have selectively damaged the fibers in pathway 3 that connect his fusiform gyrus, partly via the STS, to his amygdala while leaving both those structures, as well as pathway 2, completely intact. Because pathway 2 (meaning and language) is unaffected, he still knows his mother's face by sight and remembers everything about her. And because his amygdala and the rest of his limbic system are unaffected, he can still feel laughter and loss like any normal person. But the *link* between perception and emotion has been severed, so his mother's face doesn't evoke the expected feelings of warmth. In other words, there is recognition but without the expected emotional jolt. Perhaps the only way David's brain can cope with this dilemma is to rationalize it away by concluding that she is an imposter.[6] This seems an extreme rationalization, but as we shall see in the final chapter the brain abhors discrepancies of any kind and an absurdly far-fetched delusion is sometimes the only way out.

The advantage of our neurological theory over the Freudian view is that it can be tested experimentally. As we saw earlier, when you look at something that's emotionally evocative—a tiger, your lover, or indeed, your mother—your amygdala signals your hypothalamus to prepare your body for action. This fight-or-flight reaction is not all or nothing; it operates on a continuum. A mildly, moderately, or profoundly emotional experience elicits a mild, moderate, or profound autonomic reaction, respectively. And part of these continuous autonomic reactions to experience is microsweating: Your whole body, including your palms, becomes damper or dryer in proportion to any upticks or downticks in your level of emotional arousal at any given moment.

This is good news for us scientists because it means we can measure your emotional reaction to the things you see by simply monitoring the degree of your microsweating. This can be done simply by taping two passive electrodes to your skin and routing them through a device called an ohmmeter to monitor your galvanic skin response (GSR), the moment-to-moment fluctuations in the electrical resistance of your skin. (GSR is also called the skin conductance response, or SCR.) Thus when you see a foxy pinup or a gruesome medical picture, your body sweats, your skin resistance drops, and you get a big GSR. On the other hand, if you see something completely neutral, like a doorknob or an unfamiliar face, you get no GSR (although the doorknob may very well produce a GSR in a Freudian psychoanalyst).

Now you may well wonder why we should go through the elaborate process of measuring GSR to monitor emotional arousal. Why not simply ask people how something made them feel? The answer is that between the stage of emotional reaction and the verbal report, there are many complex layers of processing, so what you often get is an intellectualized or censored story. For instance, if a subject is a closet homosexual, he may in fact deny his arousal when he sees a Chippendales dancer. But his GSR can't lie because he has no control over it. (GSR is one of the physiological signals that is used in polygraph, or so-called lie-detector tests.) It's a foolproof test to see if emotions are genuine as opposed to verbally faked. And believe it or not, all normal people get huge GSR jolts when they are shown a picture of their mothers—they don't even have to be Jewish!

Based on this reasoning we measured David's GSR. When we flashed neutral pictures of things like a table and chairs, there was no GSR. Nor did his GSR change when he was shown unfamiliar faces, since there was no jolt of familiarity. So far, nothing unusual. But when we showed him his mother's picture, there was no GSR either. This never occurs in normal people. This observation provides striking confirmation of our theory.

But if this is true, why doesn't David call, say, his mailman an imposter, assuming he used to know his mailman prior to the accident? After all, the disconnection between vision and emotion should apply equally to the mailman—not just his mother. Shouldn't this lead to the same symptom? The answer is that his brain doesn't expect an emotional jolt when he sees the mailman. Your mother is your life; your mail carrier is just some person.

Another paradox was that David did not have the imposter delusion when his mother spoke to him on the phone from the adjacent room.

"Oh Mom, it's so good to hear from you. How are you?" he would say.

How does my theory account for this? How can someone be delusional about his mother when she shows up in person but not when she phones him? There is in fact an elegantly simple explanation. It turns out that there is a separate anatomical pathway from the hearing centers of the brain (the auditory cortex) to your amygdala. This pathway was not destroyed in David, so his mother's voice evoked the strong positive emotions he expected to feel. This time there was no need for delusion.

Soon after our findings on David were published in the journal *Proceedings of the Royal Society of London*, I received a letter from a patient named Mr. Turner, who lived in Georgia. He claimed to have developed Capgras syndrome after a head injury. He liked my theory, he said, because he now understood he wasn't crazy or losing his mind;

there was a perfectly logical explanation for his strange symptoms, which he would now try to overcome if he could. But he then went on to add that what troubled him most was not the imposter illusion, but the fact that he no longer enjoyed visual scenes—such as beautiful landscapes and flower gardens—which had been immensely pleasing prior to the accident. Nor did he enjoy great works of art like he used to. His knowledge that this was caused by the disconnection in his brain did not restore the appeal of flowers or art. This made me wonder whether these connections might play a role in all of us when we enjoy art. Can we study these connections to explore the neural basis of our aesthetic response to beauty? I'll return to this question when we discuss the neurology of art in Chapters 7 and 8.

One last twist to this strange tale. It was late at night and I was in bed, when the phone rang. I woke up and looked at the clock: it was 4 A.M. It was an attorney. He was calling me from London and had apparently overlooked the time difference.

"Is this Dr. Ramachandran?"

"Yes it is," I mumbled, still half-asleep.

"I am Mr. Watson. We have a case we would like your opinion on. Perhaps you could fly over and examine the patient?"

"What's this all about?" I said, trying not to sound irritated.

"My client, Mr. Dobbs, was in a car accident," he said. "He was unconscious for several days. When he came out of it he was quite normal except for a slight difficulty finding the right word when he talks."

"Well, I'm happy to hear that," I said. "Some slight word-finding difficulty is extremely common after brain injury—no matter where the injury is." There was a pause. So I asked, "What can I do for you?"

"Mr. Dobbs—Jonathan—wants to file a lawsuit against the people whose car collided with his. This fault was clearly the other party's, so their insurance company is going to compensate Jonathan financially for the damage to his car. But the legal system is very conservative here in England. The physicians here have found him to be physically normal—his MRI is normal and there are no neurological symptoms or other injuries anywhere in his body. So the insurance company will only pay for the car damage, not for any health-related issues."

"Well."

"The problem, Dr. Ramachandran, is that he claims to have developed the Capgras syndrome. Even though he knows that he is looking at his wife, she often seems like a stranger, a new person. This is extremely troubling to him, and he wants to sue the other party for a million dollars for having caused a permanent neuropsychiatric disturbance."

"Pray continue."

"Soon after the accident someone found your book *Phantoms in the Brain* lying on my client's coffee table. He admitted to reading it, which is when he realized he might have the Capgras syndrome. But this bit of self-diagnosis didn't help him in any way. The symptoms remained just the same. So he and I want to sue the other party for a million dollars for having produced this permanent neurological symptom. He fears he may even end up divorcing his wife.

"The trouble is, Dr. Ramachandran, the other attorney is claiming that my client has simply fabricated the whole thing after reading your book. Because if you think about it, it's very easy to fake the Capgras syndrome. Mr. Dobbs and I would like to fly you out to London so you can administer the GSR test and prove to the court that he does indeed have the Capgras syndrome, that he isn't malingering. I understand you cannot fake this test."

The attorney had done his homework. But I had no intention of flying to London just to administer this test.

"Mr. Watson, what's the problem? If Mr. Dobbs finds that his wife looks like a new woman every time he sees her, he should find her perpetually attractive. This is a good thing—not bad at all. We should all be so lucky!" My only excuse for this tasteless joke is that I was still only barely awake.

There was a long pause at the other end and a click as he hung up on me. I never heard from him again. My sense of humor is not always well received.

Even though my remark may have sounded frivolous, it wasn't entirely off the mark. There's a well-known psychological phenomenon called the Coolidge effect, named after President Calvin Coolidge. It's based on a little-known experiment performed by rat psychologists decades ago. Start with a sex-deprived male rat in a cage. Put a female rat in the cage. The male mounts the female, consummating the relationship several times until he collapses from sheer sexual exhaustion. Or so it would seem. The fun begins if you now introduce a new female into the cage. He gets going again and performs several times until he is once again thoroughly exhausted. Now introduce a third new female rat, and our apparently exhausted male rat starts all over again. This voyeuristic experiment is a striking demonstration of the potent effect of novelty on sexual attraction and performance. I have often wondered whether the effect is also true for female rats courting males, but to my knowledge that hasn't been tried—probably because for many years most psychologists were men.

The story is told that President Coolidge and his wife were on a state visit to Oklahoma, and they were invited to a chicken coop—apparently one of their major tourist attractions. The president had to first give a speech, but since Mrs. Coolidge had already heard the speech many times she decided to go to the coop an hour earlier. She was being shown around by the farmer. She was surprised to see that the coop had dozens of hens but only one majestic rooster. When she asked the guide about this, he replied, "Well, he is a fine rooster. He goes on and on all night and day servicing the hens."

"All night?" said Mrs. Coolidge. "Will you do me a big favor? When the president gets here, tell him in exactly the same words—what you just told me."

An hour later when the president showed up, the farmer repeated the story.

The president asked, "Tell me something: Does the rooster go on all night with the same hen or different hens?"

"Why, different hens of course," replied the farmer.

"Well, do me a favor," said the president. "Tell the First Lady what you just told me."

This story may be apocryphal, but it does raise a fascinating question. Would a patient with Capgras syndrome never get bored with his wife? Would she remain perpetually novel and attractive? If the syndrome could somehow be evoked temporarily with transcranial magnetic stimulation . . . one could make a fortune.

CHAPTER 3

Loud Colors and Hot Babes: Synesthesia

---

*"My life is spent in one long effort to escape from the commonplaces of
existence. These little problems help me to do so."*

—SHERLOCK HOLMES

WHENEVER FRANCESCA CLOSES HER EYES AND TOUCHES A PAR-
ticular texture, she experiences a vivid emotion: Denim, extreme sadness. Silk, peace and calm. Orange peel, shock. Wax, embarrassment.
She sometimes feels subtle nuances of emotions. Grade 60 sandpaper
produces guilt, and grade 120 evokes "the feeling of telling a white lie."

Mirabelle, on the other hand, experiences colors every time she sees
numbers, even though they are typed in black ink. When recalling a
phone number she conjures up a spectrum of the colors corresponding to
the numbers in her mind's eye and proceeds to read off the numbers one
by one, deducing them from the colors. This makes it easy to memorize
phone numbers.

When Esmeralda hears a C-sharp played on the piano, she sees blue.
Other notes evoke other distinct colors—so much so that different piano
keys are actually color coded for her, making it easier to remember and
play musical scales.

These women are not crazy, nor are they suffering from a neurological disorder. They and millions of otherwise normal people have
synesthesia, a surreal blending of sensation, perception, and emotion.
Synesthetes (as such people are called) experience the ordinary world in
extraordinary ways, seeming to inhabit a strange no-man's-land between
reality and fantasy. They taste colors, see sounds, hear shapes, or touch
emotions in myriad combinations.

When my lab colleagues and I first came across synesthesia in 1997, we didn't know what to make of it. But in the years since, it has proven to be an unexpected key for unlocking the mysteries of what makes us distinctly human. It turns out this little quirky phenomenon not only sheds light on normal sensory processing, but it takes us on a meandering path to confront some of the most intriguing aspects of our minds—such as abstract thinking and metaphor. It may illuminate attributes of human brain architecture and genetics that might underlie important aspects of creativity and imagination.

When I embarked on this journey nearly twelve years ago, I had four goals in mind. First, to show that synesthesia is real: These people aren't just making it up. Second, to propose a theory of exactly what is going on in their brains that sets them apart from nonsynesthetes. Third, to explore the genetics of the condition. And fourth, and most important, to explore the possibility that, far from being a mere curiosity, synesthesia may give us valuable clues to understanding some of the most mysterious aspects of the human mind—abilities such as language, creativity, and abstract thought that come to us so effortlessly that we take them for granted. Finally, as an additional bonus, synesthesia may also shed light on age-old philosophical questions of qualia—the ineffable raw qualities of experience—and consciousness.

Overall I am happy with the way our research has proceeded since then. We have come up with partial answers to all four questions. More important, we have galvanized an unprecedented interest in this phenomenon; there is now virtually a synesthesia industry, with over a dozen books published on the topic.

WE DON'T KNOW when synesthesia was first recognized as a human trait, but there are hints that Isaac Newton could have experienced it. Aware that the pitch of a sound depends on its wavelength, Newton invented a toy—a musical keyboard—that flashed up different colors on a screen for different notes. Thus every song was accompanied by a kaleidoscopic display of colors. One wonders if sound-color synesthesia inspired his invention. Could a mixing of senses in his brain have provided the original impetus for his wavelength theory of color? (Newton proved that white light is composed of a mixture of colors which can be separated by a prism, with each color corresponding to a particular wavelength of light.)

Francis Galton, a cousin of Charles Darwin and one of the most colorful and eccentric scientists of the Victorian era, conducted the first systematic study of synesthesia in the 1890s. Galton made many valuable contributions to psychology, especially the measurement of intelligence. Unfortunately, he was also an extreme racist; he helped usher in the pseudoscience of eugenics, whose goal was to "improve" mankind by selective breeding of the kind practiced with domesticated livestock. Galton was convinced that the poor were poor because of inferior genes, and that they must be forbidden from breeding too much, lest they overwhelm and contaminate the gene pool of the landed gentry and rich folk like him. It isn't clear why an otherwise intelligent man should hold such views, but my hunch is that he had an unconscious need to attribute his own fame and success to innate genius rather than acknowledging the role of opportunity and circumstance. (Ironically, he himself was childless.)

Galton's ideas about eugenics seem almost comical in hindsight, yet there is no denying his genius. In 1892 Galton published a short article on synesthesia in the journal *Nature*. This was one of his lesser-known papers, but about a century later it piqued my interest. Although Galton wasn't the first to notice the phenomenon, he was the first to document it systematically and encourage people to explore it further. His paper focused on the two most common types of synesthesia: the kind in which sounds evoke colors (auditory-visual synesthesia) and the kind in which printed numbers always seem tinged with inherent color (grapheme-color synesthesia). He pointed out that even though a specific number always produces the same color for any given synesthete, the number-color associations are different for different synesthetes. In other words, it's not as though all synesthetes see a 5 as red or a 6 as green. To Mary, 5 always looks blue, 6 is magenta, and 7 is chartreuse. To Susan, 5 is vermillion, 6 is light green, and 4 is yellow.

How to explain these people's experiences? Are they crazy? Do they simply have vivid associations from childhood memories? Are they just speaking poetically? When scientists encounter anomalous oddities such as synesthetes, their initial reaction is usually to brush them under the carpet and ignore them. This attitude—which many of my colleagues are very vulnerable to—is not as silly as it seems. Because a majority of anomalies—spoon bending, alien abduction, Elvis sightings—turn out to be false alarms, it's not a bad idea for a scientist to play it safe and ignore them. Whole careers, even lifetimes, have been wasted on the pursuit of oddities, such as polywater (a hypothetical form of water based on crackpot science), telepathy, or cold fusion. So I wasn't surprised that even though we had known about synesthesia for over a century, it has generally been sidelined as a curiosity because it didn't make "sense."

Even now, the phenomenon is often dismissed as bogus. When I bring it up in casual conversation, I often hear it shot down on the spot. I've heard, "So you study acid junkies?" and "Whoa! Cuckoo!" and a dozen other dismissals. Unfortunately even physicians are not immune—and ignorance in a physician can be quite hazardous to people's health. I know of at least one case in which a synesthete was misdiagnosed as having schizophrenia and was prescribed antipsychotic medication to rid her of hallucinations. Fortunately her parents took it upon themselves to get informed, and in the course of their reading came across an article on synesthesia. They drew this to the doctor's attention, and their daughter was quickly taken off the drugs.

Synesthesia as a real phenomenon did have a few supporters, including the neurologist Dr. Richard Cytowic, who wrote two books about it: *Synesthesia: A Union of the Senses* (1989) and *The Man Who Tasted Shapes* (1993/2003). Cytowic was a pioneer, but he was a prophet preaching in the wilderness and was largely ignored by the establishment. It didn't help matters that the theories he put forward to explain synesthesia were a bit vague. He suggested that the phenomenon was a kind of evolutionary throwback to a more primitive brain state in which the senses hadn't quite separated and were being mingled in the emotional core of the brain.

This idea of an undifferentiated primitive brain didn't make sense to me. If the synesthete's brain was reverting to an earlier state, then how would you explain the distinctive and specific nature of the synesthete's experiences? Why, for example, does Esmeralda "see" C-sharp as being invariably blue? If Cytowic was correct, you would expect the senses to just blend into each other to create a blurry mess.

A second explanation that is sometimes posed is that synesthetes are just remembering childhood memories and associations. Maybe they played with refrigerator magnets, and the 5 was red and the 6 was green.

Maybe they remember this association vividly, just as you might recall the smell of a rose, the taste of Marmite or curry, or the trill of a robin in the spring. Of course, this theory doesn't explain why only some people remain stuck with such vivid sensory memories. I certainly don't see colors when looking at numbers or listening to tones, and I doubt whether you do either. While I might think of cold when I look at a picture of an ice cube, I certainly don't feel it, no matter how many childhood experiences I may have had with ice and snow. I might say that I feel warm and fuzzy when stroking a cat, but I would never say touching metal makes me feel jealous.

A third hypothesis is that synesthetes are using vague tangential speech or metaphors when they speak of C-major being red or chicken tasting pointy, just as you and I speak of a "loud" shirt or "sharp" cheddar cheese. Cheese is, after all, soft to touch, so what do you mean when you say it is sharp? Sharp and dull are tactile adjectives, so why do you apply them without hesitation to the taste of cheese? Our ordinary language is replete with synesthetic metaphors—hot babe, flat taste, tastefully dressed—so maybe synesthetes are just especially gifted in this regard. But there is a serious problem with this explanation. We don't have the foggiest idea of how metaphors work or how they are represented in the brain. The notion that synesthesia is just metaphor illustrates one of the classic pitfalls in science—trying to explain one mystery (synesthesia) in terms of another (metaphor).

What I propose, instead, is to turn the problem on its head and suggest the very opposite. I suggest that synesthesia is a concrete sensory process whose neural basis we can uncover, and that the explanation might in turn provide clues for solving the deeper question of how metaphors are represented in the brain and how we evolved the capacity to entertain them in the first place. This doesn't imply that metaphor is just a form of synesthesia; only that understanding the neural basis of the latter can help illuminate the former. So when I resolved to do my own investigation of synesthesia, my first goal was to establish whether it was a genuine sensory experience.

IN 1997 A doctoral student in my lab, Ed Hubbard, and I set out to find some synesthetes to begin our investigations. But how? According to most published surveys, the incidence was anywhere from one in a thousand to one in ten thousand. That fall I was lecturing to an undergraduate class of three hundred students. Maybe we'd get lucky. So we made an announcement:

"Certain otherwise normal people claim they see sounds, or that certain numbers always evoke certain colors," we told the class. "If any one of you experiences this, please raise your hands."

To our disappointment, not a single hand went up. But later that day, as I was chatting with Ed in my office, two students knocked on the door. One of them, Susan, had striking blue eyes, streaks of red dye in her blonde ringlets, a silver ring in her belly button and an enormous skateboard. She said to us, "I'm one of those people you talked about in class, Dr. Ramachandran. I didn't raise my hand because I didn't want people to think I was weird or something. I didn't even know that there were others like me or that the condition had a name."

Ed and I looked at each other, pleasantly surprised. We asked the other student to come back later, and waved Susan into a chair. She leaned the skateboard against the wall and sat down.

"How long have you experienced this?" I asked.

"Oh, from early childhood. But I didn't really pay much attention to it at that time, I suppose. But then it gradually dawned on me that it was really odd, and I didn't discuss it with anyone . . . I didn't want people thinking I was crazy or something. Until you mentioned it in class, I didn't know that it had a name. What did you call it, syn . . . es . . . something that rhymes with anesthesia?"

"It's called synesthesia," I said. "Susan, I want you to describe your experiences to me in detail. Our lab has a special interest in it. What exactly do you experience?"

"When I see certain numbers, I always see specific colors. The number 5 is always a specific shade of dull red, 3 is blue, 7 is bright blood red, 8 is yellow, and 9 is chartreuse."

I grabbed a felt pen and pad that were on the table and drew a big 7. "What do you see?"

"Well, it's not a very clean 7. But it looks red . . . I told you that."

"Now I want you to think carefully before you answer this question. Do you actually see the red? Or does it just make you think of red or make you visualize red . . . like a memory image. For example, when

I hear the word 'Cinderella,' I think of a young girl or of pumpkins or coaches. Is it like that? Or do you literally see the color?"

"That's a tough one. It's something I have often asked myself. I guess I do really see it. That number you drew looks distinctly red to me. But I can also see that it's really black—or I should say, I know it's black. So in some sense it is a memory image of sorts . . . I must be seeing it in my mind's eye or something. But it certainly doesn't feel like that. It feels like I am actually seeing it. It's very hard to describe, Doctor."

"You are doing very well, Susan. You are a good observer and that makes everything you say valuable."

"Well, one thing I can tell you for sure is that it isn't like imagining a pumpkin when looking at a picture of Cinderella or listening to the word 'Cinderella.' I do actually see the color."

One of the first things we teach medical students is to listen to the patient by taking a careful history. Ninety percent of the time you can arrive at an uncannily accurate diagnosis by paying close attention, using physical examination and sophisticated lab tests to confirm your hunch (and to increase the bill to the insurance company). I started to wonder whether this dictum might be true not just for patients but for synesthetes as well.

I decided to give Susan some simple tests and questions. For example, was it the actual visual appearance of the numeral that evoked the color? Or was it the numerical concept—the idea of sequence, or even of quantity? If the latter, then would Roman numerals do the trick or only Arabic ones? (I should call them Indian numerals really; they were invented in India in the first millennium B.C.E. and exported to Europe via Arabs.)

I drew a big *VII* on the pad and showed it to her.

"What do you see?"

"I see it's a seven, but it looks black—no trace of red. I have always known that. Roman numerals don't work. Hey, Doctor, doesn't that prove it can't be a memory thing? Because I do know it's a seven but it still doesn't generate the red!"

Ed and I realized that we were dealing with a very bright student. It was starting to look like synesthesia was indeed a genuine sensory phenomenon, brought on by the actual visual appearance of the numeral— not by the numerical concept. But this was still well short of proof. Could

we be absolutely sure that this wasn't happening because early in kindergarten she had repeatedly seen a red seven on her refrigerator door?
I wondered what would happen if I showed her black-and-white halftone photos of fruits and vegetables which (for most of us) have strong
memory-color associations. I drew pictures of a carrot, a tomato, a pumpkin, and a banana, and showed them to her.

"What do you see?"

"Well, I don't see any colors, if that's what you're asking. I know the
carrot is orange and can imagine it to be so, or visualize it to be orange.
But I don't actually see the orange color the way I see red when you show
me the 7. It's hard to explain, Doctor, but it's like this: When I see the
black-and-white carrot, I kinda know it's orange, but I can visualize it
as being any bizarre color I want, like a blue carrot. It's very hard for me
to do that with 7; it keeps screaming red at me! Is all of this making any
sense to you guys?"

"Okay," I told her, "now I want you to close your eyes and show me
your hands."

She seemed slightly startled by my request but followed my instructions. I then drew the numeral 7 on the palm of her hand.

"What did I draw? Here, let me do it again."

"It's a 7!"

"Is it colored?"

"No, absolutely not. Well, let me rephrase that; I don't initially see red
even though I 'feel' it's 7. But then I start visualizing the 7, and it's sort of
tinged red."

"Okay, Susan, what if I say 'seven'? Here, let's try it: Seven, seven,
seven."

"It wasn't red initially, but then I started to experience red . . . Once I
start visualizing the appearance of the shape of 7, then I see the red—but
not before that."

On a whim I said, "Seven, five, three, two, eight. What did you see
then, Susan?"

"My God . . . that's very interesting. I see a rainbow!"

"What do you mean?"

"Well, I see the corresponding colors spread out in front of me as in a
rainbow, with the colors matching the number sequence you read aloud.
It's a very pretty rainbow."

"One more question, Susan. Here is that drawing of 7 again. Do you see the color directly on the number, or does it spread around it?"

"I see it directly on the number."

"What about a white number on black paper? Here is one. What do you see?"

"It's even more clearly red than the black one. Dunno why."

"What about double-digit numbers?" I drew a bold 75 on the pad and showed it to her. Would her brain start blending the colors? Or see a totally new color?

"I see each number with its appropriate color. But I have often noticed this myself. Unless the numbers are too close."

"Okay, let's try that. Here, the 7 and 5 are much closer together. What do you see?"

"I still see them in the correct colors, but they seem to 'fight' or cancel each other; they seem dimmer."

"And what if I draw the number seven in the wrong-color ink?"

I drew a green 7 on the pad and showed it to her.

"Ugh! It looks hideous. It jars, like there is something wrong with it. I certainly don't mix the real color with the mental color. I see both colors simultaneously, but it looks hideous."

Susan's remark reminded me of what I had read in the older papers on synesthesia, that the experience of color was often emotionally tinged for them and that incorrect colors could produce a strong aversion. Of course, we all experience emotions with certain colors. Blue seems calming, and red is passionate. Could it be that the same process is, for some odd reason, exaggerated in synesthetes? What can synesthesia tell us about the link between color and emotion that artists like Van Gogh and Monet have long been fascinated by?

There was a hesitant knock on the door. We hadn't noticed that almost an hour had passed and that the other student, a girl named Becky, was still outside my office. Fortunately, she was cheerful despite having waited so long. We asked Susan to come back the following week and invited Becky in. It turned out that she too was a synesthete. We repeated the same questions and conducted the same tests on her as we had on Susan. Her answers were uncannily similar with a few minor differences.

Becky saw colored numbers, but hers were not the same as Susan's.

For Becky, 7 was blue and 5 was green. Unlike Susan, she saw letters of the alphabet in vivid colors. Roman numerals and numbers drawn on her hand were ineffective, which suggested that, as in Susan, the colors were driven by the visual appearance of the number and not by the numerical concept. And lastly, she saw the same rainbow-like effect that Susan saw when we recited a string of random numbers.

I realized then and there that we were hot on the trail of a genuine phenomenon. All my doubts were dispelled. Susan and Becky had never met each other before, and the high level of similarity between their reports couldn't possibly be a coincidence. (We later learned that there's a lot of variation among synesthetes, so we were very lucky to have stumbled on two very similar cases.) But even though I was convinced, we still had a lot of work to do to produce evidence strong enough to publish. People's verbal commentaries and introspective reports are notoriously unreliable. Subjects in a laboratory setting are often highly suggestible and may unconsciously pick up what you want to hear and oblige by telling you that. Furthermore, they sometimes speak ambiguously or vaguely. What was I to make of Susan's perplexing remark? "I really do see red, but I also know it's not—so I guess I must be seeing it in my mind's eye or something."

SENSATION IS INHERENTLY subjective and ineffable: You know what it "feels" like to experience the vibrant redness of a ladybug's shell, for instance, but you could never *describe* that redness to a blind person, or even to a color-blind person who cannot distinguish red from green. And for that matter, you can never truly know whether other people's inner mental experience of redness is the same as yours. This makes it somewhat tricky (to put it mildly) to study the perception of other people. Science traffics in objective evidence, so any "observations" we make about people's subjective sensory experience are necessarily indirect or secondhand. I would point out though that subjective impressions and single-subject case studies can often provide strong clues toward designing more formal experiments. Indeed, most of the great discoveries in neurology were initially based on simple clinical testing of single cases (and their subjective reports) before being confirmed in other patients.

One of the first "patients" with whom we launched a systematic study in search of hard proof of the reality of synesthesia was Francesca, a mild-mannered woman in her midforties who had been seeing a psychiatrist because she had been experiencing a mild low-grade depression. He prescribed lorazepam and Prozac, but not knowing what to make of her synesthetic experiences, referred her to my lab. She was the same woman I mentioned earlier who claimed that right from very early childhood she experienced vivid emotions when she touched different textures. But how could we test the truth of her claim? Perhaps she was just a highly emotional person and simply enjoyed speaking about the emotions that various objects triggered in her. Perhaps she was "mentally disturbed" and just wanted attention or to feel special.

Francesca came into the lab one day, having seen an ad in the *San Diego Reader*. After tea and the usual pleasantries my student David Brang and I hooked her up to our ohmmeter to measure her GSR. As we saw in the Chapter 2, this device measures the moment-to-moment microsweating produced by fluctuating levels of emotional arousal. Unlike a person, who can verbally dissemble or even be subconsciously deluded about how something makes her feel, GSR is instantaneous and automatic. When we measured GSR in normal subjects who touched various mundane textures such as corduroy or linoleum, it was clear they experienced no emotions. But Francesca was different. For the textures that she reported gave her strong emotional reactions, such as fear or anxiety or disgust, her body produced a strong GSR signal. But when she touched textures that she said gave her warm, relaxed feelings, there was no change in the electrical resistance of her skin. Since you cannot fake GSR responses, this provided strong evidence that Francesca was telling us the truth.

But to be absolutely sure that Francesca was experiencing specific emotions, we used an added procedure. Again we took her into a room and hooked her up to the ohmmeter. We asked her to follow instructions on a computer screen that would tell her which of several objects that were laid out on the table in front of her she was to touch and for how long. We said she would be alone in the room since noises from our presence might interfere with the GSR monitoring. Unbeknownst to her, we had a hidden video camera behind the monitor to record all her facial expressions. The reason we did this secretively was to ensure that her expressions were genuine and spontaneous. After the experiment, we had independent student evaluators rate the magnitude and quality of the expressions on her face, such as fear or calm. Of course we made sure that the evaluators didn't know the purpose of the experiment and didn't know what object Francesca had been touching on any given trial. Once again we found that there was clear correlation between Francesca's subjective ratings of various textures and her spontaneous facial expressions. It seemed quite clear, therefore, that the emotions she claimed to experience were authentic.

MIRABELLE, AN EBULLIENT, dark-haired young lady, had been eavesdropping on a conversation I had been having with Ed Hubbard at the Espresso Roma Cafe on campus, a stone's throw away from my office. She arched her eyebrows—whether from amusement or skepticism, I couldn't tell.

She came to our lab shortly thereafter to volunteer as a subject. Like Susan and Becky, every number appeared to Mirabelle to be tinged with a particular color. Susan and Becky had convinced us informally that they were reporting their experience accurately and truthfully, but with Mirabelle we wanted to see if we could scare up some hard proof that she was really seeing color (as when you see an apple) rather than just experiencing a vague mental picture of color (as when you imagine an apple). This boundary between seeing and imagining has always proved elusive in neurology. Perhaps synesthesia would help resolve the distinction between them.

I waved her toward a chair in my office, but she was reluctant to sit. Her eyes darted all around the room looking at the various antique scientific instruments and fossils lying on the table and on the floor. She was like the proverbial kid in a candy store as she crawled all around the floor looking at a collection of fossil fishes from Brazil. Her jeans were sliding down her hips, and I tried not to gaze directly at the tattoo on her waist. Mirabelle's eyes lit up when she saw a long, polished fossilized bone which looked a bit like a humerus (upper arm bone). I asked her to guess what it was. She tried rib, shin bone, and thigh bone. In fact, it was the baculum (penis bone) of an extinct Pleistocene walrus. This particular one had obviously been fractured in the middle and had rehealed at an angle while the animal was alive, as evidenced by a callus formation.

There was also a healed, callused tooth mark on the fracture line, suggesting the fracture had been caused by a sexual or predatory bite. There is a detective aspect to paleontology just as there is in neurology, and we could have gone on with all this for another two hours. But we were running out of time. We needed to get back to her synesthesia.

We began with a simple experiment. We showed Mirabelle a white number 5 on a black computer screen. As expected, she saw it in color—in her case, bright red. We had her fix her gaze on a small white dot in the middle of the screen. (This is called a fixation spot; it keeps the eyes from wandering). We then gradually moved the number farther and farther away from the central spot to see if this did anything to the color that was evoked. Mirabelle pointed out that the red color became progressively less vivid as the number was moved away, eventually becoming a pale desaturated pink. This in itself may not seem very surprising; a number seen off-axis prompts a weaker color. But it was nonetheless telling us something important. Even seen off to the side the number itself was still perfectly identifiable, yet the color was much weaker. In one stroke this result showed that synesthesia can't be just a childhood memory or a metaphorical association.[1] If the number were merely evoking the memory or the idea of a color, why should it matter where it was placed in the visual field, so long as it is still clearly recognizable?

We then used a second, more direct test called popout, which psychologists employ to determine whether an effect is truly perceptual (or only conceptual). If you look at Figure 3.1 you will see a set of tilted lines scattered amid a forest of vertical lines. The tilted lines stick out like a sore thumb—they "pop out." Indeed, you can not only pick them out of the crowd almost instantly but can also group them mentally to form a separate plane or cluster. If you do this, you can easily see that the cluster of tilted lines forms the global shape of an X. Similarly in Figure 3.2, red dots scattered among green dots (pictured here as black dots among gray dots) pop out vividly and form the global shape of a triangle.

In contrast, look at Figure 3.3. You see a set of Ts scattered amid the Ls, but unlike the tilted lines and colored dots of the previous two figures, the Ts don't give you the same vivid, automatic "here I am!" popout effect, in spite of the fact that Ls and Ts are as different from each other as vertical and tilted lines. You also cannot group the Ts nearly as easily, and must instead engage in an item-by-item inspection. We may

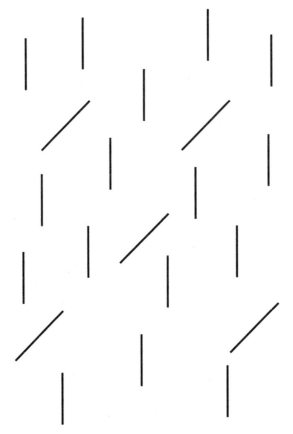

FIGURE 3.1 Tilted lines embedded in a matrix of vertical lines can be readily detected, grouped, and segregated from the straight lines by your visual system. This type of segregation can occur only with features extracted early in visual processing. (Recall from Chapter 2 that three-dimensional shape from shading can also lead to grouping.)

conclude from this that only certain "primitive," or elementary, perceptual features such as color and line orientation can provide a basis for grouping and popout. More complex perceptual tokens such as graphemes (letters and numbers) cannot do so, however different they might be from each other.

To take an extreme example, if I showed you a sheet of paper with the word *love* typed all over it and a few *hate*s scattered about, you could not find the *hate*s very easily. You would have to search for them in a more

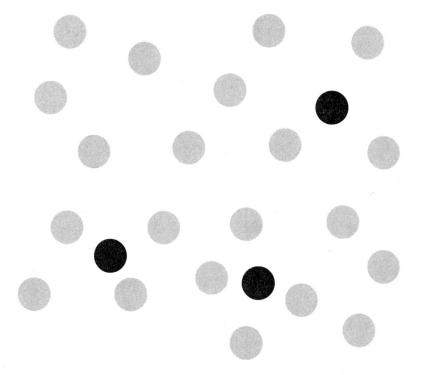

FIGURE 3.2   Dots of similar colors or shading can also be grouped effortlessly. Color is a feature detected early in visual processing.

or less serial fashion. And even as you found them, one by one, they still wouldn't segregate from the background the way the tilted lines or colors do. Again, this is because linguistic concepts like love and hate cannot serve as a basis for grouping, however dissimilar they might be conceptually.

Your ability to group and segregate similar features probably evolved mainly to defeat camouflage and discover hidden objects in the world. For instance, if a lion hides behind a mottling of green foliage, the raw image that enters your eye and hits your retina is nothing but a bunch of yellowish fragments broken up by intervals of green. However, this is not what you *see*. Your brain knits together the fragments of tawny fur to discern the global shape, and activates your visual category for lion. (And from there, it's straight on to the amygdala!) Your brain treats the probability that all those yellow patches could be truly isolated and independent from each other as essentially zero. (This is why a painting or a photograph of a lion hiding behind foliage, in which the patches of

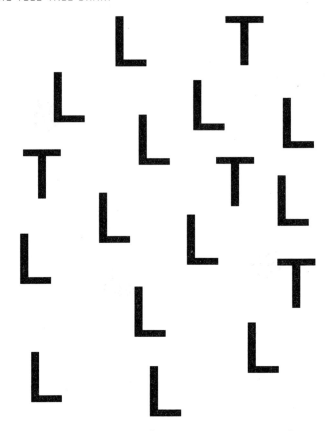

FIGURE 3.3   *T*s scattered among *L*s are not easy to detect or group, perhaps because both are made up of the same low-level features: vertical and horizontal lines. Only the arrangement of the lines is different (producing corners versus T-junctions), and this is not extracted early in visual processing.

color actually *are* independent and unrelated, still makes you "see" the lion.) Your brain automatically tries to group low-level perceptual features together to see if they add up to something important. Like lions.

Perceptual psychologists routinely exploit these effects to determine whether a particular visual feature is elementary. If the feature gives you popout and grouping, the brain must be extracting it early in sensory processing. If popout and grouping are muted or absent, higher-order sensory or even conceptual processing must be involved in representing the objects in question. L and T share the same elementary features in common (one short short horizontal and one short vertical line touching at right angles); the main things that distinguish them in our minds are linguistic and conceptual factors.

So let's get back to Mirabelle. We know that real colors can lead to grouping and popout. Would her "private" colors be able to elicit the same effects?

To answer this question I devised patterns similar to the one shown in Figure 3.4: a forest of blocky 5s with a few blocky 2s scattered among them. Since the 5s are just mirror images of the 2s, they are composed of identical features: two vertical lines and three horizontal ones. When you look at this image, you manifestly do not get popout; you can only spot the 2s through item-by-item inspection. And you can't easily discern the global shape—the big triangle—by mentally grouping the 2s; they simply don't segregate from the background. Although you can eventually deduce logically that the 2s form a triangle, you don't *see* a big triangle the way you see the one in Figure 3.5, where the 2s have been rendered in black and the 5s in gray. Now, what if you were to show Figure 3.4 to a synesthete who claims to experience 2s as red and 5s as green? If she were merely thinking of red (and green) then, just like you and me, she wouldn't instantly see the triangle. On the other hand if synesthesia were a genuinely low-level sensory effect, she might literally see the triangle the way you and I do in Figure 3.5.

For this experiment we first showed images much like Figure 3.4 to twenty normal students and told them to look for a global shape (made of little 2s) among the clutter. Some of the figures contained a triangle, others showed a circle. We flashed these figures in a random sequence on a computer monitor for about half a second each, too short a time for detailed visual inspection. After seeing each figure the subjects had to press one of two buttons to indicate whether they had just been shown a circle or a triangle. Not surprisingly, the students' hit rate was about 50 percent; in other words, they were just guessing, since they couldn't spontaneously discern the shape. But if we colored all the 5s green and all the 2s red (in Figure 3.5 this is simulated with gray and black), their performance went up to 80 or 90 percent. They could now see the shape instantly without a pause or a thought.

The surprise came when we showed the black-and-white displays to Mirabelle. Unlike the nonsynesthetes, she was able to identify the shape correctly on 80 to 90 percent of trials—just as if the numbers were

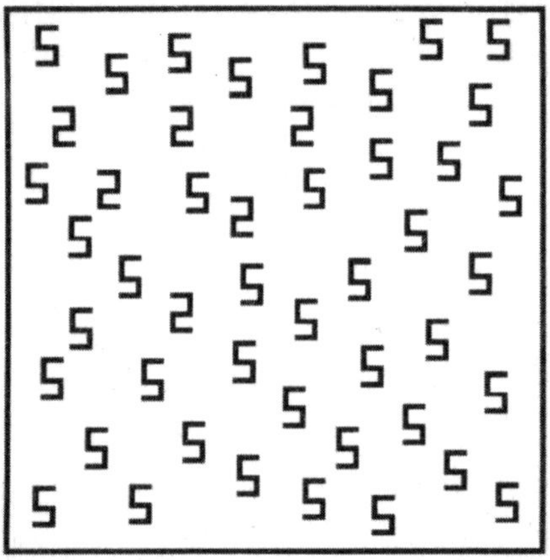

FIGURE 3.4   A cluster of 2s scattered among 5s. It is difficult for normal subjects to detect the shape formed by the 2s, but lower synesthetes as a group perform much better. The effect has been confirmed by Jamie Ward and his colleagues.

FIGURE 3.5   The same display as Figure 3.4 except that the numbers are shaded differently, allowing normal people to see the triangle instantly. Lower synesthetes ("projectors") presumably see something like this.

actually colored differently! The synesthetically induced colors were just as effective as real colors in allowing her to discover and report the global shape.[2] This experiment provides unassailable proof that Mirabelle's induced colors are genuinely sensory. There is simply no way she could fake it, and no way it could be the result of childhood memories or any of the other alternative explanations that have been proposed.

Ed and I realized that, for the first time since Francis Galton, we had clear, unambiguous proof from our experiments (grouping and popout) that synesthesia was indeed a real sensory phenomenon—proof that had eluded researchers for over a century. Indeed, our displays could not only be used to distinguish fakes from genuine synesthetes, but also to ferret out closet synesthetes, people who might have the ability but not realize it or not be willing to admit it.

ED AND I sat back in the café discussing our findings. Between our experiments with Francesca and Mirabelle, we had established that synesthesia exists. The next question was, why *does* it exist? Could a glitch in brain wiring explain it? What did we know that could help us figure this out? First, we knew that the most common type of synesthesia is apparently number-color. Second, we knew that one of the main color centers in the brain is an area called V4 in the fusiform gyrus of the temporal lobes. (V4 was discovered by Semir Zeki, professor of neuroesthetics at University College of London, and a world authority on the organization of the primate visual system.) Third, we knew that there may be areas in roughly the same part of the brain that are specialized for numbers. (We know this because small lesions to this part of the brain cause patients to lose arithmetic skills.) I thought, wouldn't it be wonderful if number-color synesthesia were simply caused by some accidental "cross-wiring" between the number and color centers in the brain? This seemed almost too obvious to be true—but why not? I suggested we look at some brain atlases to see exactly how close these two areas really are in relation to each other.

"Hey, maybe we can ask Tim," Ed responded. He was referring to Tim Rickard, a colleague of ours at the center. Tim had used sophisticated brain-imaging techniques like fMRI to map out the brain area where visual number recognition occurs. Later that afternoon, Ed and

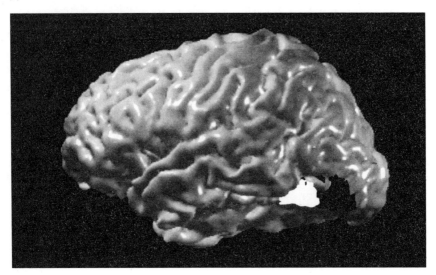

FIGURE 3.6   The left side of the brain showing the approximate location of the fusiform area: black, a number area; white, a color area (shown schematically on the surface).

I compared the exact location of V4 and the number area in an atlas of the human brain. To our amazement, we saw that the number area and V4 were right next to each other in the fusiform gyrus (Figure 3.6). This was strong support for the cross-wiring hypothesis. Can it really be a coincidence that the most common type of synesthesia is the number-color type, and the number and color areas are immediate neighbors in the brain?

This was starting to look too much like nineteenth-century phrenology, but maybe it was true! Since the nineteenth century a debate has raged between phrenology—the notion that different functions are sharply localized in different brain areas—versus holism, which holds that functions are emergent properties of the entire brain whose parts are in constant interaction. It turns out this is an artificial polarization to some degree, because the answer depends on the particular function one is talking about. It would be ludicrous to say that gambling or cooking are localized (although there may be aspects of them that are) but it would be equally silly to say that the cough reflex or the pupils' reflex to light is not localized. What's surprising, though, is that even some non-stereotyped functions, such as seeing colors or numbers (as shapes or even as numerical ideas), are in fact mediated by specialized brain regions.

Even high-level perceptions such as tools or vegetables or fruits—which border on being concepts rather than mere perceptions—can be lost selectively depending on the particular small region of the brain that is damaged by stroke or accident.

So what do we know about brain localization? How many special- ized regions are there, and how are they arranged? Just as the CEO of a corporation delegates different tasks to different people occupying differ- ent offices, your brain parcels out different jobs to different regions. The process begins when neural signals from your retina travel to an area in the back of your brain where the image gets categorized into different simple attributes such as color, motion, form, and depth. After that, infor- mation about separate features gets divvied up and distributed to several far-flung regions in your temporal and parietal lobes. For example, infor- mation about the direction of moving targets goes to V5 in your parietal lobes. Color information gets sent mainly to V4 in your temporal lobes.

The reason for this division of labor is not hard to divine. The kinds of computation you need for extracting information about wavelength (color) is very different from the computations required for extracting information about motion. It may be simpler to accomplish this if you have separate areas for each task, keeping the neural machinery distinct for economy of wiring and ease of computation.

It also makes sense to organize specialized regions into hierarchies. In a hierarchical system, each "higher" level carries out more sophisticated tasks but, just like in a corporation, there is an enormous amount of feed- back and crosstalk. For example, color information processed in V4 gets relayed to higher color areas that lie farther up in the temporal lobes, near the angular gyrus. These higher areas may be concerned with more complex aspects of color processing. The eucalyptus leaves I see all over campus appear to be the same shade of green at dusk as they do midday, even though the wavelength composition of light reflected is very dif- ferent in the two cases. (Light at dusk is red, but you don't suddenly see leaves as reddish green; they still look green because your higher color areas compensate.)

Numerical computation, too, seems to occur in stages: an early stage in the fusiform gyrus where the actual shapes of numbers are repre- sented, and a later stage in the angular gyrus concerned with numerical concepts such as ordinality (sequence) and cardinality (quantity). When the angular gyrus is damaged by a stroke or a tumor, a patient may still be able to identify numbers but can no longer divide or subtract. (Multiplication often survives because it is learned by rote.) It was this aspect of brain anatomy—the close proximity of colors and numbers in the brain in both the fusiform gyrus and near the angular gyrus—that made me suspect that number-color synesthesia was caused by crosstalk between these specialized brain areas.

But if such neural cross-wiring is the correct explanation, why does it occur at all? Galton observed that synesthesia runs in families, a finding that has been repeatedly confirmed by other researchers. Thus it is fair to ask whether there is a genetic basis for synesthesia. Perhaps synesthetes harbor a mutation that causes some abnormal connections to exist between adjacent brain areas that are normally well segregated from each other. If this mutation is useless or deleterious, why hasn't it been weeded out by natural selection?

Furthermore, if the mutation were to be expressed in a patchy manner, it might explain why some synesthetes "cross-wire" colors and numbers whereas others, like a synesthete I once saw named Esmerelda, see colors in response to musical notes. Consistent with Esmerelda's case, hearing centers in the temporal lobes are close to the brain areas that receive color signals from V4 and higher color centers. I felt the pieces were starting to fall into place.

The fact that we see various types of synesthesia provides additional evidence for cross-wiring. Perhaps the mutant gene expresses itself to a greater degree, in more brain regions, in some synesthetes than in others. But how exactly does the mutation cause cross-wiring? We know that the normal brain does not come ready-made with neatly packaged areas that are clearly delineated from each other. In the fetus there is an initial dense overproliferation of connections that get pruned back as development proceeds. One reason for this extensive pruning process is presumably to avoid leakage (signal spread) between adjacent areas, just as Michelangelo whittled away excess marble to produce *David*. This pruning is largely under genetic control. It's possible that the synesthesia mutation leads to incomplete pruning between some areas that lie close to each other. The net result would be the same: cross-wiring.

However, it is important to note that anatomical cross-wiring between brain areas cannot be the complete explanation for synesthesia.

If it were, how could you account for the commonly reported emergence of synesthesia during the use of hallucinogenic drugs such as LSD? A drug can't suddenly induce sprouting of new axon connections, and such connections would not magically vanish after the drug wore off. Thus it must be enhancing the activity of preexisting connections in some way—which is not inconsistent with the possibility that synesthetes have more of these connections than the rest of us. David Brang and I also encountered two synesthetes who temporarily lost their synesthesia when they started taking antidepressant drugs called selective serotonin reuptake inhibitors (SSRIs), a drug family that famously includes Prozac. While subjective reports cannot entirely be relied on, they do provide valuable clues for future studies. One person was able to switch her synesthesia on or off by starting or stopping her drug regimen. She detested the antidepressant Wellbutrin because it deprived her of the sensory magic that synesthesia provided; the world looked drab without it.

I have been using the word "cross-wiring" somewhat loosely, but until we know exactly what's going on at the cellular level, the more neutral term "cross-activation" might be better. We know, for instance, that adjacent brain regions often inhibit each other's activity. This inhibition serves to minimize crosstalk and keeps areas insulated from one other. What if there were a chemical imbalance of some kind that reduces this inhibition—say, the blocking of an inhibitory neurotransmitter, or a failure to produce it? In this scenario there would not be any extra "wires" in the brain, but the synesthete's wires would not be properly insulated. The result would be the same: synesthesia. We know that, even in a normal brain, extensive neural connections exist between regions that lie far apart. The normal function of these is unknown (as with most brain connections!), but a mere strengthening of these connections or a loss of inhibition might lead to the kind of cross-activation I suggest.

In light of the cross-activation hypothesis we can now also start to guess why Francesca had such powerful emotional reactions to mundane textures. All of us have a primary touch map in the brain called the primary somatosensory cortex, or S1. When I touch you on the shoulder, touch receptors in your skin detect the pressure and send a message to your S1. You feel the touch. Similarly when you touch different textures, a neighboring touch map, S2, is activated. You feel the textures: the dry grain of a wooden deck, the slippery wetness of a bar of soap. Such tactile sensations are fundamentally external, originating from the world outside your body.

Another brain region, the insula, maps internal feelings from your body. Your insula receives continuous streams of sensation from receptor cells in your heart, lungs, liver, viscera, bones, joints, ligaments, fascia, and muscles, as well as from specialized receptors in your skin that sense heat, cold, pain, sensual touch, and perhaps tickle and itch as well. Your insula uses this information to represent how you feel in relation to the outside world and your immediate environment. Such sensations are fundamentally *internal*, and comprise the primary ingredients of your emotional state. As a central player in your emotional life, your insula sends signals to and receives signals from other emotional centers in your brain including the amygdala, the autonomic nervous system (powered by the hypothalamus), and the orbitofrontal cortex, which is involved in nuanced emotional judgments. In normal people these circuits are activated when they touch certain emotionally charged objects. Caressing, say, a lover, could generate complex feelings of ardor, intimacy, and pleasure. Squeezing a lump of feces, in contrast, likely leads to strong feelings of disgust and revulsion. Now think of what would happen if there were an extreme exaggeration of these very connections linking S2, the insula, the amygdala, and the orbitofrontal cortex. You would expect to see precisely the sort of touch-triggered complex emotions that Francesca experiences when she touches denim, silver, silk, or paper—things that would leave most of us unmoved.

Incidentally, Francesca's mother also has synesthesia. But in addition to emotions, she reports taste sensations in response to touch. For example, caressing a wrought-iron fence evokes an intense salty flavor in her mouth. This too makes sense: The insula receives strong taste input from the tongue.

WITH THE IDEA of cross-activation we seemed to be homing in on a neurological explanation for number-color and textural synesthesia.[3] But as other synesthetes showed up in my office, we realized there are many more forms of the condition. In some people, days of the week or months of the year produced colors: Monday might be green, Wednesday pink, and December yellow. No wonder many scientists thought they were crazy! But, as I said earlier, I've learned over the years to listen to what people say. In this particular case, I realized that the only thing days of the week, months, and numbers have in common is the concept of numerical sequence or ordinality. So in these individuals, unlike Becky and Susan, perhaps it *is* the abstract concept of numerical sequence that evokes the color, rather than the visual appearance of the number. Why the difference between the two types of synesthetes? To answer this, we have to return to brain anatomy.

After the shape of a number is recognized in your fusiform, the message is relayed further on to your angular gyrus, a region in your parietal lobes involved, among other things, in higher color processing. The idea that some types of synesthesia might involve the angular gyrus is consistent with an old clinical observation that this structure is involved in cross-sensory synthesis. In other words, it is thought that this is a grand junction where information about touch, hearing, and vision flow together to enable the construction of high-level percepts. For example, a cat purrs and is fluffy (touch), it purrs and meows (hearing), and it has a certain appearance (vision) and fishy breath (smell)—all of which are evoked by the memory of a cat or the sound of the word "cat." No wonder patients with damage here lose the ability to name things (anomia) even though they can recognize them. They have difficulty with arithmetic, which, if you think about it, also involves cross-sensory integration: in kindergarten you learn to count with your fingers, after all. (Indeed, if you touch the patient's finger and ask her which one it is, she often can't tell you.) All of these bits of clinical evidence strongly suggest that the angular gyrus is a great center in the brain for sensory convergence and integration. So perhaps it's not so outlandish, after all, that a flaw in the circuitry could lead to colors being quite literally evoked by certain sounds.

According to clinical neurologists, the left angular gyrus in particular may be involved in juggling numerical quantity, sequences, and arithmetic. When this region is damaged by stroke, the patient can recognize numbers and can still think reasonably clearly, but he has difficulty with even the simplest arithmetic. He can't subtract 7 from 12. I have seen patients who cannot tell you which of two numbers—3 or 5—is larger.

Here we have the perfect arrangement for another type of cross-wiring. The angular gyrus is involved in color processing and numerical sequences. Could it be that, in some synesthetes, the crosstalk occurs between these two higher areas near the angular gyrus rather than lower down in the fusiform? If so, that would explain why, in them, even abstract number representations or the idea of a number prompted by days of the week or months will strongly manifest color. In other words, depending on which part of the brain the abnormal synesthesia gene is expressed, you get different types of synesthetes: "higher" synesthetes driven by numerical concept, and "lower" synesthetes driven by visual appearance alone. Given the multiple back-and-forth connections between brain areas, it is also possible that numerical ideas about sequentiality are sent back down to the fusiform gyrus to evoke colors.

In 2003 I began a collaboration with Ed Hubbard and Geoff Boynton from the Salk Institute for Biological Studies to test these ideas with brain imaging. The experiment took four years, but we were finally able to show that, in grapheme-color synesthetes, the color area V4 lights up even when you present colorless numbers. This cross-activation could never happen in you or me. In recent experiments carried out in Holland, researchers Romke Rouw and Steven Scholte found that there were substantially more axons ("wires") linking V4 and the grapheme area in lower synesthetes compared to the general population. And even more remarkably, in higher synesthetes, they found a greater number of fibers in the general vicinity of the angular gyrus. This all is precisely what we had proposed. The fit between prediction and subsequent confirmation rarely proceeds so smoothly in science.

The observations we had made so far broadly support the cross-activation theory and provide an elegant explanation of the different perceptions of "higher" and "lower" synesthetes.[4] But there are many other tantalizing questions we can ask about the condition. What if a letter synesthete were bilingual and knew two languages with different alphabets, such as Russian and English? The English *P* and the Cyrillic *Π* represent more or less the same phoneme (sound) but look completely dissimilar. Would they evoke the same or different colors? Is the grapheme alone critical, or is it the phoneme? Maybe in lower synesthetes it's the visual appearance that drives it whereas in higher synesthetes it's the sound. And what about uppercase versus lowercase letters? Or letters depicted in cursive writing? Do the colors of two adjacent graphemes run or flow into each other, or do they cancel each other out? To my

knowledge none of these questions have been adequately answered yet—
which means we have many exciting years of synesthesia research ahead
of us. Fortunately, many new researchers have joined us in the enterprise
including Jamie Ward, Julia Simner, and Jason Mattingley. There is now
a whole thriving industry on the subject.

Let me tell you about one last patient. In Chapter 2 we noted that
the fusiform gyrus represents not only shapes like letters of the alphabet
but faces as well. Thus, shouldn't we expect there to be cases in which a
synesthete sees different *faces* as possessing intrinsic colors? We recently
came across a student, Robert, who reported experiencing exactly that.
He usually saw the color as a halo around the face, but when he was
inebriated the color would become much more intense and spread into
the face itself.[5] To find out if Robert was being truthful we did a simple
experiment. I asked him to stare at the nose of a photograph of another
college student and asked Robert what color he saw around the face.
Robert said the student's halo was red. I then briefly flashed either red or
green dots on different locations in the halo. Robert's gaze immediately
darted toward a green spot but only rarely toward a red one; in fact,
he claimed not to have seen the red spots at all. This provides compelling evidence that Robert really was seeing halos: On a red background,
green would be conspicuous while red would be almost imperceptible.

To add to the mystery, Robert also had Asperger syndrome, a high-
functioning form of autism. This made it difficult for him to understand
and "read" people's emotions. He could do so through intellectual deduction from the context, but not with the intuitive ease most of us enjoy.
Yet for Robert, every emotion also evoked a specific color. For example,
anger was blue and pride was red. So his parents taught him very early
in life to use his colors to develop a taxonomy of emotions to compensate for his deficit. Interestingly, when we showed him an arrogant face,
he said it was "purple and therefore arrogant." (It later dawned on all
three of us that purple is a blend or red and blue, evoked by pride and
aggression, and the latter two, if combined, would yield arrogance. Robert hadn't made this connection before.) Could it be that Robert's whole
subjective color spectrum was being mapped in some systematic manner onto his "spectrum" of social emotions? If so, could we potentially
use him as a subject to understand how emotions—and complex blends
of them—are represented in the brain? For example, are pride and

arrogance differentiated solely on the basis of the surrounding social context, or are they inherently distinct subjective qualities? Is a deep-seated insecurity also an ingredient of arrogance? Are the whole spectrum of subtle emotions based on various combinations, in different ratios, of a small number of basic emotions?

Recall from Chapter 2 that color vision in primates has an intrinsically rewarding aspect that most other components of visual experience do not elicit. As we saw, the evolutionary rationale for neurally linking color with emotion was probably initially to attract us to ripe fruits and/ or tender new shoots and leaves, and later to attract males to swollen female rumps. I suspect that these effects arise through interactions between the insula and higher brain regions devoted to color. If the same connections are abnormally strengthened—and perhaps slightly scrambled—in Robert, this would explain why he saw many colors as strongly tinged with arbitrary emotional associations.

BY NOW I was intrigued by another question. What's the connection—if any—between synesthesia and creativity? The only thing they seem to have in common is that both are equally mysterious. Is there truth to the folklore that synesthesia is more common in artists, poets, and novelists, and perhaps in creative people in general? Could synesthesia explain creativity? Wassily Kandinsky and Jackson Pollock were synesthetes, and so was Vladimir Nabokov. Perhaps the higher incidence of synesthesia in artists is rooted deep in the architecture of their brains.

Nabokov was very curious about his synesthesia and wrote about it in some of his books. For example:

> . . . In the green group, there are alder-leaf f, the unripe apple of
> p, and pistachio t. Dull green, combined somehow with violet,
> is the best I can do for w. The yellows comprise various e's and
> i's, creamy d, bright-golden y, and u, whose alphabetical value I
> can express only by "brassy with an olive sheen." In the brown
> group, there are the rich rubbery tone of soft g, paler j, and the
> drab shoelace of h. Finally, among the reds, b has the tone called
> burnt sienna by painters, m is a fold of pink flannel, and today I
> have at last perfectly matched v with "Rose Quartz" in Maerz and

Paul's *Dictionary of Color.* (From *Speak, Memory: An Autobiography Revisited,* 1966)

He also pointed out that both his parents were synesthetes and seemed intrigued that his father saw K as yellow, his mother saw it as red, and he saw it as orange—a blend of the two! It isn't clear from his writings whether he regarded this blending as a coincidence (which it almost certainly is) or thought of it as a genuine hybridization of synesthesia.

Poets and musicians also seem to enjoy a higher incidence of synesthesia. On his website the psychologist Sean Day provides his translation of a passage from an 1895 German article that quotes the great musician Franz Liszt:

> When Liszt first began as Kapellmeister in Weimar (1842), it astonished the orchestra that he said: "O please, gentlemen, a little bluer, if you please! This tone type requires it!" Or: "That is a deep violet, please, depend on it! Not so rose!" First the orchestra believed Liszt just joked; . . . later they got accustomed to the fact that the great musician seemed to see colors there, where there were only tones.

The French poet and synesthete Arthur Rimbaud wrote the poem, "Vowels," which begins:

> *A black, E white, I red, U green, O blue: vowels,*
> *I shall tell, one day, of your mysterious origins:*
> *A, black velvety jacket of brilliant flies*
> *which buzz around cruel smells, . . .*

According to one recent survey, as many as a third of all poets, novelists, and artists claim to have had synesthetic experiences of one sort or another, though a more conservative estimate would be one in six. But is this simply because artists have vivid imaginations and are more apt to express themselves in metaphorical language? Or maybe they are just less inhibited about admitting having had such experiences? Or are they simply claiming to be synesthetes because it is "sexy" for an artist to be a synesthete? If the incidence is genuinely higher, why?

One thing that poets and novelists have in common is that they are especially good at using metaphor. ("It is the East, and Juliet is the sun!") It's as if their brains are better set up than the rest of ours to forge links between seemingly unrelated domains—like the sun and a beautiful young woman. When you hear "Juliet is the sun," you don't say, "Oh, does that mean she is an enormous, glowing ball of fire?" If asked to explain the metaphor, you instead say things like, "She is warm like the sun, nurturing like the sun, radiant like the sun, dispels darkness like the sun." Your brain instantly finds the right links highlighting the most salient and beautiful aspects of Juliet. In other words, just as synesthesia involves making arbitrary links between seemingly unrelated perceptual entities like colors and numbers, metaphor involves making nonarbitrary links between seemingly unrelated conceptual realms. Perhaps this isn't just a coincidence.

The key to this puzzle is the observation that at least some high-level concepts are anchored, as we have seen, in specific brain regions. If you think about it, there is nothing more abstract than a number. Warren McCulloch, a founder of the cybernetics movement in the mid-twentieth century, once asked the rhetorical question, "What is a number that Man may know it? And what is Man that he may know number?" Yet there it is, number, neatly packaged in the small, tidy confines of the angular gyrus. When it is damaged, the patient can no longer do simple arithmetic.

Brain damage can make a person lose the ability to name tools but not fruits and vegetables, or only fruits and not tools, or only fruits but not vegetables. All of these concepts are stored close to one other in the upper parts of the temporal lobes, but clearly they are sufficiently separated so that a small stroke can knock out one but leave the others intact. You might be tempted to think of fruits and tools as perceptions rather than concepts, but in fact two tools—say, a hammer and saw—can be visually as dissimilar from each other as they are from a banana; what unites them is a semantic understanding about their purpose and use.

If ideas and concepts exist in the form of brain maps, perhaps we have the answer to our question about metaphor and creativity. If a mutation were to cause excess connections (or alternatively, to permit excess cross-leakage) between different brain areas, then depending on where and how widely in the brain the trait was expressed, it could lead to both synesthesia *and* a heightened facility for linking seemingly unrelated concepts, words, images, or ideas. Gifted writers and poets may have excess connections between word and language areas. Gifted painters and graphic artists may have excess connections between high-level visual areas. Even a single word like "Juliet" or "sun" can be thought of as the center of a semantic whirlpool, or of a rich swirl of associations. In the brain of a gifted wordsmith, excess connections would mean larger whirlpools and therefore larger regions of overlap and a concomitantly higher propensity toward metaphor. This could explain the higher incidence of synesthesia in creative people in general. These ideas take us back full circle. Instead of saying "Synesthesia is more common among artists because they are being metaphorical," we should say, "They are better at metaphors because they are synesthetes."

If you listen to your own conversations, you will be amazed to see how frequently metaphors pop up in ordinary speech. ("Pop up"—see?) Indeed, far from being mere decoration, the use of metaphor and our ability to uncover hidden analogies is the basis of all creative thought. Yet we know almost nothing about why metaphors are so evocative and how they are represented in the brain. Why is "Juliet is the sun" more effective than "Juliet is a warm, radiantly beautiful woman"? Is it simply economy of expression, or is it because the mention of the sun automatically evokes a visceral feeling of warmth and light, making the description more vivid and in some sense real? Maybe metaphors allow you to carry out a sort of virtual reality in the brain. (Bear in mind also that even "warm" and "radiant" are metaphors! Only "beautiful" isn't.)

There is no simple answer to this question, but we do know that some very specific brain mechanisms—even specific brain regions—might be critical, because the ability to use metaphors can be selectively lost in certain neurological and psychiatric disorders. For instance, in addition to experiencing difficulty using words and numbers, there are hints that people with damage to the left inferior parietal lobule (IPL) often also lose the ability to interpret metaphors and become extremely literal minded. This hasn't been "nailed down" yet, but the evidence is compelling.

If asked, "What does 'a stitch in time saves nine' mean?" a patient with an IPL stroke might say, "It's good to stitch up a hole in your shirt before it gets too large." He will completely miss the metaphorical meaning of the proverb even when told explicitly that it is a proverb. This leads me to wonder whether the angular gyrus may have originally evolved for mediating cross-sensory associations and abstractions but then, in humans, was coopted for making all kinds of associations, including metaphorical ones. Metaphors seem paradoxical: On the one hand, a metaphor isn't literally true, and yet on the other hand a well-turned metaphor seems to strike like lightning, revealing the truth more deeply or directly than a drab, literal statement.

I get chills whenever I hear Macbeth's immortal soliloquy from Act 5, Scene 5:

Out, out, brief candle!
Life's but a walking shadow, a poor player
That struts and frets his hour upon the stage,
And then is heard no more. It is a tale
Told by an idiot, full of sound and fury,
Signifying nothing.

Nothing he says is literal. He is not actually talking about candles or stagecraft or idiots. If taken literally, these lines really would be the ravings of an idiot. And yet these words are one of the most profound and deeply moving remarks about life that anyone has ever made!

Puns, on the other hand, are based on superficial associations. Schizophrenics, who have miswired brains, are terrible at interpreting metaphors and proverbs. Yet according to clinical folklore, they are very good at puns. This seems paradoxical because, after all, both metaphors and puns involve linking seemingly unrelated concepts. So why should schizophrenics be bad at the former but good with the latter? The answer is that even though the two appear similar, puns are actually the opposite of metaphor. A metaphor exploits a surface-level similarity to reveal a deep hidden connection. A pun is a surface-level similarity that masquerades as a deep one—hence its comic appeal. ("What fun do monks have on Christmas?" Answer: "Nun.") Perhaps a preoccupation with "easy" surface similarities erases or deflects attention from deeper connections. When I asked a schizophrenic what an elephant had in common with a man, he answered "They both carry a trunk"; alluding maybe to the man's penis (or maybe to an actual trunk used for storage).

Leaving puns aside, if my ideas about the link between synesthesia and metaphor are correct, then why isn't every synesthete highly gifted or every great artist or poet a synesthete? The reason may be that synesthesia might merely predispose you to be creative, but this does not mean other factors (both genetic and environmental) aren't involved in the full flowering of creativity. Even so, I would suggest that similar—though not identical—brain mechanisms might be involved in both phenomena, and so understanding one might help us understand the other.

An analogy might be helpful. A rare blood disorder called sickle cell anemia is caused by a defective recessive gene that causes red blood cells to assume an abnormal "sickle" shape, making them unable to transport oxygen. This can be fatal. If you happen to inherit two copies of this gene (in the unlikely event that both your parents had either the trait or the disease itself), then you develop the full-blown disease. However, if you inherit just one copy of this gene, you do not come down with the disease, though you can still pass it on to your children. Now it turns out that, although sickle-cell anemia is extremely rare in most parts of the world, where natural selection has effectively weeded it out, its incidence is ten times higher in certain parts of Africa. Why should this be? The surprising answer is that the sickle-cell trait actually seems to protect the affected individual from malaria, a disease caused by a mosquito-borne parasite that infects and destroys blood cells. This protection conferred on the population as a whole from malaria outweighs the reproductive disadvantage caused by the occasional rare appearance of an individual with double copies of the sickle-cell gene. Thus the apparently maladaptive gene has actually been selected for by evolution, but only in geographic locations where malaria is endemic.

A similar argument has been proposed for the relatively high incidence of schizophrenia and bipolar disorder in humans. The reason these disorders have not been weeded out may be because having *some* of the genes that lead to the full-blown disorder are advantageous—perhaps boosting creativity, intelligence, or subtle social-emotional faculties. Thus humanity as a whole benefits from keeping these genes in its gene pool, but the unfortunate side effect is a sizable minority who get bad combinations of them.

Carrying this logic forward, the same could well be true for synesthesia. We have seen how, by dint of anatomy, genes that lead to enhanced cross-activation between brain areas could have been highly advantageous by making us creative as a species. Certain uncommon variants or combinations of these genes might have the benign side effect of producing synesthesia. I hasten to emphasize the part about benign: Synesthesia is not deleterious like sickle-cell disease and mental illness, and in fact most synesthetes seem to really enjoy their abilities and would not opt to have them "cured" even if they could. This is only to say that the general mechanism might be the same. This idea is important because it makes clear that synesthesia and metaphor are not synonymous, and yet they share a deep connection that might give us deep insights into our marvelous uniqueness.[6]

Thus synesthesia is best thought of as an example of subpathological cross-modal interactions that could be a signature or marker for creativity. (A modality is a sensory faculty, such as smell, touch, or hearing. "Cross-modal" refers to sharing information between senses, as when your vision and hearing together tell you that you're watching a badly dubbed foreign film.) But as often happens in science, it got me thinking about the fact that even in those of us who are nonsynesthetes a great deal of what goes on in our mind depends on entirely normal cross-modal interactions that are not arbitrary. So there is a sense in which at some level we are all "synesthetes." For example, look at the two shapes in Figure 3.7. The one on the left looks like a paint splat. The one on the right resembles a jagged piece of shattered glass. Now let me ask you, if you had to guess, which of these is a "bouba" and which is a "kiki"? There is no right answer, but odds are you picked the splat as "bouba" and the glass as "kiki." I tried this in a large classroom recently, and 98 percent of the students made this choice. Now you might think this has something to do with the blob resembling the physical form of the letter *B* (for "bouba") and the jagged thing resembling a *K* (as in "kiki"). But if you try the experiment on non-English-speaking people in India or China, where the writing systems are completely different, you find exactly the same thing.

Why does this happen? The reason is that the gentle curves and undulations of contour on the amoeba-like figure metaphorically (one might say) mimic the gentle undulations of the sound *bouba*, as represented in the hearing centers in the brain and in the smooth rounding and relaxing of the lips for producing the curved *booo-baaa* sound. On

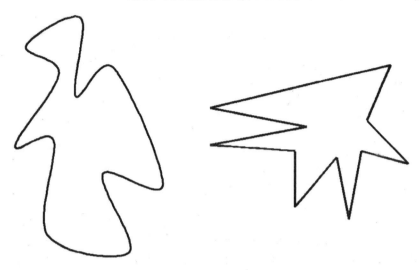

FIGURE 3.7   Which of these shapes is "bouba" and which is "kiki"? Such stimuli were originally used by Heinz Werner to explore interactions between hearing and vision.

the other hand, the sharp wave forms of the sound *kee-kee* and the sharp inflection of the tongue on the palate mimic the sudden changes in the jagged visual shape. We will return to this demonstration in Chapter 6 and see how it might hold the key to understanding many of the most mysterious aspects of our minds, such as the evolution of metaphor, language, and abstract thought.[7]

I HAVE ARGUED so far that synesthesia, and in particular the existence of "higher" forms of synesthesia (involving abstract concepts rather than concrete sensory qualities) can provide clues to understanding some of the high-level thought processes that humans alone are capable of.[8] Can we apply these ideas to what is arguably the loftiest of our mental traits, mathematics? Mathematicians often speak of seeing numbers laid out in space, roaming this abstract realm to discover hidden relationships that others might have missed, such as Fermat's Last Theorem or Goldbach's conjecture. Numbers and space? Are they being metaphorical?

One day in 1997, after I had consumed a glass of sherry, I had a flash of insight—or at least thought I had. (Most of the "insights" I have when inebriated turn out to be false alarms.) In his original *Nature* paper,

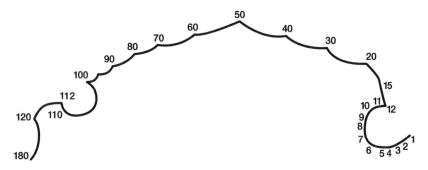

FIGURE 3.8   Galton's number line. Notice that *12* is a tiny bit closer to *1* than it is to *6*.

Galton described a second type of synesthesia that is even more intriguing than the number-color condition. He called it "number forms." Other researchers use the phrase "number line." If I asked you to visualize the numbers 1 to 10 in your mind's eye, you will probably report a vague tendency to see them mapped in space sequentially, left to right, as you were taught in grade school. But number-line synesthetes are different. They able to visualize numbers vividly and do not see the numbers arranged sequentially from left to right, but on a snaking, twisting line that can even double back on itself, so that 36 might be closer to 23, say, than it is to 38 (Figure 3.8). One could think of this as "number-space" synesthesia, in which every number is always in a particular location in space. The number line for any individual remains constant even if tested on intervals separated by months.

As with all experiments in psychology, we needed a method to prove Galton's observation experimentally. I called upon my students Ed Hubbard and Shai Azoulai to help set up the procedures. We first decided to look at the well-known "number distance" effect seen in normal people. (Cognitive psychologists have examined every conceivable variation of the effect on hapless student volunteers, but its relevance to number-space synesthesia was missed until we came along.) Ask anyone which of two numbers is larger, 5 or 7? 12 or 50? Anyone who has been through grade school will get it right every time. The interesting part comes when you clock how long it takes people to spit out each of their answers. This latency between showing them a number pair and their verbal response is their reaction time (RT). It turns out that the greater the distance between two numbers the shorter the RT, and contrariwise, the closer two numbers are, the longer it takes to form an answer. This suggests that your brain represents numbers in some sort of an actual mental number line which you consult "visually" to determine which is greater. Numbers that are far apart can be easily eyeballed, while numbers that are close together need closer inspection, which takes a few extra milliseconds.

We realized we could exploit this paradigm to see if the convoluted number-line phenomenon really existed or not. We could ask a number-space synesthete to compare number pairs and see if her RTs corresponded to the real conceptual distance between numbers or would reflect the idiosyncratic geometry of her own personal number line. In 2001 we managed to recruit an Austrian student named Petra who was a number-space synesthete. Her highly convoluted number line doubled back on itself so that, for example, 21 was spatially closer to 36 than it was to 18. Ed and I were very excited. As of that time there had not been any study on the number-space phenomenon since the time when Galton discovered it in 1867. No attempt had been made to establish its authenticity or to suggest what causes it. So any new information, we realized, would be valuable. At least we could set the ball rolling.

We hooked Petra up to a machine that measured her RT to questions such as "Which is bigger, 36 or 38?" or (on a different trial) "36 or 23?" As often happens in science, the result wasn't entirely clear one way or the other. Petra's RT seemed to depend partially on the numerical distance and partially on spatial distance. This wasn't the conclusive result we had hoped for, but it did suggest that her number-line representation wasn't entirely left-to-right and linear as it is in normal brains. Some aspects of number representation in her brain were clearly messed up.

We published our finding in 2003 in a volume devoted to synesthesia, and it inspired much subsequent research. The results have been mixed, but at the very least we revived interest in an old problem that had been largely ignored by the pundits, and we suggested ways of testing it objectively.

Shai Azoulai and I followed up with a second experiment on two new number-space synesthetes that was designed to prove the same point. This time we used a memory test. We asked each synesthete to remember sets of nine numbers (for example, 13, 6, 8, 18, 22, 10, 15, 2, 24) displayed randomly on various spatial locations on the screen. The experiment contained two conditions. In condition A, nine random numbers were scattered randomly about the two-dimensional screen. In condition B, each number was placed where it "should" be on each synesthete's personal convoluted line as if it had been projected, or "flattened," onto the screen. (We had initially interviewed each subject to find out the geometry of his or her personal number line and determined which numbers the subject placed close to each other within that idiosyncratic coordinate system.) In each condition the subjects were asked to view the display for 30 seconds in order to memorize the numbers. After a few minutes they were simply asked to report all the numbers they could recall having seen. The result was striking: The most accurate recall was for the numbers they had seen in condition B. Again we had shown that these people's personal number lines were real. If they weren't, or if their shapes varied across time, why should it matter where the numbers had been placed? Putting the numbers where they "should" be in each synesthete's personal number line apparently facilitated that person's memory for the numbers—something you wouldn't see in a normal person.

One more observation deserves special mention. Some of our number-space synesthetes told us spontaneously that the shape of their personal number lines strongly influenced their ability to do arithmetic. In particular, subtraction or division (but not multiplication, which, again, is memorized by rote) was much more difficult across sudden sharp kinks in their lines than it was along relatively straight portions of it. On the other hand, some creative mathematicians have told me that their twisted number lines enable them to see hidden relationships between numbers that elude us lesser mortals. This observation convinced me that both mathematical savants and creative mathematicians are not being merely metaphorical when they speak of wandering a spatial landscape of numbers. They are seeing relationships that are not obvious to us less-gifted mortals.

As for how these convoluted number lines come to exist in the first place, that is still hard to explain. A number represents many things—eleven apples, eleven minutes, the eleventh day of Christmas—but what they have in common are the semiseparate notions of order and quantity. These are very abstract qualities, and our apish brains surely were not under selective pressure to handle mathematics per se. Studies of hunter-gatherer societies suggest that our prehistoric ancestors probably had names for a few small numbers—perhaps up to ten, the number of our fingers—but more advanced and flexible counting systems are cultural inventions of historical times; there simply wouldn't have been enough for the brain to evolve a "lookup table" or number module starting from scratch. On the other hand (no pun intended), the brain's representation of space is almost as ancient as mental faculties come. Given the opportunistic nature of evolution, it is possible that the most convenient way to represent abstract numerical ideas, including sequentiality, is to map them onto a preexisting map of visual space. Given that the parietal lobe originally evolved to represent space, is it a surprise that numerical calculations are also computed there, especially in the angular gyrus? This is a prime example of what might have been a unique step in human evolution.

In the spirit of taking a speculative leap, I would like to argue that further specialization might have occurred in our space-mapping parietal lobes. The left angular gyrus might be involved in representing ordinality. The right angular gyrus might be specialized for quantity. The simplest way to spatially map out a numerical sequence in the brain would be a straight line from left to right. This in turn might be mapped onto notions of quantity represented in the right hemisphere. But now let's assume that the gene that allows such remapping of sequence on visual space is mutated. The result might be a convoluted number line of the kind you see in number-space synesthetes. If I were to guess, I'd say other types of sequence—such as months or weeks—are also housed in the left angular gyrus. If this is correct, we should expect that a patient with a stroke in this area might have difficulty in quickly telling you whether, for example, Wednesday comes after or before Tuesday. Someday I hope to meet such a patient.

ABOUT THREE MONTHS after I had embarked on synesthesia research, I encountered a strange twist. I received an email from one of my undergraduate students, Spike Jahan. I opened it expecting to find the usual "please reconsider my grade" request, but it turned out that he's a number-color synesthete who had read about our work and wanted to be tested. Nothing strange so far, but then he dropped a bombshell: He's color-blind. A color-blind synesthete! My mind began to reel. If he experiences colors, are they anything like the colors you or I experience? Could synesthesia shed light on that ultimate human mystery, conscious awareness?

Color vision is a remarkable thing. Even though most of us can experience millions of subtly different hues, it turns out our eyes use only three kinds of color photoreceptors, called cones, to represent all of them. As we saw in Chapter 2, each cone contains a pigment that responds optimally to just one color: red, green, or blue. Although each type of cone responds optimally only to one specific wavelength, it will also respond to a lesser extent to other wavelengths that are close to the optimum. For example, red cones respond vigorously to red light, fairly well to orange, weakly to yellow, and hardly at all to green or blue. Green cones respond best to green, less well to yellowish green, and even less to yellow. Thus every specific wavelength of (visible) light stimulates your red, green, and blue cones by a specific amount. There are literally millions of possible three-way combinations, and your brain knows to interpret each one as a separate color.

Color blindness is a congenital condition in which one or more of these pigments is deficient or absent. A color-blind person's vision works perfectly normally in nearly every respect, but she can see only a limited range of hues. Depending on which cone pigment is lost and on the extent of loss, she may be red-green color-blind or blue-yellow color-blind. In rare cases two pigments are deficient, and the person sees purely in black and white.

Spike had the red-green variety. He experienced far fewer colors in the world than most of us do. What was truly bizarre, though, was that he often saw numbers tinged with colors that he had never seen in the real world. He referred to them, quite charmingly and appropriately, as "Martian colors" that were "weird" and seemed quite "unreal." He could only see these when looking at numbers.

Ordinarily one would be tempted to ignore such remarks as being crazy, but in this case the explanation was staring me in the face. I realized that my theory about cross-activation of brain maps provides a neat explanation for this bizarre phenomenon. Remember, Spike's cone receptors are deficient, but the problem is entirely in his eyes. His retinas are unable to send the full normal range of color signals up to the brain, but in all likelihood his cortical color-processing areas, such as V4 in the fusiform, are perfectly normal. At the same time, he is a number-color synesthete. Thus number shapes are processed normally all the way up to his fusiform and then, due to cross-wiring, produce cross-activation of cells in his V4 color area. Since Spike has never experienced his missing colors in the real world and can do so only by looking at numbers, he finds them incredibly strange. Incidentally, this observation also demolishes the idea that synesthesia arises from early-childhood memory associations such as having played with colored magnets. For how can someone "remember" a color he has never seen? After all, there are no magnets painted with Martian colors!

It is worth pointing out that non-color-blind synesthetes may also see "Martian" colors. Some describe letters of the alphabet as being composed of multiple colors simultaneously "layered on top of each other" making them not quite fit the standard taxonomy of colors. This phenomenon probably arises from mechanisms similar to those observed in Spike; the colors look weird because the connections in his visual pathways are weird and thus uninterpretable.

What is it like to experience colors that don't appear anywhere in the rainbow, colors from another dimension? Imagine how frustrating it must be to sense something you cannot describe. Could you explain what it feels like to see blue to a person who has been blind from birth? Or the smell of Marmite to an Indian, or saffron to an Englishman? It raises the old philosophical conundrum of whether we can ever really know what someone else is experiencing. Many a student has asked the seemingly naïve question, "How do I know that your red isn't my blue?" Synesthesia reminds us that this question may not be that naïve after all. As you may recall from earlier, the term for referring to the ineffable subjective quality of conscious experience is "qualia." These questions about whether other people's qualia are similar to our own, or different, or possibly absent, may seem as pointless as asking how many angels can dance on the head of a pin—but I remain hopeful. Philosophers have struggled with these questions for centuries, but here at last, with our blooming knowledge about synesthesia, a tiny crack in the door of this mystery may be opening. This is the way science works: Begin with simple, clearly formulated, tractable questions that can pave the way for eventually answering the Big Questions, such as "What are qualia," "What is the self," and even "What is consciousness?"

Synesthesia might be able to give us some clues to these abiding mysteries[9,10] because it provides a way of selectively activating some visual areas while skipping or bypassing others. It is not ordinarily possible to do this. So instead of asking the somewhat nebulous questions "What is consciousness?" and "What is the self?" we can refine our approach to the problem by focusing on just one aspect of consciousness—our awareness of visual sensations—and ask ourselves, Does conscious awareness of redness require activation of all or most of the thirty areas in the visual cortex? Or only a small subset of them? What about the whole cascade of activity from the retina to the thalamus to the primary visual cortex before the messages get relayed to the thirty higher visual areas? Is their activity also required for conscious experience, or can you skip them and directly activate V4 and experience an equally vivid red? If you look at a red apple, you would ordinarily activate the visual area for both color (red) and form (apple-like). But what if you could artificially stimulate the color area without stimulating cells concerned with form? Would you experience disembodied red color floating out there in front of you like a mass of amorphous ectoplasm or other spooky stuff? And lastly, we also know that there are many more neural projections going backward from each level in the hierarchy of visual processing to earlier areas than there are going forward. The function of these back-projections is completely unknown. Is their activity required for conscious awareness of red? What if you could selectively silence them with a chemical while you looked at a red apple—would you lose awareness? These questions come perilously close to being the kind of impossible-to-do armchair thought experiments that philosophers revel in. The key difference is that such experiments really can be done—maybe within our lifetimes.

And then we may finally understand why apes care about nothing beyond ripe fruit and red rumps, while we are drawn to the stars.

CHAPTER 4

The Neurons That Shaped Civilization

———

*Even when we are alone, how often do we think with pain and plea-
sure of what others think of us, or their imagined approbation or disap-
probation; and this all follows from sympathy, a fundamental element
of the social instincts.*

—CHARLES DARWIN

A FISH KNOWS HOW TO SWIM THE INSTANT IT HATCHES, AND OFF
it darts to fend for itself. When a duckling hatches, it can follow its
mother over land and across the water within moments. Foals, still drip-
ping with amniotic fluid, spend a few minutes bucking around to get the
feel of their legs, then join the herd. Not so with humans. We come out
limp and squalling and utterly dependent on round-the-clock care and
supervision. We mature glacially, and do not approach anything resembling adult competence for many, many years. Obviously we must gain
some very large advantage from this costly, not to mention risky up-front
investment, and we do: It's called culture.

In this chapter I explore how a specific class of brain cells, called mirror neurons, may have played a pivotal role in our becoming the one
and only species that veritably lives and breathes culture. Culture consists of massive collections of complex skills and knowledge which are
transferred from person to person through two core mediums, language
and imitation. We would be nothing without our savant-like ability to
imitate others. Accurate imitation, in turn, may depend on the uniquely
human ability to "adopt another's point of view"—both visually and

metaphorically—and may have required a more sophisticated deployment of these neurons compared with how they are organized in the
brains of monkeys. The ability to see the world from another person's
vantage point is also essential for constructing a mental model of another
person's complex thoughts and intentions in order to predict and manipulate his behavior. ("Sam thinks I don't realize that Martha hurt him.")
This capacity, called theory of mind, is unique to humans. Finally, certain aspects of language itself—that vital medium of cultural transmission—was probably built at least partly on our facility for imitation.

Darwin's theory of evolution is one of the most important scientific
discoveries of all time. Unfortunately, however, the theory makes no pro-
vision for an afterlife. Consequently it has provoked more acrimonious
debate than any other topic in science—so much so that some school
districts in the United States have insisted on giving the "theory" of intelligent design (which is really just a fig leaf for creationism) equal status
in textbooks. As has been pointed out repeatedly by the British scientist and social critic Richard Dawkins, this is little different from giving
equal status to the idea that the sun goes around Earth. At the time evolutionary theory was proposed—long before the discovery of DNA and
the molecular machinery of life, back when paleontology had just barely
begun to piece together the fossil record—the gaps in our knowledge
were sufficiently large to leave room for honest doubt. That point is long
past, but that doesn't mean we have solved the entire puzzle. It would
be arrogant for a scientist to deny that there are still many important
questions about the evolution of the human mind and brain that remain
unanswered. At the top of my list would be the following:

1.   The hominin brain reached nearly its present size, and
perhaps even its present intellectual capacity, about 300,000 years
ago. Yet many of the attributes we regard as uniquely human—
such as toolmaking, fire building, art, music, and perhaps even
full-blown language—appeared only much later, around 75,000
years ago. Why? What was the brain doing during that long incu-
bation period? Why did it take so long for all this latent potential
to blossom, and then why did it blossom so suddenly? Given that
natural selection can only select expressed abilities, not latent ones,
how did all this latent potential get built up in the first place? I

shall call this "Wallace's problem" after the Victorian naturalist Alfred Russel Wallace, who first proposed it when discussing the origins of language:

> The lowest savages with the least copious vocabularies [have] the capacity of uttering a variety of distinct articulate sounds and of applying them to an almost infinite amount of modulation and inflection [which] is not in any way inferior to that of the higher [European] races. An instrument has been developed in advance of the needs of its possessor.

2. Crude Oldowan tools—made by just a few blows to a core stone to create an irregular edge—emerged 2.4 million years ago and were probably made by *Homo habilis*, whose brain size was halfway between that of chimps and modern humans. After another million years of evolutionary stasis, aesthetically pleasing symmetrical tools began to appear which reflected a standardization of production technique. These required switching from a hard hammer to a soft, perhaps wooden, hammer while the tool was being made, so as to ensure a smooth rather than a jagged, irregular edge. And lastly, the invention of stereotyped assembly-line tools—sophisticated symmetrical bifacial tools that were hafted to a handle—took place only two hundred thousand years ago. Why was the evolution of the human mind punctuated by these relatively sudden upheavals of technological change? What was the role of tool use in shaping human cognition?

3. Why was there a sudden explosion—what Jared Diamond, in his book *Guns, Germs, and Steel*, calls the "great leap"—in mental sophistication around sixty thousand years ago? This is when widespread cave art, clothing, and constructed dwellings appeared. Why did these advances come along only then, even though the brain had achieved its modern size almost a million years earlier? It's the Wallace problem again.

4. Humans are often called the "Machiavellian primate," referring to our ability to predict other people's behavior and outsmart them. Why are we humans so good at reading one another's intentions? Do we have a specialized brain module, or circuit,

for generating a theory of other minds, as proposed by the British cognitive neuroscientists Nicholas Humphrey, Uta Frith, Marc Hauser, and Simon Baron-Cohen? Where is this circuit and when did it evolve? Is it present in some rudimentary form in monkeys and apes, and if so, what makes ours so much more sophisticated than theirs?

5.   How did language evolve? Unlike many other human traits such as humor, art, dancing, and music, the survival value of language is obvious: It lets us communicate our thoughts and intentions. But the question of how such an extraordinary ability actually came into being has puzzled biologists, psychologists, and philosophers since at least Darwin's time. One problem is that the human vocal apparatus is vastly more sophisticated than that of any other ape, but without the correspondingly sophisticated language areas in the human brain, such exquisite articulatory equipment alone would be useless. So how did these two mechanisms with so many elegant interlocking parts evolve in tandem? Following Darwin's lead, I suggest that our vocal equipment and our remarkable ability to modulate our voice evolved mainly for producing emotional calls and musical sounds during courtship in early primates, including our hominin ancestors. Once that evolved, the brain—especially the left hemisphere—could start using it for language.

But an even bigger puzzle remains. Is language mediated by a sophisticated and highly specialized mental "language organ" that is unique to humans and that emerged completely out of the blue, as suggested by the famous MIT linguist Noam Chomsky? Or was there a more primitive gestural communication system already in place that provided scaffolding for the emergence of vocal language? A major piece of the solution to this riddle comes from the discovery of mirror neurons.

I HAVE ALREADY alluded to mirror neurons in earlier chapters and will return to them again in Chapter 6, but here in the context of evolution let's take a closer look. In the frontal lobes of a monkey's brain, there are certain cells that fire when the monkey performs a very specific action.

For instance, one cell fires during the pulling of a lever, a second for grabbing a peanut, a third for putting a peanut in the mouth, and yet a fourth for pushing something. (Bear in mind, these neurons are part of a small *circuit* performing a highly specific task; a single neuron by itself doesn't move a hand, but its response allows you to eavesdrop on the circuit.) Nothing new so far. Such motor-command neurons were discovered by the renowned Johns Hopkins University neuroscientist Vernon Mountcastle several decades ago.

While studying these motor-command neurons in the late 1990s, another neuroscientist, Giacomo Rizzolatti, and his colleagues Giuseppe Di Pellegrino, Luciano Fadiga, and Vittorio Gallese, from the University of Parma in Italy, noticed something very peculiar. Some of the neurons fired not only when the monkey performed an action, but also when it watched another monkey performing the same action! When I heard Rizzolatti deliver this news during a lecture one day, I nearly jumped off my seat. These were not mere motor-command neurons; they were adopting the other animal's point of view (Figure 4.1). These neurons (again, actually the neural circuit to which they belong; from now on I'll use the word "neuron" for "the circuit") were for all intents and purposes reading the other monkey's mind, figuring out what it was up to. This is an indispensable trait for intensely social creatures like primates.

It isn't clear how exactly the mirror neuron is wired up to allow this predictive power. It is as if higher brain regions are reading the output from it and saying (in effect), "The same neuron is now firing in my brain as would be firing if I were reaching out for a banana; so the other monkey must be intending to reach for that banana now." It is as if mirror neurons are nature's own virtual-reality simulations of the intentions of other beings.

In monkeys these mirror neurons enable the prediction of simple goal-directed actions of other monkeys. But in humans, and in humans alone, they have become sophisticated enough to interpret even complex intentions. How this increase in complexity took place will be hotly debated for some time to come. As we will see later, mirror neurons also enable you to imitate the movements of others, thereby setting the stage for the cultural "inheritance" of skills developed and honed by others. They may have also propelled a self-amplifying feedback loop that kicked in at one point to accelerate brain evolution in our species.

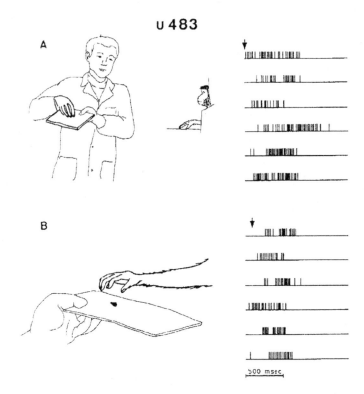

FIGURE 4.1 Mirror neurons: Recordings of nerve impulses (shown on the right) from the brain of a rhesus monkey (a) watching another being reach for a peanut, and (b) reaching out for the peanut. Thus each mirror neuron (there are six) fires both when the monkey observes the action and when the monkey executes the action itself.

As Rizzolatti noted, mirror neurons may also enable you to mime the lip and tongue movements of others, which in turn could provide the evolutionary basis for verbal utterances. Once these two abilities are in place—the ability to read someone's intentions and the ability to mimic their vocalizations—you have set in motion two of the many foundational events that shaped the evolution of language. You need no longer speak of a unique "language organ," and the problem doesn't seem quite so mysterious anymore. These arguments do not in any way negate the idea that there are specialized brain areas for language in humans. We are dealing here with the question of how such areas may have evolved, not whether they exist or not. An important piece of the puzzle is

Rizzolatti's observation that one of the chief areas where mirror neurons abound, the ventral premotor area in monkeys, may be the precursor of our celebrated Broca's area, a brain center associated with the expressive aspects of human language.

Language is not confined to any single brain area, but the left inferior parietal lobe is certainly one of the areas that are crucially involved, especially in the representation of word meaning. Not coincidentally, this area is also rich in mirror neurons in the monkey. But how do we actually know that mirror neurons exist in the human brain? It is one thing to saw open the skull of a monkey and spend days or weeks probing around with a microelectrode, but people do not seem interested in volunteering for such procedures.

One unexpected hint comes from patients with a strange disorder called anosognosia, a condition in which people seem unaware of or deny their disability. Most patients with a right-hemisphere stroke have complete paralysis of the left side of their body and, as you might expect, complain about it. But about one in twenty of them will vehemently deny their paralysis even though they are mentally otherwise lucid and intelligent. For example, President Woodrow Wilson, whose left side was paralyzed by a stroke in 1919, insisted that he was perfectly fine. Despite the clouding of his thought processes and against all advice, he remained in office, making elaborate travel plans and major decisions pertaining to American involvement in the League of Nations.

In 1996 some colleagues and I made our own little investigation of anosognosia and noticed something new and amazing: Some of these patients not only denied their own paralysis, but also denied the paralysis of another patient—and let me assure you, the second patient's inability to move was as clear as day. Denying one's own paralysis is odd enough, but why deny another patient's paralysis? We suggest that this bizarre observation is best understood in terms of damage to Rizzolatti's mirror neurons. It's as if anytime you want to make a judgment about someone else's movements, you have to run a virtual-reality simulation of the corresponding movements in your own brain. And without mirror neurons you cannot do this.

The second piece of evidence for mirror neurons in humans comes from studying certain brain waves in humans. When people perform volitional actions with their hands, the so-called mu wave disappears completely. My colleagues Eric Altschuler, Jaime Pineda, and I found that mu-wave suppression also occurs when a person watches someone *else* moving his hand, but not if he watches a similar movement by an inanimate object, such as a ball bouncing up and down. We suggested at the Society for Neuroscience meeting in 1998 that this suppression was caused by Rizzolatti's mirror-neuron system.

Since Rizzolatti's discovery, other types of mirror neurons have been found. Researchers at the University of Toronto were recording from cells in the anterior cingulate in conscious patients who were undergoing neurosurgery. Neurons in this area have long been known to respond to physical pain. On the assumption that such neurons respond to pain receptors in the skin, they are often called sensory pain neurons. Imagine the head surgeon's astonishment when he found that the sensory pain neuron he was monitoring responded equally vigorously when the patient watched another patient being poked! It was as though the neuron was empathizing with someone else. Neuroimaging experiments on human volunteers conducted by Tania Singer also supported this conclusion. I like calling these cells "Gandhi neurons" because they blur the boundary between self and others—not just metaphorically, but quite literally, since the neuron can't tell the difference. Similar neurons for touch have since been discovered in the parietal lobe by a group headed by Christian Keysers using brain-imaging techniques.

Think of what this means. Anytime you watch someone doing something, the neurons that your brain would use to do the same thing become active—as if you yourself were doing it. If you see a person being poked with a needle, your pain neurons fire away as though you were being poked. It is utterly fascinating, and it raises some interesting questions. What prevents you from blindly imitating every action you see? Or from literally feeling someone else's pain?

In the case of motor mirror neurons, one answer is that there may be frontal inhibitory circuits that suppress the automatic mimicry when it is inappropriate. In a delicious paradox, this need to inhibit unwanted or impulsive actions may have been a major reason for the evolution of free will. Your left inferior parietal lobe constantly conjures up vivid images of multiple options for action that are available in any given context, and your frontal cortex suppresses all but one of them. Thus it has been suggested that "free won't" may be a better term than free will. When these frontal inhibitory circuits are damaged, as in frontal lobe syndrome, the patient sometimes mimics gestures uncontrollably, a symptom called echopraxia. I would predict, too, that some of these patients might literally experience pain if you poke someone else, but to my knowledge this has never been looked for. Some degree of leakage from the mirror-neuron system can occur even in normal individuals. Charles Darwin pointed out that, even as adults, we feel ourselves unconsciously flexing our knee when watching an athlete getting ready to throw a javelin, and clench and unclench our jaws when we watch someone using a pair of scissors.[1]

Turning now to the *sensory* mirror neurons for touch and pain, why doesn't their firing automatically make us feel everything we witness? It occurred to me that perhaps the null signal ("I am not being touched") from skin and joint receptors in your own hand block the signals from your mirror neurons from reaching conscious awareness. The overlapping presence of the null signals and the mirror-neuron activity is interpreted by higher brain centers to mean, "Empathize, by all means, but don't literally feel that other guy's sensations." Speaking in more general terms, it is the dynamic interplay of signals from frontal inhibitory circuits, mirror neurons (both frontal and parietal), and null signals from receptors that allow you to enjoy reciprocity with others while simultaneously preserving your individuality.

At first this explanation was an idle speculation on my part, but then I met a patient named Humphrey. Humphrey had lost his hand in the first Gulf War and now had a phantom hand. As is true in other patients, whenever he was touched on his face, he felt sensations in his missing hand. No surprises so far. But with ideas about mirror neurons brewing in my mind, I decided to try a new experiment. I simply had him watch another person—my student Julie—while I stroked and tapped her hand. Imagine our amazement when he exclaimed with considerable surprise that he could not merely see but actually feel the things being done to Julie's hand on his phantom. I suggest this happens because his mirror neurons were being activated in the normal fashion but there was no longer a null signal from the hand to veto them. Humphrey's mirror neuron activity was emerging fully into conscious experience. Imagine: The only thing separating your consciousnesses from another's might be your skin! After seeing this phenomenon in Humphrey we tested three other patients and found the same effect, which we dubbed "acquired hyperempathy." Amazingly, it turns out that some of these patients get relief from phantom limb pain by merely watching another person being massaged. This might prove useful clinically because, obviously, you can't directly massage a phantom.

These surprising results raise another fascinating question. Instead of amputation, what if a patient's brachial plexus (the nerves connecting the arm to the spinal cord) were to be anesthetized? Would the patient then experience touch sensations in his anesthetized hand when merely watching an accomplice being touched? The surprising answer is yes. This result has radical implications, for it suggests that no major structural reorganization in the brain is required for the hyperempathy effect; merely numbing the arm is adequate. (I did this experiment with my student Laura Case.) Once again, the picture that emerges is a much more dynamic view of brain connections than what you would be led to believe from the static picture implied by textbook diagrams. Sure enough, brains are made up of modules, but the modules are not fixed entities; they are constantly being updated through powerful interactions with each other, with the body, the environment, and indeed with other brains.

MANY NEW QUESTIONS have emerged since mirror neurons were discovered. First, are mirror-neuron functions present innately, or learned, or perhaps a little of both? Second, how are mirror neurons wired up, and how do they perform their functions? Third, why did they evolve (if they did)? Fourth, do they serve any purpose beyond the obvious one for which they were named? (I will argue that they do.)

I have already hinted at possible answers but let me expand. One skeptical view of mirror neurons is that they are just a result of associative learning, as when a dog salivates in anticipation of dinner when she hears her master's key in the front door lock each evening. The argument is that every time a monkey moves his hand toward the peanut, not only does the "peanut grabbing" command neuron fire, but so does the visual neuron that is activated by the appearance of his own hand reaching for a peanut. Since neurons that "fire together wire together," as the old mnemonic goes, eventually even the mere sight of a moving hand (its own or another monkey's) triggers a response from the command neurons. But if this is the correct explanation, why do only a subset of the command neurons fire? Why aren't all the command neurons for this action mirror neurons? Furthermore, the visual appearance of another person reaching toward a peanut is very different from your view of your own hand. So how does the mirror neuron apply the appropriate correction for vantage point? No simple straightforward associationist model can account for this. And finally, so what if learning plays a role in constructing mirror neurons? Even if it does, that doesn't make them any less interesting or important for understanding brain function. The question of what mirror neurons are doing and how they work is quite independent of the question of whether they are wired up by genes or by the environment.

Highly relevant to this discussion is an important discovery made by Andrew Meltzoff, a cognitive psychologist at the University of Washington's Institute for Learning and Brain Sciences in Seattle. He found that a newborn infant will often protrude its tongue when watching its mother do it. And when I say newborn I mean it—just a few hours old. The neural circuitry involved must be hardwired and not based on associative learning. The child's smile echoing the mother's smile appears a little later, but again it can't be based on learning since the baby can't see its own face. It has to be innate.

It has not been proven whether mirror neurons are responsible for these earliest imitative behaviors, but it's a fair bet. The ability would depend on mapping the visual appearance of the mother's protruding tongue or smile onto the child's own motor maps, controlling a finely adjusted sequence of facial muscle twitches. As I noted in my BBC Radio Reith Lectures in 2003, entitled "The Emerging Mind," this sort of translation between maps is precisely what mirror neurons are thought to do, and if this ability is innate, it is truly astonishing. I'll call it the "sexy" version of the mirror-neuron function.

Some people argue that the complex computational ability for true imitation—based on mirror neurons—emerges only later in development, whereas the tongue protrusion and first smile are merely hardwired reflexes in response to simple "triggers" from mom, the same way a cat's claws come out when it sees a dog. The only way to distinguish the sexy from the mundane explanation would be to see whether a baby can imitate a nonstereotyped movement it is unlikely to ever encounter in nature, such as an asymmetrical smile, a wink, or a curious distortion of the mouth. This couldn't be done by a simple hardwired reflex. The experiment would settle the issue once and for all.

INDEPENDENT OF THE question of whether mirror neurons are innate or acquired, let us now take a closer look at what they actually do. Many functions were proposed when they were first reported, and I'd like to build on these earlier speculations.[2] Let's make a list of things they might be doing. Bear in mind they may have originally evolved for purposes other than the ones listed here. These secondary functions may simply be a bonus, but that doesn't make them any less useful.

First, and most obvious, they allow you to figure out someone else's intentions. When you see your friend Josh's hand moves toward the ball, your own ball-reaching neurons start firing. By running this virtual simulation of being Josh, you get the immediate impression that he is intending to reach for the ball. This ability to entertain a theory of mind may exist in the great apes in rudimentary form, but we humans are exceptionally good at it.

Second, in addition to allowing us to see the world from another person's *visual* vantage point, mirror neurons may have evolved further, enabling us to adopt the other person's *conceptual* vantage point. It may not be entirely coincidental that we use metaphors like "I see what you mean" or "Try to see it from my point of view." How this magic step from literal to conceptual viewpoint occurred in evolution—if indeed it occurred—is of fundamental importance. But it is not an easy proposition to test experimentally.

As a corollary to adopting the other's point of view, you can also see yourself as others see you—an essential ingredient of self-awareness. This is seen in common language: When we speak of someone being "self-conscious," what we really mean is that she is conscious of someone else being conscious of her. Much the same can be said for a word like "self-pity." I will return to this idea in the concluding chapter on consciousness and mental illness. There I will argue that other-awareness and self-awareness coevolved in tandem, leading to the I-you reciprocity that characterizes humans.

A less obvious function of mirror neurons is abstraction—again, something humans are especially good at. This is well illuminated by the bouba-kiki experiment discussed discussed in Chapter 3 in the context of synesthesia. To reiterate, over 95 percent of people identify the jagged form as the "kiki" and the curvy one as "bouba." The explanation I gave is that the sharp inflections of the jagged shape mimic the inflection of the sound *ki-ki*, not to mention the sudden deflection of the tongue from the palate. The gentle curves of bulbous shape, on the other hand, mimic the *boooooo-baaaaaa* contour of the sound and the tongue's undulation on the palate. Similarly, the sound *shhhhhhhh* (as in "shall") is linked to a blurred, smudged line, whereas *rrrrrrrrrrrrrrrr* is linked to a sawtooth-shaped line, and an *sssssssss* (as in "sip") to a fine silk thread—which shows that it's not the mere similarity of the jagged shape to the letter *K* that produces the effect, but genuine cross-sensory abstraction. The link between the bouba-kiki effect and mirror neurons may not be immediately evident, but there is a fundamental similarity. The main computation done by mirror neurons is to transform a map in one dimension, such as the visual appearance of someone else's movement, into another dimension, such as the motor maps in the observer's brain, which contain programs for muscle movements (including tongue and lip movements).

This is exactly what's going on in the bouba-kiki effect: Your brain is performing an impressive feat of abstraction in linking your visual and auditory maps. The two inputs are entirely dissimilar in every way except one—the abstract properties of jaggedness or curviness—and your brain homes in on this common denominator very swiftly when you are asked to pair them up. I call this process "cross-modal abstraction." This ability to compute similarities despite surface differences may have paved the way for more complex types of abstraction that our species takes great delight in. Mirror neurons may be the evolutionary conduit that allowed this to happen.

Why did a seemingly esoteric ability like cross-modal abstraction evolve in the first place? As I suggested in a previous chapter, it may have emerged in ancestral arboreal primates to allow them to negotiate and grasp tree branches. The vertical visual inputs of tree limbs and branches reaching the eye had to be matched with totally dissimilar inputs from joints and muscles and the body's felt sense of where it is in space—an ability that would have favored the development of both canonical neurons and mirror neurons. The readjustments that were required in order to establish a congruence between sensory and motor maps may have initially been based on feedback, both at the genetic level of the species and at the experiential level of the individual. But once the rules of congruence were in place, the cross-modal abstraction could occur for novel inputs. For instance, picking up a shape that is visually perceived to be tiny would result in a spontaneous movement of almost-opposed thumb and forefingers, and if this were mimicked by the lips to produce a correspondingly diminutive orifice (through which you blow air), you would produce sounds (words) that sound small (such as "teeny weeny," "diminutive," or in French "*un peu*," and so on). These small "sounds" would in turn feed back via the ears to be linked to tiny shapes. (This, as we shall see in Chapter 6, may have been how the first words evolved in our ancestral hominins.) The resulting three-way resonance between vision, touch, and hearing may have progressively amplified itself as in an echo chamber, culminating in the full-fledged sophistication of cross-sensory and other more complex types of abstraction.

If this formulation is correct, some aspects of mirror-neuron function may indeed be acquired through learning, building on a genetically specified scaffolding unique to humans. Of course, many monkeys and even lower vertebrates may have mirror neurons, but the neurons may need to develop a certain minimum sophistication and number of connections with other brain areas before they can engage in the kinds of abstractions that humans are good at.

What parts of the brain are involved in such abstractions? I already hinted (about language) that the inferior parietal lobule (IPL) may have played a pivotal role, but let's take a closer look. In lower mammals the IPL isn't very large, but it becomes more conspicuous in primates. Even within primates it is disproportionately large in the great apes, reaching a climax in humans. Finally, only in humans do we see a major portion of this lobule splitting further into two, the angular gyrus and the supramarginal gyrus, suggesting that something important was going on in this region of the brain during human evolution. Lying at the crossroads between vision (occipital lobes), touch (parietal lobes), and hearing (temporal lobes), the IPL is strategically located to receive information from all sensory modalities. At a fundamental level, cross-modal abstraction involves the dissolution of barriers to create modality-free representations (as exemplified by the bouba-kiki effect). The evidence for this is that when we tested three patients who had damage to the left angular gyrus, they performed poorly on the bouba-kiki task. As I already noted, this ability to map one dimension onto another is one of the things that mirror neurons are thought to be doing, and not coincidentally such neurons are plentiful in the general vicinity of the IPL. The fact that this region in the human brain is disproportionately large and differentiated suggests an evolutionary leap.

The upper part of the IPL, the supramarginal gyrus, is another structure unique to humans. Damage here leads to a disorder called ideomotor apraxia: a failure to perform skilled actions in response to the doctor's commands. Asked to pretend he is combing his hair, an apraxic will raise his arm, look at it, and flail it around his head. Asked to mime hammering a nail, he will make a fist and bang it on the table. This happens even though his hand isn't paralyzed (he will spontaneously scratch an itch) and he knows what "combing" means ("It means I am using a comb to tidy up my hair, Doctor"). What he lacks is the ability to conjure up a mental picture of the required action—in this case combing—which must precede and orchestrate the actual execution of the action. These are functions one would normally associate with mirror neurons, and indeed the supramarginal gyrus has mirror neurons. If our speculations are on the right track, then one would expect patients with apraxia to be terrible at understanding and imitating other people's movements. Although we have seen some hints of this, the matter requires careful investigation.

One also wonders about the evolutionary origin of metaphors. Once the cross-modal abstraction mechanism was set up between vision and touch in the IPL (originally for grasping branches), this mechanism could have paved the way for cross-sensory metaphors ("stinging rebuke," "loud shirt") and eventually for metaphors in general. This is supported by our recent observations that patients with angular gyrus lesions not only have difficulty with bouba-kiki, but also with understanding simple proverbs, interpreting them literally rather than metaphorically. Obviously these observations need to be confirmed on a larger sample of patients. It is easy to imagine how cross-modal abstraction might work for bouba-kiki, but how do you explain metaphors that combine very abstract concepts like "it is the east, and Juliet is the sun" given the seemingly infinite number of such concepts in the brain? The surprising answer to this question is that the number of concepts is *not* infinite, nor is the number of words that represent them. For all practical purposes, most English speakers have a vocabulary of about ten thousand words (although you can get by with far fewer if you are a surfer). There may be only some mappings that make sense. As the eminent cognitive scientist and polymath Jaron Lanier pointed out to me, Juliet can be the sun, but it makes little sense to say she is a stone or an orange juice carton. Bear in mind that the metaphors that get repeated and become immortal are the apt ones, the resonant ones. In doggerel, comically bad metaphors abound.

Mirror neurons play another important role in the uniqueness of the human condition: They allow us to imitate. You already know about tongue protrusion mimicry in infants, but once we reach a certain age, we can mime very complex motor skills, such as your mom's baseball swing or a thumbs-up gesture. No ape can match our imitative talents. However, I will note as an interesting aside here, the ape that comes closest to us in this regard is not our nearest cousin, the chimpanzee, but the orangutan. Orangutans can even open locks or use an oar to row, once they have seen someone else do it. They are also the most arboreal and prehensile of the great apes, so their brains may be jam-packed with mirror neurons for allowing their babies to watch mom in order to learn how to negotiate trees without the penalties of trial and error. If by some miracle an isolated pocket of orangs in Borneo survives the environmental holocaust that *Homo sapiens* seems hell-bent on bringing about, these meek apes may well inherit the earth.

Miming may not seem like an important skill—after all, "aping" someone is a derogatory term, which is ironic given that most apes are actually not very good at imitation. But as I have previously argued, miming may have been the key step in hominin evolution, resulting in our ability to transmit knowledge through example. When this step was taken, our species suddenly made the transition from gene-based Darwinian evolution through natural selection—which can take millions of years—to cultural evolution. A complex skill initially acquired through trial and error (or by accident, as when some ancestral hominid first saw a shrub catching fire from lava) could be transmitted rapidly to every member of a tribe, both young and old. Other researchers including

Merlin Donald have made the same point, although not in relation to mirror neurons.[3]

THIS LIBERATION FROM the constraints of a strictly gene-based Darwinian evolution was a giant step in human evolution. One of the big puzzles in human evolution is what we earlier referred to as the "great leap forward," the relatively sudden emergence between sixty thousand and a hundred thousand years ago of a number of traits we regard as uniquely human: fire, art, constructed shelters, body adornment, multicomponent tools, and more complex use of language. Anthropologists often assume this explosive development of cultural sophistication must have resulted from a set of new mutations affecting the brain in equally complex ways, but that doesn't explain why all of these marvelous abilities should have emerged at roughly the same time.

One possible explanation is that the so-called great leap is just a statistical illusion. The arrival of these traits may in fact have been smeared out over a much longer period of time than the physical evidence depicts. But surely the traits don't have to emerge at exactly the same time for the question to still be valid. Even spread out, thirty thousand years is just a blip compared to the millions of years of small, gradual behavioral changes that took place prior to that. A second possibility is that the new brain mutations simply increased our general intelligence, the capacity for abstract reasoning as measured by IQ tests. This idea is on the right track, but it doesn't tell us much—even leaving aside the very legitimate criticism that intelligence is a complex, multifaceted ability which can't be meaningfully averaged into a single general ability.

That leaves a third possibility, one that brings us back full circle to mirror neurons. I suggest that there was indeed a genetic change in the brain, but ironically the change *freed* us from genetics by enhancing our ability to learn from one another. This unique ability liberated our brain from its Darwinian shackles, allowing the rapid spread of unique inventions—such as making cowry-shell necklaces, using fire, constructing tools and shelter, or indeed even inventing new words. After 6 billion years of evolution, culture finally took off, and with culture the seeds of civilization were sown. The advantage of this argument is that you don't need to postulate separate mutations arriving nearly simultaneously to account for the coemergence of our many and various unique mental abilities. Instead, increased sophistication of a single mechanism—such as imitation and intention reading—could explain the huge behavioral gap between us and apes.

I'll illustrate with an analogy. Imagine a Martian naturalist watching human evolution over the last five hundred thousand years. She would of course be puzzled by the great leap forward that occurred fifty thousand years ago, but would be even more puzzled by a second great leap which occurred between 500 B.C.E. and the present. Thanks to certain innovations such as those in mathematics—in particular, the zero, place value, and numerical symbols (in India in the first millennium B.C.E.), and geometry (in Greece during the same period)—and, more recently, in experimental science (by Galileo)—the behavior of a modern civilized person is vastly more complex than that of humans ten thousand to fifty thousand years ago.

This second leap forward in culture was even more dramatic than the first. There is a greater behavioral gap between pre– and post–500 B.C.E. humans than between, say, *Homo erectus* and early *Homo sapiens*. Our Martian scientist might conclude that a new set of mutations made this possible. Yet given the time scale, that's just not possible. The revolution stemmed from a set of purely environmental factors which happened fortuitously at the same time. (Let's not forget the invention of the printing press, which allowed the extraordinary spread and near universal availability of knowledge that usually remained confined to the elite.) But if we admit this, then why doesn't the same argument apply to the first great leap? Maybe there was a lucky set of environmental circumstances and a few accidental inventions by a gifted few which could tap into a preexisting ability to learn and propagate information quickly— the basis of culture. And in case you haven't guessed by now, that ability might hinge on a sophisticated mirror-neuron system.

A caveat is in order. I am not arguing that mirror neurons are sufficient for the great leap or for culture in general. I'm only saying that they played a crucial role. Someone has to discover or invent something— like noticing the spark when two rocks are struck together—before the discovery can spread. My argument is that even if such accidental innovations were hit upon by chance by individual early hominins, they would have fizzled out were it not for a sophisticated mirror-neuron system. After all, even monkeys have mirror neurons, but they are not bearers of a proud culture. Their mirror-neuron system is either not advanced enough or is not adequately connected to other brain structures to allow the rapid propagation of culture. Furthermore, once the propagation mechanism was in place, it would have exerted selective pressure to make some outliers in the population more innovative. This is because innovations would only be valuable if they spread rapidly. In this respect, we could say mirror neurons served the same role in early hominin evolution as the Internet, Wikipedia, and blogging do today. Once the cascade was set in motion, there was no turning back from the path to humanity.

CHAPTER 5

Where Is Steven? The Riddle of Autism

———

You must always be puzzled by mental illness. The thing I would dread most, if I became mentally ill, would be your adopting a common sense attitude; that you could take it for granted that I was deluded.

—LUDWIG WITTGENSTEIN

"I KNOW STEVEN IS TRAPPED IN THERE SOMEWHERE, DR. RAMA-chandran. If only you could find a way to tell our son how dearly we love him, perhaps you could bring him out."

How often have physicians heard that heartbreaking lament from parents of children with autism? This devastating developmental disorder was discovered independently by two physicians, Leo Kanner in Baltimore and Hans Asperger in Vienna, in the 1940s. Neither doctor had any knowledge of the other, and yet by an uncanny coincidence they gave the syndrome the same name: autism. The word comes from the Greek *autos* meaning "self," a perfect description because the most striking feature of autism is a complete withdrawal from the social world and a marked reluctance or inability to interact with people.

Take Steven, for instance. He is six years old, with freckled cheeks and sandy-brown hair. He is sitting at a play table drawing pictures, his brow lightly furrowed in concentration. He is producing some beautiful drawings of animals. There's one of a galloping horse that is so wonderfully animated that it seems to leap out of the paper. You might be tempted to walk over and praise him for his talent. The possibility that he might be profoundly incapacitated would never cross your mind. But the moment you try to talk to him, you realize that there's a sense in which Steven the person simply isn't there. He is incapable of anything remotely resembling the two-way exchange of normal conversation. He refuses to make eye contact. Your attempts to engage him make him extremely anxious. He fidgets and rocks his body to and fro. All attempts to communicate with him meaningfully have been, and will be, in vain.

Since the time of Kanner and Asperger, there have been hundreds of case studies in the medical literature documenting, in detail, the various seemingly unrelated symptoms that characterize autism. These fall into two major groups: social-cognitive and sensorimotor. In the first group we have the single most important diagnostic symptom: mental aloneness and a lack of contact with the world, particularly the social world, as well as a profound inability to engage in normal conversation. Going hand in hand with this is an absence of emotional empathy for others. Even more surprising, autistic children express no outward sense of play, and they do not engage in the untrammeled make-believe with which normal children fill their waking hours. Humans, it has been pointed out, are the only animals that carry our sense of whimsy and playfulness into adulthood. How sad it must for parents to see their autistic sons and daughters impervious to the enchantment of childhood. Yet despite this social withdrawal, autistic children have a heightened interest in their inanimate surroundings, often to the point of being obsessive. This can lead to the emergence of odd, narrow preoccupations and a fascinations with things that seem utterly trivial to most of us, like memorizing all the phone numbers in a directory.

Let us turn now to the second cluster of symptoms: sensorimotor. On the sensory side, autistic children may find specific sensory stimuli highly distressing. Certain sounds, for example, can set off a violent temper tantrum. There is also a fear of novelty and change, and an obsessive insistence on sameness, routine, and monotony. The motor symptoms include a to-and-fro rocking of the body (such as we saw with Steven), repetitive hand movements including flapping motions and self-slapping, and sometimes elaborate, repetitive rituals. These sensorimotor symptoms are not quite as definitive or as devastating as the social-emotional ones, but they co-occur so frequently that they must be connected somehow. Our picture of what causes autism would be incomplete if we failed to account for them.

There is one more motor symptom to mention, one that I think holds the key to unraveling the mystery: Many autistic children have difficulty with miming and imitating other people's actions. This simple observation suggested to me a deficiency in the mirror-neuron system. Much of the remainder of this chapter chronicles my pursuit of this hypothesis and the fruit it has borne so far.

Not surprisingly, there have been dozens of theories of what causes autism. These can be broadly divided into psychological explanations and physiological explanations—the latter emphasizing innate abnormalities in brain wiring or neurochemistry. One ingenious psychological explanation, put forward by Uta Frith of University College of London and Simon Baron-Cohen of Cambridge University, is the notion that children with autism have a deficient theory of other minds. Less credible is the psychodynamic view that blames bad parenting, an idea that is so absurd that I won't consider it further.

We encountered the term "theory of mind" in passing in the previous chapter in relation to apes. Now let me explain it more fully. It is a technical term that is widely used in the cognitive sciences, from philosophy to primatology to clinical psychology. It refers to your ability to attribute intelligent mental beingness to other people: to understand that your fellow humans behave the way they do because (you assume) they have thoughts, emotions, ideas, and motivations of more or less the same kind as you yourself possess. In other words, even though you cannot actually feel what it is like to be another individual, you use your theory of mind to automatically project intentions, perceptions, and beliefs into the minds of others. In so doing you are able to infer their feelings and intentions and to predict and influence their behavior. Calling it a theory can be a little misleading, since the word "theory" is normally used to refer to an intellectual system of statements and predictions, rather than in this sense, where it refers to an innate, intuitive mental faculty. But that is the term my field uses, so that is the term I will use here. Most people do not appreciate just how complex and, frankly, miraculous it is that they possess a theory of mind. It seems as natural, as immediate, and as simple as looking and seeing. But as we saw in Chapter 2, the ability to see is actually a very complicated process that engages a widespread network of brain regions. Our species' highly sophisticated theory of mind is one of the most unique and powerful faculties of the human brain.

Our theory-of-mind ability apparently does not rely on our general intelligence—the rational intelligence you use to reason, to draw inferences, to combine facts, and so forth—but on a specialized set of brain mechanisms that evolved to endow us with our equally important degree of *social* intelligence. The idea that there might be specialized circuitry for social cognition was first suggested by psychologist Nick Humphrey and primatologist David Premack in the 1970s, and it now has a great deal of empirical support. So Frith's hunch about autism and theory of mind was compelling: Perhaps autistic children's profound deficits in social interactions stem from their theory-of-mind circuitry being somehow compromised. This idea is undoubtedly on the right track, but if you think about it, saying that autistic children cannot interact socially because they have a deficient theory of mind doesn't go very far beyond restating the observed symptoms. It's a good starting point, but what is really needed is to identify brain systems whose known functions match those that are deranged in autism.

Many brain-imaging studies have been conducted on children with autism, some pioneered by Eric Courchesne. It has been noted, for example, that children with autism have larger brains with enlarged ventricles (cavities in the brain). The same group of researchers has also noted striking changes in the cerebellum. These are intriguing observations that will surely have to be accounted for when we have a clearer understanding of autism. But they do not explain the symptoms that characterize the disorder. In children with damage to the cerebellum due to other organic diseases, one sees very characteristic symptoms, such as intention tremor (when the patient attempts to touch his nose, the hand begins to oscillate wildly), nystagmus (jerky eye movements), and ataxia (swaggering gait). None of these symptoms are typical of autism. Conversely, symptoms typical of autism (such as lack of empathy and social skills) are never seen in cerebellar disease. One reason for this might be that the cerebellar changes observed in autistic children may be the unrelated side effects of abnormal genes whose other effects are the true causes of autism. If so, what might these other effects be? What's needed, if we wish to explain autism, is candidate neural structures in the brain whose specific functions precisely match the particular symptoms that are unique to autism.

The clue comes from mirror neurons. In the late 1990s it occurred to my colleagues and me that these neurons provided precisely the candidate neural mechanism we were looking for. You can refer back to the previous chapter if you want a refresher, but suffice it to say, the discovery of mirror neurons was significant because they are essentially a network of mind-reading cells within the brain. They provided the missing physiological basis for certain high-level abilities that had long been challenging for neuroscientists to explain. We were struck by the fact that it is precisely these presumed functions of mirror neurons—such as empathy, intention-reading, mimicry, pretend play, and language learning—that are dysfunctional in autism.[1] (All of these activities require adopting the other's point of view—even if the other is imaginary—as in pretend play or enjoying action figures.) You can make two columns side by side, one for the known characteristics of mirror neurons and one for the clinical symptoms of autism, and there is an almost precise match. It seemed reasonable, therefore, to suggest that the main cause of autism is a dysfunctional mirror-neuron system. The hypothesis has the advantage of explaining many seemingly unrelated symptoms in terms of a single cause.

It might seem quixotic to suppose that there could be a single cause behind such a complex disorder, but we have to bear in mind that multiple effects do not necessarily imply multiple causes. Consider diabetes. Its manifestations are numerous and varied: polyuria (excessive urination), polydypsia (incessant thirst), polyphagia (increased appetite), weight loss, kidney disorders, ocular changes, nerve damage, gangrene, plus quite a few others. But underlying this miscellany is something relatively simple: either insulin deficiency or fewer insulin receptors on cell surfaces. Of course the disease is not simple at all. There are a lot of complex ins and outs; there are numerous environmental, genetic, and behavioral effects in play. But in the big picture, it comes down to insulin or insulin receptors. Analogously, our suggestion was that in the big picture the main cause of autism is a disturbed mirror-neuron system.

ANDREW WHITTEN'S GROUP in Scotland made this proposal at about the same time ours did, but the first experimental evidence for it came from our lab working in collaboration with researchers Eric Altschuler and Jaime Pineda here at UC San Diego. We needed a way to eavesdrop on mirror-neuron activity noninvasively, without opening the children's skulls and inserting electrodes. Fortunately, we found there was an easy way to do this using EEG (electroencephalography), which uses a grid of electrodes placed on the scalp to pick up brain waves. Long before CT scans and MRIs, EEG was the very first brain-imaging technology invented by humans. It was pioneered in the early twentieth century, and has been in clinical use since the 1940s. As the brain hums along in various states—awake, asleep, alert, drowsy, daydreaming, focused, and so on—it generates tell-tale patterns of electrical brain waves at different frequencies. It had been known for over half a century that, as mentioned in Chapter 4, one particular brain wave, the mu wave, is suppressed anytime a person makes a volitional movement, even a simple movement like opening and closing the fingers. It was subsequently discovered that mu-wave suppression also occurs when a person *watches* another person performing the same movement. We therefore suggested that mu-wave suppression might provide a simple, inexpensive, and noninvasive probe for monitoring mirror-neuron activity.

We ran a pilot experiment with a medium-functioning autistic child, Justin, to see if it would work. (Very young low-functioning children did not participate in this pilot study as we wanted to confirm that any difference between normal and autistic mirror-neuron activity that we found was not due to problems in attention, understanding instructions, or a general effect of mental retardation.) Justin had been referred to us by a local support group created to promote the welfare of local children with autism. Like Steven, he displayed many of the characteristic symptoms of autism but was able to follow simple instructions such as "look at the screen" and was not reluctant to have electrodes placed on his scalp.

As in normal children, Justin exhibited robust a mu wave while he sat around idly, and the mu wave was suppressed whenever we asked him to make simple voluntary movements. But remarkably, when he watched someone else perform the action, the suppression did not occur as it ought to. This observation provided a striking vindication of our hypothesis. We concluded that the child's motor-command system was intact—he could, after all, open doors, eat potato chips, draw pictures, climb stairs, and so on—but his mirror-neuron system was deficient. We presented this single-subject case study at the 2000 annual meeting of the Society for Neuroscience, and we followed it up with ten additional children in 2004. Our results were identical. This observation has since received extensive confirmation over the years from many different groups, using a variety of techniques.[2]

For example, a group of researchers led by Riitta Hari at the Aalto University of Science and Technology corroborated our conjecture using

MEG (magnetoencephalography), which is to EEG what jets are to biplanes. More recently, Michele Villalobos and her colleagues at San Diego State University used fMRI to show a reduction in functional connectivity between the visual cortex and the prefrontal mirror-neuron region in autistic patients.

Other researchers have tested our hypothesis using TMS (transcranial magnetic stimulation). TMS is, in one sense, the opposite of EEG: Rather than passively eavesdropping on the electrical signals emanating from the brain, TMS *creates* electrical currents in the brain using a powerful magnet held over the scalp. Thus with TMS you can induce neural activity artificially in any brain region that happens to be near the scalp. (Unfortunately, many brain regions are tucked away in the brain's deep folds, but plenty of other regions, including the motor cortex, are conveniently located directly beneath the skull where TMS can "zap" them easily.) The researchers used TMS to stimulate the motor cortex, then recorded electromuscular activation while the subjects watched other people performing actions. When a normal subject watches another person performing an action—say, squeezing a tennis ball with the right hand—the muscles in the subject's own right hand will register a tiny uptick in their electrical "chatter." Even though the subject doesn't perform a squeezing action herself, the mere act of watching the action leads to a tiny but measurable increase in the action-readiness of the muscles that *would* contract if she were performing it. The subject's own motor system automatically simulates the perceived action, but at the same time it automatically suppresses the spinal motor signal to prevent it from being carried out—and yet a tiny trickle of the suppressed motor command still manages to leak through and down to reach the muscles. That's what happens in normal subjects. But the autistic subjects showed no sign of increased muscle potentials while watching actions being performed. Their mirror neurons were missing in action. These results, taken together with our own, provide conclusive evidence that the hypothesis is correct.

THE MIRROR-NEURON HYPOTHESIS can explain several of the more quirky manifestations of autism. For instance, it has been known for some time that autistic children often have problems interpreting proverbs and metaphors. When asked to "get a grip on yourself," the autistic child may literally start grabbing his own body. When asked to explain the meaning of "all that glitters is not gold," we have noticed that some high-functioning autistics provide literal answers: "It means it's just some yellow metal—doesn't have to be gold." Although seen in only a subset of autistic children, this difficulty with metaphor cries out for an explanation.

There is a branch of cognitive science known as embodied cognition, which holds that human thought is deeply shaped by its interconnection with the body and by the inherent nature of human sensory and motor processes. This view stands in contrast to what we might call the classical view, which dominated cognitive science from the mid- through late twentieth century, and held that the brain was essentially the same thing as a general-purpose "universal computer" that just happened to be connected to a body. While it is possible to overstate the view of embodied cognition, it now has a lot of support; whole books have been written on the subject, Let me just give you one specific example of an experiment I did in collaboration with Lindsay Oberman and Piotr Winkielman. We showed that if you bite into a pencil (as if it were a bridle bit) to stretch your mouth into a wide, fake smile, you will have difficulty detecting another person's smile (but not a frown). This is because biting the pencil activates many of the same muscles as a smile, and this floods your brain's mirror-neuron system, creating a confusion between action and perception. (Certain mirror neurons fire when you make a facial expression and when you observe the same expression on another person's face.) The experiment shows that action and perception are much more closely intertwined in the brain than is usually assumed.

So what has this got to do with autism and metaphor? We recently noticed that patients with lesions in the left supramarginal gyrus who have apraxia—an inability to mime skilled voluntary actions, such as stirring a cup of tea or hammering a nail—also have difficulty interpreting action-based metaphors such as "reach for the stars." Since the supramarginal gyrus also has mirror neurons, our evidence suggests that the mirror-neuron system in humans is involved not only in interpreting skilled actions but in understanding action metaphors and, indeed, in other aspects of embodied cognition. Monkeys also have mirror neurons, but for their mirror neurons to play a role in metaphor monkeys may have to reach a higher level of sophistication—of the kind seen only in humans.

The mirror-neuron hypothesis also lends insight into autistic language difficulties. Mirror neurons are almost certainly involved when an infant first repeats a sound or word that she hears. It may require internal translation: the mapping of sound patterns onto corresponding motor patterns and vice versa. There are two ways such a system could be set up. First, as soon as the word is heard, a memory trace of the phonemes (speech sounds) is set up in the auditory cortex. The baby then tries various random utterances and, using error feedback from the memory trace, progressively refines the output to match memory. (We all do this when we internally hum a recently heard tune and then sing it out loud, progressively refining the output to match the internal humming.) Second, the networks for translating heard sounds into spoken words may have been innately specified through natural selection. In either case the net result would be a system of neurons with properties of the kind we ascribe to mirror neurons. If the child could, without delay and opportunity for feedback from rehearsal, repeat a phoneme cluster it has just heard for the first time, that would argue for a hardwired translational mechanism. Thus there is a variety of ways this unique mechanism could be set up. But whatever the mechanism, our results suggest that a flaw in its initial setup might cause the fundamental deficit in autism. Our empirical results with mu-wave suppression support this and also allow us to provide a unitary explanation for an array of seemingly unrelated symptoms.

Finally, although the mirror-neuron system evolved initially to create an internal model of other people's actions and intentions, in humans it may have evolved further—turning inward to represent (or re-represent) one's own mind to itself. A theory of mind is not only useful for intuiting what is happening in the minds of friends, strangers, and enemies; but in the unique case of *Homo sapiens*, it may also have dramatically increased the insight we have into our own minds' workings. This probably happened during the mental phase transition we underwent just a couple hundred millennia ago, and would have been the dawn of full-fledged self awareness. If the mirror-neuron system underlies theory of mind and if theory of mind in normal humans is supercharged by being applied inward, toward the self, this would explain why autistic individuals find social interaction and strong self-identification so difficult, and why so many autistic children have a hard time correctly using the pronouns "I" and "you" in conversation: They may lack a mature-enough mental self-representation to understand the distinction. This hypothesis would predict that even otherwise high-functioning autistics who can talk normally (highly verbal autistics are said to have Asperger syndrome, a subtype among autistic spectrum disorders) would have difficulty with such conceptual distinctions between words such as "self-esteem," "pity," "mercy," "forgiveness," and "embarrassment," not to mention "self-pity," which would make little sense without a full-fledged sense of self. Such predictions have never been tested on a systematic basis, but my student Laura Case is doing so. And we will return to these questions about self-representation and self-awareness, and derangements of these elusive faculties, in the last chapter.

This may be a good place to add three qualifying remarks. First, small groups of cells with mirror-neuron-like properties are found in many parts of the brain, and should really be thought of as parts of a large, interconnected circuit—a "mirror network," if you will. Second, as I noted earlier, we must be careful not to attribute all puzzling aspects about the brain to mirror neurons. They don't do everything! Nonetheless, they seem to have been key players in our transcendence of apehood, and they keep turning up in study after study of various mental functions that go far beyond our original "monkey see, monkey do" conception of them. Third, ascribing certain cognitive capacities to certain neurons (in this case, mirror neurons) or brain regions is only a beginning; we still need to understand how the neurons carry out their computations. However, understanding the anatomy can substantially guide the way and help reduce the complexity of the problem. In particular anatomical data can constrain our theoretical speculations and help eliminate many initially promising hypotheses. On the other hand, saying that "mental capacities emerge in a homogeneous network" gets you nowhere and flies in the face of empirical evidence of the exquisite anatomical specialization in the brain. Diffuse networks capable of learning exist in pigs and apes as well, but only humans are capable of language and self-reflection.

AUTISM IS STILL very difficult to treat, but the discovery of mirror-neuron dysfunction opens up some novel therapeutic approaches. For example, the lack of mu-wave suppression could become an invaluable diagnostic tool for screening for the disorder in early infancy, so that currently available behavioral therapies can be instituted long before other, more "florid" symptoms appear. Unfortunately, in most cases it is the unfolding of the florid symptoms, during the second or third year of life, that tips parents and doctors off. The earlier autism is caught, the better.

A second, more intriguing possibility would be to use biofeedback to treat the disorder. In biofeedback, a physiological signal from a subject's body or brain is tracked by a machine and represented back to the subject through some sort of external display. The goal is for the subject to concentrate on nudging that signal up or down and thereby gain some measure of conscious control over it. For example, a biofeedback system can show a person his heart rate, represented as a bouncing, beeping dot on a display screen; most people, with practice, can use this feedback to learn how to slow their hearts at will. Brain waves can also be used for biofeedback. For example, Stanford University professor Sean Mackey put chronic pain patients in a brain-imaging scanner and showed them a computer-animated image of a flame. The size of the flame at any given moment was a representation of the neural activity in each patient's anterior cingulate (a cortical region involved in pain perception), and was thus proportional to the subjective amount of pain he or she was in. By concentrating on the flame, most of the patients were able to gain some control over its size and to keep it small, and ipso facto to reduce the amount of pain they were experiencing. By the same token, one could monitor mu waves on an autistic child's scalp and display them on a screen in front of her, perhaps in the guise of a simple thought-controlled video game, to see if she can somehow learn to suppress them. Assuming her mirror-neuron function is weak or dormant rather than absent, this kind of exercise might boost her ability to see through to the intentionality of others, and bring her a step closer to joining the social world that swirls invisibly around her. As this book went to press, this approach was being pursued by our colleague Jaime Pineda at UC San Diego.

A third possibility—one that I suggested in an article for *Scientific American* that I coauthored with my graduate student Lindsay Oberman—would be to try certain drugs. There is a great deal of anecdotal evidence that MDMA (the party drug ecstasy) enhances empathy, which it may do by increasing the abundance of neurotransmitters called empathogens, which naturally occur in the brains of highly social creatures such as primates. Could a deficiency in such transmitters contribute to the symptoms of autism? If so, could MDMA (with its molecule suitably modified) ameliorate some of the most troubling symptoms of the disorder? It is also known that prolactin and oxytocin—so-called affiliation hormones—promote social bonding. Perhaps this connection, too, could be exploited therapeutically. If administered sufficiently early, cocktails of such drugs might help tide over some early symptom manifestations enough to minimize the subsequent cascade of events that lead to the full spectrum of autistic symptoms.

Speaking of prolactin and oxytocin, we recently encountered an autistic child whose brain MRI showed a substantial reduction in the size of the olfactory bulb, which receives smell signals from the nose. Given that smell is a major factor in the regulation of social behavior in most mammals, we wondered, Is it conceivable that olfactory-bulb malfunction plays a major role in the genesis of autism? Reduced olfactory-bulb activity would diminish oxytocin and prolactin, which in turn might reduce empathy and compassion. Needless to say, this is all pure speculation on my part, but in science, fancy is often the mother of fact—at least often enough that premature censorship of speculation is never a good idea.

One final option for reviving dormant mirror neurons in autism would be to take advantage of the great delight that all humans—including autistics—take in dancing to a rhythm. Although such dance therapy using rhythmic music has been tried with autistic children, no attempt has been made to directly tap into the known properties of the mirror-neuron system. One way to do this might be, for example, to have several model dancers moving simultaneously to rhythm and having the child mime the same dance in synchrony. Immersing all of them in a hall of multiply reflecting mirrors might also help by multiplying the impact on the mirror-neuron system. It seems like a far-fetched possibility, but then so was the idea of using vaccines to prevent rabies or diphtheria.[3]

THE MIRROR-NEURON HYPOTHESIS does a good job of accounting for the defining features of autism: lack of empathy, pretend play, imitation, and a theory of mind.[4] However, it is not a complete account, because there are some other common (though not defining) symptoms of autism that mirror neurons do not have any apparent bearing on. For example, some autistics display a rocking to-and-fro movement, avoid eye contact, show hypersensitivity and aversion to certain sounds, and often engage in tactile self-stimulation—sometimes even beating themselves—which seems intended to dampen this hypersensitivity. These symptoms are common enough that they too need to be explained in any full account of autism. Perhaps beating themselves is a way of enhancing the salience of the body, thereby helping anchor the self and reaffirming its existence. But can we put this idea in the context of the rest of what we have said so far about autism?

In the early 1990s our group (in collaboration with Bill Hirstein, my postdoctoral colleague; and Portia Iversen, cofounder of Cure Autism Now, an organization devoted to autism) thought a lot about how to account for these other symptoms of autism. We came up with what we called the "salience landscape theory": When a person looks at the world, she is confronted with a potentially bewildering sensory overload. As we saw in Chapter 2 when we considered the two branches of the "what" stream in the visual cortex, information about the world is first discriminated in the brain's sensory areas and then relayed to the amygdala. As the gateway to the emotional core of your brain, the amygdala performs an emotional surveillance of the world you inhabit, gauges the emotional significance of everything you see, and decides whether it is trivial and humdrum or something worth getting emotional over. If the latter, the amygdala tells the hypothalamus to activate the autonomic nervous system in proportion to the arousal worthiness of the triggering sight—it could be anything from mildly interesting to downright terrifying. Thus the amygdala is able to create a "salience landscape" of your world, with hills and valleys corresponding to high and low salience.

It is sometimes possible for this circuit to go haywire. Your autonomic response to something arousing manifests as increased sweating, heart rate, muscular readiness, and so on, to prepare your body for action. In extreme cases this surge of physiological arousal can feed back into your brain and prompt your amygdala to say, in effect, "Wow, it's even more dangerous than I thought. We'll need more arousal to get out of this!" The result is an autonomic blitzkrieg. Many adults are prone to such panic attacks, but most of us, most of the time, are not in danger of getting swept away by such autonomic maelstroms.

With all this in mind, our group explored the possibility that children with autism have a distorted salience landscape. This may be partially due to indiscriminately enhanced (or reduced) connections between sensory cortices and the amygdala, and possibly between limbic structures and the frontal lobes. As a result of these abnormal connections, every trivial event or object sets off an uncontrollable autonomic storm, which would explain autistics' preference for sameness and routine. If the emotional arousal is less florid, on the other hand, the child might attach abnormally high significance to certain unusual stimuli, which could account for their strange preoccupations, including their sometimes savant-like skills. Conversely, if some of the connections from the sensory cortex to the amygdala are partially effaced by the distortions in salience landscape, the child might ignore things, like eyes, that most normal children find highly attention grabbing.

To test the salience landscape hypothesis we measured galvanic skin response (GSR) in a group of 37 autistic and 25 normal children. The normal children showed arousal for certain categories of stimuli as expected but not for others. For example, they had GSR responses to photos of parents but not of pencils. The children with autism, on the other hand, showed a more generally heightened autonomic arousal that was further amplified by the most trivial objects and events, whereas some highly salient stimuli such as eyes were completely ineffective.

If salience landscape theory is on the right track, one would expect to find abnormalities in visual pathway 3 of autistic brains. Pathway 3 not only projects to the amygdala, but it routes through the superior temporal sulcus, which—along with its neighboring region, the insula—is rich in mirror neurons. In the insula, mirror neurons have been shown to be involved in perceiving as well as expressing certain emotions—like disgust, including social and moral disgust—in an empathetic manner. Thus damage to these areas, or perhaps a deficiency of mirror neurons within them, might not only distort the salience landscape, but also diminish empathy, social interaction, imitation, and pretend play.

As an added bonus, salience landscape theory may also explain two other quirky aspects of autism that have always been puzzling. First, some parents report that their child's autistic symptoms are temporarily relieved by a bout of high fever. Fever is ordinarily caused by certain bacterial toxins that act on temperature-regulating mechanisms in the hypothalamus in the base of your brain. Again, this is part of pathway

3. I realized that it may not be coincidental that certain dysfunctional behaviors such as tantrums originate in networks that neighbor the hypothalamus. Thus the fever might have a "spillover" effect that happens to dampen activity at one of the bottlenecks of the feedback loop that generates those autonomic-arousal storms and their associated tantrums. This is a highly speculative explanation but it's better than none at all, and if it pans out it could provide another basis for intervention. For example, there might be some way to safely dampen the feedback loop artificially. A damped circuit might be better than a malfunctioning one, especially if it could get a kid like Steven to engage even just a little bit more with his mother. For example, one could give him high fever harmlessly by injecting denatured malarial parasites; repeated injections of such pyrogens (fever-inducing substances) might help "reset" the circuit and alleviate symptoms permanently.

Second, children with autism often repeatedly bang and beat themselves. This behavior is called somatic self-stimulation. In terms of our theory, we would suggest that this leads to a damping of the autonomic-arousal storms that the child suffers from. Indeed, our research team has found that such self-stimulation not only has a calming effect but leads to a measurable reduction in GSR. This suggests a possible symptomatic therapy for autism: One could have a portable device for monitoring GSR that then feeds back to a body stimulation device which the child wears under his clothing. Whether such a device would prove practical in a day-to-day setting remains to be seen; it is being tested by my postdoctoral colleague Bill Hirstein.

The to-and-fro rocking behavior of some autistic children may serve a similar purpose. We know it likely stimulates the vestibular system (sense of balance), and we know that balance-related information splits at some point to travel down pathway 3, especially to the insula. Thus repetitive rocking might provide the same kind of damping that self-beating does. More speculatively, it might help anchor the self in the body, providing coherence to an otherwise chaotic world, as I'll describe in a moment.

Aside from possible mirror-neuron deficiency, what other factors might account for the distorted salience landscapes through which many autistic people seem to view the world? It is well documented that there are genetic predispositions to autism. But less well known is the fact that nearly a third of children with autism have had temporal lobe epilepsy (TLE) in infancy. (The proportion could be much higher if we include clinically undetected complex partial seizures.) In adults TLE manifests as florid emotional disturbances, but because their brains are fully mature, it does not appear to lead to deep-seated cognitive distortions. But less is known about what TLE does to a developing brain. TLE seizures are caused by repeated random volleys of nerve impulses coursing through the limbic system. If they occur frequently in a very young brain, they might lead, through a process of synapse enhancement called kindling, to selective but widespread, indiscriminate enhancement (or sometimes effacement) of the connections between the amygdala and the high-level visual, auditory, and somatosensory cortices. This could account both for the frequent false alarms set off by trivial or mundane sights and otherwise neutral sounds, and conversely for the failure to react to socially salient information, which are so characteristic of autism.

In more general terms, our sense of being an integrated, embodied self seems to depend crucially on back-and-forth, echo-like "reverberation" between the brain and the rest of the body—and indeed, thanks to empathy, between the self and others. Indiscriminate scramblings of the connections between high-level sensory areas and the amygdala, and the resulting distortions to one's salience landscape, could as part of the same process cause a disturbing loss of this sense of embodiment—of being a distinct, autonomous self anchored in a body and embedded in a society. Perhaps somatic self-stimulation is some children's attempt to regain their embodiment by reviving and enhancing body-brain interactions while at the same time damping spuriously amplified autonomic signals. A subtle balance of such interactions may be crucial for the normal development of an integrated self, something we ordinarily take for granted as the axiomatic foundation of being a person. No wonder, then, that this very sense of being a person is profoundly disturbed in autism.

We have so far considered two candidate theories for explaining the bizarre symptoms of autism: the mirror-neuron dysfunction hypothesis and the idea of a distorted salience landscape. The rationale for proposing these theories is to provide unitary mechanisms for the bewildering array of seemingly unrelated symptoms that characterize the disorder. Of course, the two hypotheses are not necessarily mutually exclusive. Indeed, there are known connections between the mirror-neuron system and the limbic system. It is possible that distortions in limbic-sensory connections are what lead ultimately to a deranged mirror-neuron system. Clearly, we need more experiments to resolve these issues. Whatever the underlying mechanisms turn out to be, our results strongly suggest that children with autism have a dysfunctional mirror-neuron system that may help explain many features of the syndrome. Whether this dysfunction is caused by genes concerned with brain development or by genes that predispose to certain viruses (that in turn might predispose to seizures), or is due to something else entirely remains to be seen. Meanwhile, it might provide a useful jumping off point for future research into autism, so that someday we may find a way to "bring Steven back."

Autism reminds us that the uniquely human sense of self is not an "airy nothing" without "habitation and a name." Despite its vehement tendency to assert its privacy and independence, the self actually emerges from a reciprocity of interactions with others and with the body it is embedded in. When it withdraws from society and retreats from its own body it barely exists; at least not in the sense of a mature self that defines our existence as human beings. Indeed, autism could be regarded fundamentally as a disorder of self-consciousness, and if so, research on this disorder may help us understand the nature of consciousness itself.

CHAPTER 6

The Power of Babble:
The Evolution of Language

. . . Thoughtful men, once escaped from the blinding influences of traditional prejudice, will find in the lowly stock whence Man has sprung, the best evidence of the splendor of his capacities; and will discern in his long progress through the past, a reasonable ground of faith in his attainment of a nobler future.

—THOMAS HENRY HUXLEY

ON THE LONG FOURTH OF JULY WEEKEND OF 1999, I RECEIVED A phone call from John Hamdi, who had been a colleague of mine at Trinity College, Cambridge, nearly fifteen years earlier. We hadn't been in contact and it was a pleasant surprise to hear his voice after such a long time. As we exchanged greetings, I smiled to myself, reminded of the many adventures we had shared during our student days. He was now a professor of orthopedic surgery in Bristol, he said. He had noticed a book I'd recently published.

"I know you are mainly involved in research these days," he said, "but my father, who lives in La Jolla, has had a head injury from a skiing accident followed by a stroke. His right side is paralyzed, and I'd be grateful if you could take a look at him. I want to make sure he's getting the best treatment available. I heard there's a new rehab procedure which employs mirrors to help patients recover the use of a paralyzed arm. Do you know anything about this?"

A week later John's father, Dr. Hamdi, was brought to my office by his wife. He had been a world-renowned professor of chemistry here at UC San Diego until his retirement three years earlier. About six months prior to my seeing him he sustained a skull fracture. In the emergency room at Scripps Clinic he was informed that a stroke, caused by a blood clot in his middle cerebral artery, had cut off the blood supply to the left hemisphere of his brain. Since the left hemisphere controls the right side of the body, Dr. Hamdi's right arm and leg were paralyzed. Much more alarming than the paralysis, though, was the fact that he could no longer speak fluently. Even simple requests such as "I want water" required great effort, and we had to pay careful attention to understand what he was saying.

Assisting me in examining Dr. Hamdi was Jason Alexander, a medical student on a six-month rotation in our lab. Jason and I looked at Dr. Hamdi's charts and also obtained a medical history from Mrs. Hamdi. We then conducted a routine neurological workup, testing in sequence his motor functions, sensory functions, reflexes, cranial nerves, and his higher mental functions such as memory, language, and intelligence. I took the handle of my knee hammer and, while Dr. Hamdi was lying in bed, stroked the outer border of his right foot and then the left foot, running the tip of the hammer handle from the pinky to sole. Nothing much happened in the normal foot, but when I repeated the procedure on the paralyzed right foot, the big toe instantly curled upward and all the other toes fanned out. This is Babinski's sign, arguably the most famous sign in neurology. It reliably indicates damage to the pyramidal tracts, the great motor pathway that descends from the motor cortex down into the spinal cord conveying commands for volitional movements.

"Why does the toe go up?" asked Jason.

"We don't know," I said, "but one possibility is that it's a throwback to an early stage in evolutionary history. The reflexive withdrawal tendency for the toes to fan out and curl up is seen in lower mammals. But the pyramidal tracts in primates become especially pronounced, and they inhibit this primitive reflex. Primates have a more sophisticated grasp reflex, with a tendency for the toes to curl inward as if to clutch a branch. It may be a reflex to avoid falling out of trees."

"Sounds far-fetched," said Jason skeptically.

"But when the pyramidal tracts are damaged," I said, ignoring his remark, "the grasp reflex goes away and the more primitive withdrawal reflex emerges because it's no longer inhibited. That's why you also see it in infants; their pyramidal tracts haven't fully developed yet."

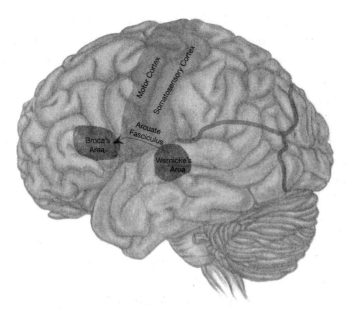

FIGURE 6.1   The two main language areas in the brain are Broca's area (in the frontal lobes) and Wernicke's area (in the temporal lobes). The two are connected by a band of fibers called the arcuate fasciculus. Another language area, the angular gyrus (not labeled in this figure), lies near the bottom of the parietal lobe, at the intersection of temporal, occipital, and parietal lobes.

The paralysis was bad enough, but Dr. Hamdi was more troubled by his speech impediment. He had developed a language deficit called Broca's aphasia, named after the French neurologist Paul Broca, who first described the syndrome in 1865. The damage is usually in the left frontal lobe in a region (Figure 6.1) that lies just in front of the large fissure, or vertical furrow, that separates the parietal and frontal lobes.

Like most patients with this disorder, Dr. Hamdi could convey the general sense of what he was trying to say, but his speech was slow and effortful, conveyed in a flat monotone, filled with pauses, and almost completely devoid of syntax (loosely speaking, grammatical structure). His utterances were also deficient in (though not devoid of) so-called function words such as "and," "but," and "if," which don't refer to anything in the world but specify relationships between different parts of a sentence.

"Dr. Hamdi, tell me about your skiing accident," I said.

"*Ummmmm* . . . Jackson, Wyoming," he began. "And skied down and *ummmmm* . . . tumbled, all right, gloves, mittens, *uhhhh* . . . poles, *uhhhh* . . . the *uhhhh* . . . but the blood drained three days pass hospital and *ummmmm* . . . coma . . . ten days . . . switch to Sharpe [memorial hospital] . . . *mmmmm* . . . four months and back . . . *ummmmmmmm* . . . it's *ummmmm* slow process and a bit of medicine *ummmmmm* . . . six medicines. One tried eight or nine months."

"Okay continue."

"And seizures."

"Oh? Where was the blood hemorrhage from?"

Dr. Hamdi pointed to the side of his neck.

"The carotid?"

"Yeah. Yeah. But . . . *uhhhh, uhhh, uhhh,* this, this and this, this . . ." he said, using his left hand to point to multiple places on his right leg and arm.

"Go on," I said, "Tell us more."

"It's *ummmmm* . . . it's difficult [referring to his paralysis], *ummm*, left side perfectly okay."

"Are you right-handed or left-handed?"

"Right-handed."

"Can you write with the left now?"

"Yeah."

"Okay. Good. What about word processing?"

"Processing *ummmm* write."

"But when you write, is it slow?"

"Yeah."

"Just like your speech?"

"Right."

"When people talk fast you have no problem understanding them?"

"Yeah, yeah."

"You can understand."

"Right."

"Very good."

"*Uhhhhhh* . . . but *uhhhh* . . . the speech, *uhhhhh, ummmmm* slowed down."

"Okay, do you think your speech is slowed down, or your thought is slowed down?"

"Okay. But *ummmm* [points to head] *uhhh* . . . words are beautiful. *Ummmmm* speech . . ."

He then made twisting motions with his mouth. Presumably he meant that his flow of thought felt intact, but the words were not coming out fluently.

"Supposing I ask you a question," I said. "Mary and Joe together have eighteen apples."

"All right."

"Joe has twice as many apples as Mary."

"Okay."

"So how many does Mary have? How many does Joe have?"

"*Ummmmm* . . . lemme think. Oh God."

"Mary and Joe together have eighteen apples . . ."

"Six, *ahhhh* twelve!" he blurted.

"Excellent!"

So Dr. Hamdi had basic conceptual algebra, was able to do simple arithmetic, and had good comprehension of language even for relatively complex sentences. I was told he had been a superb mathematician before his accident. Yet later, when Jason and I tested Dr. Hamdi on more complex algebra using symbols, he kept trying hard but failing. I was intrigued by the possibility that the Broca's area might be specialized not just for the syntax, or syntactic structure, of natural language, but also for other, more arbitrary languages that have formal rules, such as algebra or computer programming. Even though the area might have evolved for natural language, it may have the latent capacity for other functions that bear a certain resemblance to the rules of syntax.

What do I mean by "syntax"? To understand Dr. Hamdi's main problem, consider a routine sentence such as "I lent the book you gave me to Mary." Here an entire noun phrase—"the book you gave me"—is embedded in a larger sentence. That embedding process, called recursion, is facilitated by function words and is made possible by a number of unconscious rules—rules that all languages follow, no matter how different they may seem on the surface. Recursion can be repeated any number of times to make a sentence as complex as it needs to be in order to convey its ideas. With each recursion, the sentence adds a new branch to its phrase structure. Our example sentence can be expanded, for instance, to "I lent the book you gave me while I was in the hospital to Mary," and from there to "I lent the book you gave me while I was in the hospital to a nice woman I met there named Mary," and so on. Syntax allows us to create sentences as complex as our short-term memory can handle. Of course, if we go on too long, it can get silly or start to feel like a game, as in the old English nursery rhyme:

This is the man all tattered and torn
That kissed the maiden all forlorn
That milked the cow with the crumpled horn
That tossed the dog that worried the cat
That killed the rat that ate the malt
That lay in the house that Jack built.

Now, before we go on discussing language, we need to ask how we can be sure Dr. Hamdi's problem was really a disorder of language at this abstract level and not something more mundane. You might think, reasonably, that the stroke had damaged the parts of his cortex that control his lips, tongue, palate, and other small muscles required for the execution of speech. Because talking required such effort, he was economizing on words. The telegraphic nature of his speech may have been to save effort. But I did some simple tests to show Jason that this couldn't be the reason.

"Dr. Hamdi, can you write down on this pad the reason why you went to the hospital? What happened?"

Dr. Hamdi understood our request and proceeded to write, using his left hand, a long paragraph about the circumstances that brought him to our hospital. Although the handwriting wasn't good, the paragraph made sense. We could understand what he had written. Yet remarkably, his writing also had poor grammatical structure. Too few "ands," "ifs," and "buts." If his problem were related to speech muscles, why did his writing also have the same abnormal form as his speech? After all, there was nothing wrong with his left hand.

I then asked Dr. Hamdi to sing "Happy Birthday." He sang it effortlessly. Not only could he carry the tune well, but all the words were there and correctly pronounced. This was in stark contrast to his speech, which, in addition to missing important connecting words and lacking phrase structure, also contained mispronounced words and lacked the intonation, rhythm, and the melodious flow of normal speech. If his problem were poor control of his vocal apparatus, he shouldn't have been able to sing, either. To this day we don't know why Broca's patients can sing. One possibility is that language function is based mainly in the left hemisphere, which is damaged in these patients, whereas singing is done by the right hemisphere.

We had already learned a great deal after just a few minutes of testing. Dr. Hamdi's problems with expressing himself were not caused by a partial paralysis or weakness of his mouth and tongue. He had a disorder of language, not of speech, and the two are radically different. A parrot can talk—it has speech, you might say—but it doesn't have language.

HUMAN LANGUAGE SEEMS so complex, multidimensional, and richly evocative that one is tempted to think that almost the entire brain, or large chunks of it at least, must be involved. After all, even the utterance of a single word like "rose" evokes a whole host of associations and emotions: the first rose you ever got, the fragrance, rose gardens you were promised, rosy lips and cheeks, thorns, rose-colored glasses, and so on. Doesn't this imply that many far-flung regions of the brain must cooperate to generate the concept of a rose? Surely the word is just the handle, or focus, around which swirls a halo of associations, meanings, and memories.

There's probably some truth to this, but the evidence from aphasics such as Dr. Hamdi suggests the very opposite—that the brain has neural circuits specialized for language. Indeed, it may even be that separate components or stages of language processing are dealt with by different parts of the brain, although we should really think of them as parts of one large interconnected system. We are accustomed to thinking of language as a single function, but this is an illusion. Vision feels like a unitary faculty to us as well, yet as noted in Chapter 2, seeing relies on numerous quasi-independent areas. Language is similar. A sentence, loosely speaking, has three distinct components, which are normally so closely interwoven that they don't feel separate. First, there are the building blocks we call words (lexicon) that denote objects, actions, and events. Second, there is the actual meaning (semantics) conveyed by the sentence. And third, there is syntactic structure (loosely speaking, grammar), which involves the use of function words and recursion. The rules of syntax generate the complex hierarchical phrase structure of human language, which at its core allows the unambiguous communication of fine nuances of meaning and intention.

Human beings are the only creatures to have true language. Even chimps, who can be trained to sign simple sentences like "Give me fruit," can't come close to complex sentences such as "It's true that Joe is the big alpha male, but he's starting to get old and lazy, so don't worry about what he might do unless he seems to be in an especially nasty mood." The seemingly infinite flexibility and open-endedness of our language is one of the hallmarks of the human species. In ordinary speech, meaning and syntactic structure are so closely intertwined that it's hard to believe that they are really distinct. But you can have a perfectly grammatical sentence that is meaningless gibberish, as in the linguist Noam Chomsky's famous example, "Colorless green ideas sleep furiously." Conversely, a meaningful idea can be conveyed adequately by a nongrammatical sentence, as Dr. Hamdi has shown us. ("It's difficult, *ummm*, left side perfectly okay.")

It turns out that different parts of the brain are specialized for these three different aspects of language: lexicon, semantics, and syntax. But the agreement among researchers ends there. The degree of specialization is hotly debated. Language, more than any other topic, tends to polarize academics. I don't quite know why, but fortunately it isn't my field. In any case, by most accounts Broca's area seems mainly concerned with syntactic structure. So Dr. Hamdi had no better chance than a chimp of generating long sentences full of hypotheticals and subordinate clauses. Yet he had no difficulty in communicating his ideas by just stringing words together in approximately the right order, like Tarzan. (Or surfer dudes in California.)

One reason for thinking that Broca's area is specialized exclusively for syntactic structure is the observation that it seems to have a life of its own, quite independent of the meaning conveyed. It's almost as though this patch of cortex has an autonomous set of grammatical rules that are intrinsic to its networks. Some of them seem quite arbitrary and apparently nonfunctional, which is the main reason linguists assert its independence from semantics and meaning and dislike thinking of it as having evolved from anything else in the brain. The extreme view is exemplified by Chomsky, who believes that it didn't even evolve through natural selection!

The brain region concerned with semantics is located in the left temporal lobe near the back of the great horizontal cleft in the middle of the brain (see Figure 6.1). This region, called Wernicke's area, appears to be specialized for the representation of meaning. Dr. Hamdi's Wernicke's area was obviously intact. He could still comprehend what was said to him and could convey some semblance of meaning in his conversations. Conversely, Wernicke's aphasia—what you get if your Wernicke's area is damaged but your Broca's area remains intact—is in a sense the mirror image of Broca's aphasia: The patient can fluently generate elaborate, smoothly articulated, grammatically flawless sentences, but it's all meaningless gibberish. At least that's the official party line, but later I'll provide evidence that this isn't entirely true.

THESE BASIC FACTS about the major language-related brain areas have been known for more than a century. But many questions remain. How complete is the specialization? How does the neural circuitry within each area actually do its job? How autonomous are these areas, and how do they interact to generate smoothly articulated, meaningful sentences? How does language interact with thought? Does language enable us to think, or does thinking enable us to talk? Can we think in a sophisticated manner without silent internal speech? And lastly, how did this extraordinarily complex, multicomponent system originally come into existence in our hominin ancestors?

This last question is the most vexing. Our journey into full-blown humanity began with nothing but the primitive growls, grunts, and groans available to our primate cousins. By 75,000 to 150,000 years ago, the human brain was brimming with complex thoughts and linguistic skills. How did this happen? Clearly, there must have been a transitional phase, yet it's hard to imagine how linguistic brain structures of intermediate complexity might have worked, or what functions they might have served along the way. The transitional phase must have been at least partially functional; otherwise it couldn't have been selected for, nor served as an evolutionary bridge for the eventual emergence of more sophisticated language functions.

To understand what this bridge might have been is the main purpose of this chapter. I should point out that by "language" I don't mean just "communication." We often use the two words interchangeably, but in fact they are very different. Consider the vervet monkey. Vervets have three alarm calls to alert each other about predators. The call for leopard prompts the troupe to bolt for the nearest trees. The call for serpent causes the monkeys to stand up on two legs and peer down into the grass. And when vervets hear the eagle call, they look up into the air and seek shelter in the underbrush. It's tempting to conclude that these calls are like words, or at least the precursors to words, and that the monkey does have a primitive vocabulary of sorts. But do the monkeys really know there's a leopard, or do they just rush for the nearest tree reflexively when an alarm call is sounded? Or perhaps the call really just means "climb" or "there's danger on the ground," rather than the much richer concept of leopard that a human brain harbors. This example tells us that mere communication isn't language. Like an air-raid siren or a fire alarm, vervets' cries are generalized alerts that refer to specific situations; they are almost nothing like words.

In fact, we can list a set of five characteristics that make human language unique and radically different from other types of communication we see in vervets or dolphins:

1. Our vocabulary (lexicon) is enormous. By the time a child is eight years old, she has almost six hundred words at her disposal—a figure that vastly exceeds the nearest runner-up, the vervet monkey, by two orders of magnitude. One could argue, though, that this is really a matter of degree than a qualitative jump; maybe we just have much better memories.

2. More important than the sheer size of our lexicon is the fact that only humans have function words that exist exclusively in the context of language. While words like "dog," "night," or "naughty" refer to actual things or events, function words have no existence independent of their linguistic function. So even though a sentence such as "If gulmpuk is buga, then gadul will be too" is meaningless, we do understand the conditional nature of the statement because of the conventional usage of "if" and "then."

3. Humans can use words "off-line," that is, to refer to things or events that are not currently visible or exist only in the past, the future, or a hypothetical reality: "I saw an apple on the tree

yesterday, and decided I will pluck it tomorrow but only if it is ripe." This type of complexity isn't found in most spontaneous forms of animal communication. (Apes who are taught sign language can, of course, use signs in the absence of the object being referred to. For example, they can sign "banana" when hungry.)

4.   Only humans, as far as we know, can use metaphor and analogy, although here we are in a gray area: the elusive boundary between thought and language. When an alpha male ape makes a genital display to intimidate a rival into submission, is this analogous to the metaphor "F—k you" that humans use to insult one another? I wonder. But even so, this limited kind of metaphor falls far short of puns and poems, or of Tagore's description of the Taj Mahal as a "tear drop on the cheek of time." Here again is that mysterious boundary between language and thought.

5.   Flexible, recursive syntax is found only in human language. Most linguists single out this feature to argue for a qualitative jump between animal and human communication, possibly because it has more regularities and can be tackled more rigorously than other, more nebulous aspects of language.

These five aspects of language are by and large unique to humans. Of these, the first four are often lumped together as protolanguage, a term invented by the linguist Derek Bickerton. As we'll see, protolanguage set the stage for the subsequent emergence and culmination of a highly sophisticated system of interacting parts that we call, as a whole system, true language.

TWO TOPICS IN brain research always seem to attract geniuses and crackpots. One is consciousness and the other is the question of how language evolved. So many zany ideas on language origins were being proposed in the nineteenth century that the Linguistic Society of Paris introduced a formal ban on all papers dealing with this topic. The society argued that, given the paucity of evolutionary intermediates or fossil languages, the whole enterprise was doomed to fail. More likely, linguists of the day were so fascinated by the intricacies of rules intrinsic to language itself that they were not curious about how it may have all started. But censorship bans and negative predictions are never a good idea in science.

A number of cognitive neuroscientists, myself included, believe that mainstream linguists have been overemphasizing the structural aspects of language. Pointing to the fact that the mind's grammatical systems are to a large extent autonomous and modular, most linguists have shunned the question of how these interact with other cognitive processes. They profess interest solely in the rules that are fundamental to the brain's grammatical circuits, not how the circuits actually work. This narrow focus removes the incentive to investigate how this mechanism interacts with other mental capacities such as semantics (which orthodox linguists don't even regard as an aspect of language!), or to ask evolutionary questions about how it might have evolved from preexisting brain structures.

The linguists can be forgiven, if not applauded, for their wariness of evolutionary questions. With so many interlocking parts working in such a coordinated manner, it's hard to figure out, or even imagine, how language could have evolved by the essentially blind process of natural selection. (By "natural selection," I mean the progressive accumulation of chance variations that enhance the organism's ability to pass on its genes to the next generation.) It's not difficult to imagine a single trait, such as a giraffe's long neck, being a product of this relatively simple adaptive process. Giraffe ancestors that had mutant genes conferring slightly longer necks had better access to tree leaves, causing them to survive longer or breed more, which caused the beneficial genes to increase in number down through the generations. The result was a progressive increase in neck length.

But how can multiple traits, each of which would be useless without the other, evolve in tandem? Many complex, interwoven systems in biology have been held up by would-be debunkers of evolutionary theory to argue for so-called intelligent design—the idea that the complexities of life could only occur through divine intervention or the hand of God. For example, how could the vertebrate eye evolve via natural selection? A lens and a retina are mutually necessary, so each would be useless without the other. Yet by definition the mechanism of natural selection has no foresight, so it couldn't have created the one in preparation for the other.

Fortunately, as Richard Dawkins has pointed out, there are numerous creatures in nature with eyes at all stages of complexity. It turns out there is a logical evolutionary sequence that leads from the simplest possible light-sensing mechanism—a patch of light-sensitive cells on the outer skin—to the exquisite optical organ we enjoy today.

Language is similarly complex, but in this case we have no idea what the intermediate steps might have been. As the French linguists pointed out, there are no fossil languages or half-human creatures around for us to study. But this hasn't stopped people from speculating on how the transition might have come about. Broadly speaking, there have been four main ideas. Some of the confusion between these ideas results from failing to define "language" clearly in the narrow sense of syntax versus the broader sense that includes semantics. I will use the term in the broader sense.

THE FIRST IDEA was advanced by Darwin's contemporary Alfred Russel Wallace, who independently discovered the principle of natural selection (though he rarely gets the credit he deserves, probably because he was Welsh rather than English). Wallace argued that while natural selection was fine for turning fins into feet or scales into hair, language was too sophisticated to have emerged in this way. His solution to the problem was simple: Language was put into our brains by God. This idea may or may not be right but as scientists we can't test it, so let's move on.

Second, there's the idea put forward by the founding father of modern linguistic science, Noam Chomsky. Like Wallace, he too was struck by the sophistication and complexity of language. Again, he couldn't conceive of natural selection being the correct explanation for how language evolved.

Chomsky's theory of language origins is based on the principle of emergence. The word simply means the whole is greater—sometimes vastly so—than the mere sum of the parts. A good example would be the production of salt—an edible white crystal—by combining the pungent, greenish, poisonous gas chlorine with the shiny, light metal sodium. Neither of these elements has anything saltlike about it, yet they combine into salt. Now if such a complex, wholly unpredictable new property can emerge from a simple interaction between two elementary substances, then who can predict what novel unforeseen properties might emerge when you pack 100 billion nerve cells into the tiny space of the human cranial cavity? Maybe language is one such property.

Chomsky's idea isn't quite as silly as some of my colleagues think. But even if it's right, there's not much one can say or do about it given the current state of brain science. There's simply no way of testing it. And although Chomsky doesn't speak of God, his idea comes perilously close to Wallace's. I don't know for sure that he is wrong, but I don't like the idea for the simple reason that one can't get very far in science by saying (in effect) something miraculous happened. I'm interested in finding a more convincing explanation that's based on the known principles of organic evolution and brain function.

The third theory, proposed by one of the most distinguished exponents of evolutionary theory in this country, the late Stephen Jay Gould, argues that contrary to what most linguists claim, language is not a specialized mechanism based on brain modules and that it did not evolve specifically for its most obvious present purpose, communication. On the contrary, it represents the specific implementation of a more general mechanism that evolved earlier for other reasons, namely thinking. In Gould's theory, language is rooted in a system that gave our ancestors a more sophisticated way to mentally represent the world and, as we shall see in the Chapter 9, a way to represent themselves within that representation. Only later did this system get repurposed or extended into a means of communication. In this view, then, thinking was an exaptation—a mechanism that originally evolved for one function and then provided the opportunity for something very different (in this case language) to evolve.

We need to bear in mind that the exaptation itself must have evolved by conventional natural selection. Failure to appreciate this has resulted in much confusion and bitter feuds. The principle of exaptation is not an alternative to natural selection, as Gould's critics believe, but actually complements and expands its scope and range of applicability. For instance, feathers originally evolved from reptilian scales as an adaptation to provide insulation (just like hair in mammals), but then were exapted for flight. Reptiles evolved a three-bone multihinged lower jaw to permit swallowing large prey, but two of these three bones became an exaptation for improved hearing. The convenient location of these bones made possible the evolution of two little sound-amplifying bones inside your middle ear. No engineer would have dreamed of such an inelegant solution, which goes to illustrate the opportunistic nature of evolution.

(As Francis Crick once said, "God is a hacker, not an engineer.") I will expand on these ideas about jawbones transforming into ear bones at the end of this chapter.

Another example of a more general-purpose adaptation is the evolution of flexible fingers. Our arboreal ancestors originally evolved them for climbing trees, but hominins adapted them for fine manipulation and tool use. Today, thanks to the power of culture, fingers are a general-purpose mechanism that can be used for rocking a cradle, wielding a scepter, pointing, or even counting for math. But no one—not even a naïve adaptationist or evolutionary psychologist—would argue that fingers evolved because they were selected for pointing and counting.

Similarly, Gould argues, thinking may have evolved first, given its obvious usefulness in dealing with the world, which then set the stage for language. I agree with Gould's general idea that language didn't originally evolve specifically for communication. But I don't like the idea that thinking evolved first and language (by which I mean all of language—not just in the Chomskian sense of emergence) was simply a byproduct. One reason I don't like it is that it merely postpones the problem rather than solving it. Since we know even less about thinking and how it might have evolved than we do about language, saying language evolved from thought doesn't tell us very much. As I have said many times before, you can't get very far in science by trying to explain one mystery with another mystery.

The fourth idea—diametrically opposed to Gould's—was proposed by the distinguished Harvard University linguist Steven Pinker, who declares language to be an instinct, as ingrained in human nature as coughing, sneezing, or yawning. By this he doesn't mean it's as simple as these other instincts, but that it is a highly specialized brain mechanism, an adaptation that is unique to humans and that evolved through conventional mechanisms of natural selection expressly for communication. So Pinker agrees with his former teacher Chomsky in asserting (correctly, I believe) that language is a highly specialized organ, but disagrees with Gould's views on the important role played by exaptation. I think there is merit to Pinker's view, but I also think his idea is far too general to be useful. It is not actually wrong, but it is incomplete. It seems a bit like saying that the digestion of food must be based on the first law of thermodynamics—which is true for sure, but it's also true for every other system on earth. The idea doesn't tell you much about the detailed mechanisms of digestion. In considering the evolution of any complex biological system (whether the ear or the language "organ"), we would like to know not merely that it was done by natural selection, but exactly how it got started and then evolved to its present level of sophistication. This isn't as important for a more straightforward problem like the giraffe's neck (although even there, one wants to know how genes selectively lengthen neck vertebrae). But it is an important part of the story when you are dealing with more complex adaptations.

So there you have it, four different theories of language. Of these we can discard the first two—not because we know for sure that they are wrong, but because they can't be tested. But of the remaining two, who's right—Gould or Pinker? I'd like to suggest that neither of them is, although there's a grain of truth in each (so if you are a Gould/Pinker fan, you could say they were both right but didn't take their arguments far enough).

I would like to propose a different framework for thinking about language evolution that incorporates some features of both but then goes well beyond them. I call it the "synesthetic bootstrapping theory." As we shall see, it provides a valuable clue to understanding the origins of not only language, but also a host of other uniquely human traits such as metaphorical thinking and abstraction. In particular, I'll argue that language and many aspects of abstract thought evolved through exaptations whose fortuitous combination yielded novel solutions. Notice that this is different from saying that language evolved from some general mechanism such as thinking, and it also differs from Pinker's idea that language evolved as a specialized mechanism exclusively for communication.

NO DISCUSSION OF the evolution of language would be complete without considering the question of nature versus nurture. To what extent are the rules of language innate, and to what extent are they absorbed from the world early in life? Arguments about the evolution of language have been fierce, and the nature-versus-nurture debate has been the most acrimonious of all. I mention it here only briefly because it has already been the subject of a number of recent books. Everyone agrees that words are not hardwired in the brain. The same object can have different names in different languages—"dog" in English, "chien" in French, "kutta" in Hindi, "maaa" in Thai, and "nai" in Tamil—which don't even sound alike. But with regard to the rules of language, there is no such agreement. Rather, three viewpoints vie for supremacy.

In the first view, the *rules themselves* are entirely hardwired. Exposure to adult speech is needed only to act as a switch to turn the mechanism on. The second view asserts that the rules of language are extracted statistically through listening. Bolstering this idea, artificial neural networks have been trained to categorize words and infer rules of syntax simply through passive exposure to language.

While these two models certainly capture some aspect of language acquisition, they cannot be the whole story. After all, apes, housecats, and iguanas have neural networks in their skulls, but they do not learn language even when raised in human households. A bonobo ape educated at Eton or Cambridge would still be an ape without language.

According to the third view, the *competence to acquire the rules* is innate, but exposure is needed to pick up the actual rules. This competence is bestowed by a still-unidentified "language acquisition device," or LAD. Humans have this LAD. Apes lack it.

I favor this third view because it is the one most compatible with my evolutionary framework, and is supported by two complementary facts. First, apes cannot acquire true language even when they are treated like human children and trained daily in hand signs. They end up being able to sign for something they need right away, but their signing lacks generativity (the ability to generate arbitrarily complex new combinations of words), function words, and recursion. Conversely, it is nearly impossible to prevent human children from acquiring language. In some areas of the world, where people from different language backgrounds must trade or work together, children and adults develop a simplified pseudo-language—one with a limited vocabulary, rudimentary syntax, and little flexibility—called a pidgin. But the first generation of children who grow up surrounded by a pidgin spontaneously turn it into a creole—a full-fledged language, with true syntax and all the flexibility and nuance needed to compose novels, songs, and poetry. The fact that creoles arise time and time again from pidgins is compelling evidence for an LAD.

These are important and obviously difficult issues, and it's unfortunate that the popular press often oversimplifies them by just asking questions like, Is language mainly innate or mainly acquired? Or similarly, Is IQ determined mainly by one's genes or mainly by one's environment? When two processes interact linearly, in ways that can be tracked with arithmetic, such questions can be meaningful. You can ask, for instance, "How much of our profits came from investments and how much from sales?" But if the relationships are complex and nonlinear—as they are for any mental attribute, be it language, IQ, or creativity—the question should be not, Which contributes more? but rather, How do they interact to create the final product? Asking whether language is mainly nurture is as silly as asking whether the saltiness of table salt comes mainly from chlorine or mainly from sodium.

The late biologist Peter Medawar provides a compelling analogy to illustrate the fallacy. An inherited disorder called phenylketonuria (PKU) is caused by a rarely occurring abnormal gene that results in a failure to metabolize the amino acid phenylalanine in the body. As the amino acid starts accumulating in the child's brain, he becomes profoundly retarded. The cure is simple. If you diagnose it early enough, all you do is withhold phenylalanine-containing foods from the diet and the child grows up with an entirely normal IQ.

Now imagine two boundary conditions. Assume there is a planet where the gene is uncommon and phenylalanine is everywhere, like oxygen or water, and is indispensable for life. On this planet, retardation caused by PKU, and therefore variance in IQ in the population, would be entirely attributable to the PKU gene. Here you would be justified in saying that retardation was a genetic disorder or that IQ was inherited. Now consider another planet in which the converse is true: Everyone has the PKU gene but phenylalanine is rare. On this planet you would say that PKU is an environmental disorder caused by a poison called phenylalanine, and most of the variance in IQ is caused by the environment. This example shows that when the interaction between two variables is labyrinthine it is meaningless to ascribe percentage values to the contribution made by either. And if this is true for just one gene interacting with one environmental variable, the argument must hold with even greater force for something as complex and multifactorial as human intelligence, since genes interact not only with the environment but with each other.

Ironically, the IQ evangelists (such as Arthur Jensen, William Shockley, Richard Herrnstein, and Charles Murray) use the heritability of IQ itself (sometimes called "general intelligence" or "little g") to argue that intelligence is a single measurable trait. This would be roughly analogous to saying that general health is one thing just because life span has a strong heritable component that can be expressed as a single number—age! No medical student who believed in "general health" as a monolithic entity would get very far in medical school or be allowed to become a physician—and rightly so—and yet whole careers in psychology and political movements have been built on the equally absurd belief in single measurable general intelligence. Their contributions have little more than shock value.

Returning to language, it should now be obvious which side of the fence I am on: neither. I straddle it proudly. Hence this chapter is not really about how language evolved—though I have been using that phrasing as shorthand—but how language *competence*, or the ability to acquire language so quickly, evolved. This competence is controlled by genes that were selected for by the evolutionary process. Our questions in the rest of this chapter are, Why were these genes selected, and how did this highly sophisticated competence evolve? Is it modular? How did it all get started? And how did we make the evolutionary transition from the grunts and howls of our apelike ancestors to the transcendent lyricism of Shakespeare?

RECALL THE SIMPLE bouba-kiki experiment. Could it hold the key to understanding how the first words evolved among a band of ancestral hominins in the African savanna between one and two hundred thousand years ago? Since words for the same object are often utterly different in different languages, one is tempted to think that the words chosen for particular objects are entirely arbitrary. This in fact is the standard view among linguists. Now, maybe one night the first band of ancestral hominins just sat around the tribal fire and said,

"Okay, let's all call this thing a bird. Now let's all say it together, *biiirrrrddddd*. Okay let's repeat again, *birrrrrrrddddddd*."

This story is downright silly, of course. But if it's not how an initial lexicon was constructed, how did it happen? The answer comes from our bouba-kiki experiment, which clearly shows that there is a built-in, nonarbitrary correspondence between the visual shape of an object and the sound (or at least, the kind of sound) that might be its "partner." This preexisting bias may be hardwired. This bias may have been very small, but it may have been sufficient to get the process started. This idea sounds very much like the now discredited "onomatopoeic theory" of language origins, but it isn't. "Onomatopoeia" refers to words that are based on an imitation of a sound—for example, "thump" and "cluck" to refer to certain sounds, or how a child might call a cat a "meow-meow." The onomatopoeic theory posited that sounds associated with an object become shorthand to refer to the objects themselves. But the theory I favor, the synesthetic theory, is different. The rounded visual shape of the bouba doesn't make a rounded sound, or indeed any sound at all. Instead, its visual profile resembles the profile of the undulating sound at an abstract level. The onomatopoeic theory held that the link between word and sound was arbitrary and merely occurred through repeated association. The synesthetic theory says the link is nonarbitrary and grounded in a true resemblance of the two in a more abstract mental space.

What's the evidence for this? The anthropologist Brent Berlin has pointed out that the Huambisa tribe of northern Peru have over thirty different names for thirty bird species in their jungle and an equal number of fish names for different Amazonian fishes. If you were to jumble up these sixty names and give them to someone from a completely different sociolinguistic background—say, a Chinese peasant—and ask him to classify the names into two groups, one for birds, one for fish, you would find that, astonishingly, he succeeds in this task well above chance level even though his language doesn't bear the slightest shred of resemblance to the South American one. I would argue that this is a manifestation of the bouba-kiki effect, in other words, of sound-shape translation.[1]

But this is only a small part of the story. In Chapter 4, I introduced some ideas about the contribution mirror neurons may have made to the evolution of language. Now, in the remainder of this chapter, we can look at the matter more deeply. To understand the next part, let's return to Broca's area in the frontal cortex. This area contains maps, or motor programs, that send signals down to the various muscles of the tongue, lips, palate, and larynx to orchestrate speech. Not coincidentally, this region is also rich in mirror neurons, providing an interface between the oral actions for sounds, listening to sounds, and (least important) watching lip movements.

Just as there is a nonarbitrary correspondence and cross-activation between brain maps for sights and sounds (the bouba-kiki effect), perhaps there is a similar correspondence—a built-in translation—between visual and auditory maps, on the one hand, and the motor maps in Broca's area on the other. If this sounds a bit cryptic, think again of words like "teeny-weeny," "un peau," and "diminutive," for which the mouth and lips and pharynx actually become small as if to echo or mime the visual small-ness, whereas words like "en*o*rmous" and "l*a*rge" entail an actual physical enlargement of the mouth. A less obvious example is "fu*dge*," "tru*dge*," "slu*dge*," "smu*dge*," and so on, in which there is a prolonged tongue press-ing on the palate before the sudden release, as if to mimic the prolonged sticking of the shoe in mud before the relatively sudden release. Here, yet again, is a built-in abstraction device that translates visual and auditory contours into vocal contours specified by muscle twitches.

Another less obvious piece of the puzzle is the link between man-ual gestures and lip and tongue movements. As mentioned in Chapter 4, Darwin noticed that when you cut with a pair of scissors, you may unconsciously echo these movements by clenching and unclenching your jaws. Since the cortical areas concerned with the mouth and hand are right next to each other, perhaps there is an actual spillover of signals from hands to mouth. As in synesthesia, there appears to be a built-in cross-activation between brain maps, except here it is between two motor maps rather than between sensory maps. We need a new name for this, so let's call it "synkinesia" (*syn* meaning "together," *kinesia* meaning "movement").

Synkinesia may have played a pivotal role in transforming an earlier gestural language (or protolanguage, if you prefer) of the hands into spo-ken language. We know that emotional growls and shrieks in primates arise mainly in the right hemisphere, especially from a part of the limbic system (the emotional core of the brain) called the anterior cingulate. If a manual gesture were being echoed by orofacial movements while the creature was simultaneously making emotional utterances, the net result would be what we call words. In short, ancient hominins had a built-in, preexisting mechanism for spontaneously translating gestures into words. This makes it easier to see how a primitive gestural language could have evolved into speech—an idea that many classical psycholin-guists find unappealing.

As a concrete example, consider the phrase "come hither." Notice that you gesture this idea by holding your palm up and flexing your fingers toward yourself as if to touch the lower part of the palm. Amazingly, your tongue makes a very similar movement as it curls back to touch the palate to utter "hither" or "here"—examples of synkinesia. "Go" involves pouting the lips outward, whereas "come" involves drawing the lips together inward. (In the Indian Dravidian language Tamil—unrelated to English—the word for go is "po").

Obviously, whatever the original language was back in the Stone Age, it has since been embellished and transformed countless times beyond reckoning, so that today we have languages as diverse as English, Japanese, !Kung, and Cherokee. Language, after all, evolves with incredible rapidity; sometimes just two hundred years is enough to alter a language to the point where a young speaker would be barely able to communicate with her great-great-grandmother. By this token, once the juggernaut of full linguistic competence arose in the human mind and culture, the original synkinetic correspondences were probably lost or blended beyond recognition. But in my account, synkinesia sowed the initial seeds of lexicon, helping to form the original vocabulary base on which subsequent linguistic elaboration was built.

Synkinesia and other allied attributes, such as mimicry of other people's movements and extraction of commonalities between vision and hearing (*bouba-kiki*), may all rely on computations analogous to what mirror neurons are supposed to do: link concepts across brain maps. These sorts of linkages remind us again of their potential role in the evolution of protolanguage. This hypothesis may seem speculative to orthodox cognitive psychologists, but it provides a window of opportunity—indeed, the only one we have to date—for exploring the actual neural mechanisms of language. And that's a big step forward. We will pick up the threads of this argument later in this chapter.

We also need to ask how gesturing evolved in the first place.[2] At least for verbs like "come" or "go," it may have emerged through the ritualization of movements that were once used for performing those actions. For instance, you may actually pull someone toward you by flexing your fingers and elbow toward you while grabbing the person. So the movement itself (even if divorced from the actual physical object) became a means of communicating intent. The result is a gesture. You can see how the same argument applies to "push," "eat," "throw," and other basic verbs. And once you have a vocabulary of gestures in place, it becomes easier for corresponding vocalizations to evolve, given the preexisting hard-wired translation produced by synkinesia. (The ritualization and reading of gestures may, in turn, have involved mirror neurons, as alluded to in previous chapters.)

So we now have three types of map-to-map resonance going on in the early hominin brain: visual-auditory mapping (*bouba-kiki*); mapping between auditory and visual sensory maps, and motor vocalization maps in Broca's area; and mapping between Broca's area and motor areas controlling manual gestures. Bear in mind that each of these biases was probably very small, but acting in conjunction they could have progressively bootstrapped each other, creating the snowball effect that culminated in modern language.

IS THERE ANY neurological evidence for the ideas discussed so far? Recall that many neurons in a monkey's frontal lobe (in the same region that appears to have become Broca's area in us) fire when the animal performs a highly specific action like reaching for a peanut, and that a subset of these neurons also fires when the monkey watches another monkey grab a peanut. To do this, the neuron (by which I really mean "the network of which the neuron is a part") has to compute the abstract similarity between the command signals specifying muscle contraction sequences and the visual appearance of peanut reaching seen from the other monkey's vantage point. So the neuron is effectively reading the other individual's intention and could, in theory, also understand a ritualized gesture that resembles the real action. It struck me that the *bouba-kiki* effect provides an effective bridge between these mirror neurons and ideas about synesthetic bootstrapping I have presented so far. I considered this argument briefly in an earlier chapter, let me elaborate the argument now to make the case for its relevance to the evolution of protolanguage.

The bouba-kiki effect requires a built-in translation between visual appearance, sound representation in the auditory cortex, and sequences of muscle twitches in Broca's area. Performing this translation almost certainly involves the activation of circuits with mirror-neuron-like properties, mapping one dimension onto another. The inferior parietal lobule

(IPL), rich in mirror neurons, is ideally suited for this role. Perhaps the IPL serves as a facilitator for all such types of abstraction. I emphasize, again, that these three features (visual shape, sound inflections, and lip and tongue contour) have absolutely nothing in common except the abstract property of, say, jaggedness or roundness. So what we are seeing here is the rudiments—and perhaps relics of the origins—of the process called abstraction that we humans excel at, namely, the ability to extract the common denominator between entities that are otherwise utterly dissimilar. From being able to extract the jaggedness of the broken glass shape and the sound *kiki* to seeing the "fiveness" of five pigs, five donkeys, or five chirps may have been a short step in evolution but a giant step for humankind.

I HAVE ARGUED, so far, that the bouba-kiki effect may have fueled the emergence of protowords and a rudimentary lexicon. This was an important step, but language isn't just words. There are two other important aspects to consider: syntax and semantics. How are these represented in the brain and how did they evolve? The fact that these two functions are at least partially autonomous is well illustrated by Broca's and Wernicke's aphasias. As we have seen, a patient with the latter syndrome produces elaborate, smoothly articulated, grammatically flawless sentences that convey no meaning whatsoever. The Chomskian "syntax box" in the intact Broca's area goes "open loop" and produces well-formed sentences, but without Wernicke's area to inform it with cultivated content, the sentences are gibberish. It's as though Broca's area on its own can juggle the words with the correct rules of grammar—just like a computer program might—without any awareness of meaning. (Whether it is capable of more complex rules such as recursion remains to be seen; it's something we are currently studying.)

We'll come back to syntax, but first let's look at semantics (again, roughly speaking, the meaning of a sentence). What exactly is meaning? It's a word that conceals vast depths of ignorance. Although we know that Wernicke's area and parts of the temporo-parieto-occipital (TPO) junction, including the angular gyrus (Figure 6.2), are critically involved, we have no idea how neurons in these areas actually do their job. Indeed, the manner in which neural circuitry embodies meaning is one of the

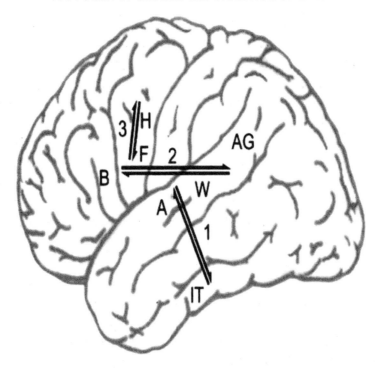

FIGURE 6.2   A schematic depiction of resonance between brain areas that may have accelerated the evolution of protolanguage. Abbreviations: B, Broca's area (for speech and syntactic structure). A, auditory cortex (hearing). W, Wernicke's area for language comprehension (semantics). AG, angular gyrus for cross-modal abstraction. H, hand area of the motor cortex, which sends motor commands to the hand (compare with Penfield's sensory cortical map in Figure 1.2). F, face area of the motor cortex (which sends command messages to the facial muscles, including lips and tongue). IT, the inferotemporal cortex/fusiform area, which represents visual shapes. Arrows depict two-way interactions that may have emerged in human evolution: 1, connections between the fusiform area (visual processing) and auditory cortex mediate the bouba-kiki effect. The cross-modal abstraction required for this probably requires initial passage through the angular gyrus. 2, interactions between the posterior language areas (including Wernicke's area) and motor areas in or near Broca's area. These connections (the arcuate fasciculus) are involved in cross-domain mapping between sound contours and motor maps (mediated partly by neurons with mirror-neuron-like properties) in Broca's area. 3, cortical motor-to-motor mappings (synkinesia) caused by links between hand gestures and tongue, lip, and mouth movements in Penfield's motor map. For example, the oral gestures for "diminutive," "little," "teeny-weeny," and the French phrase "en peau" synkinetically mimic the small pincer gesture made by opposing thumb and index finger (as opposed to "large" or "enormous"). Similarly, pouting your lips outward to say "you" or (in French) "vous" mimic pointing outward.

great unsolved mysteries of neuroscience. But if you allow that abstraction is an important step in the genesis of meaning, then our bouba-kiki example might once again provide the clue. As already noted, the sound *kiki* and the jagged drawing would seem to have nothing in common. One is a one-dimensional, time-varying pattern on the sound receptors in your ear, whereas the other is a two-dimensional pattern of light arriving on your retina all in one instant. Yet your brain has no difficulty in abstracting the property of jaggedness from both signals. As we have seen, there are strong hints that the angular gyrus is involved in this remarkable ability we call cross-modal abstraction.

There was an accelerated development of the left IPL in primate evolution culminating in humans. In addition, the front part of the lobule in humans (and humans alone), split into two gyri called the supramarginal gyrus and the angular gyrus. It doesn't require deep insight to suggest therefore that the IPL and its subsequent splitting must have played a pivotal role in the emergence of functions unique to humans. Those functions, I suggest, include high-level types of abstraction.

The IPL (including the angular gyrus)—strategically located between the touch, vision, and hearing parts of the brain—evolved originally for cross-modal abstraction. But once this happened, cross-modal abstraction served as an exaptation for more high-level abstraction of the kind we humans take great pride in. And since we have two angular gyri (one in each hemisphere), they may have evolved different styles of abstraction: the right for visuospatial and body-based metaphors and abstraction, and the left for more language-based metaphors, including puns. This evolutionary framework may give neuroscience a distinct advantage over classical cognitive psychology and linguistics because it allows us to embark on a whole new program of research on the representation of language and thought in the brain.

The upper part of the IPL, the supramarginal gyrus, is also unique to humans, and is directly involved in the production, comprehension, and imitation of complex skills. Once again, these abilities are especially well developed in us compared with the great apes. When the left supramarginal gyrus is damaged, the result is apraxia, which is a fascinating disorder. A patient with apraxia is mentally normal in most respects, including his ability to understand and produce language. Yet when you ask him to mime a simple action—"pretend you are hammering a nail"—he will make a fist and bang it on the table instead of holding a "pretend" handle as you or I might. If asked to pretend he is combing his hair, he might stroke his hair with his palm or wiggle his fingers in his hair instead of "holding" and moving an imaginary comb through his hair. If requested to pretend waving goodbye, he may stare at his hand intently trying to figure out what to do or flail it around near his face. But if questioned, "What does 'waving goodbye' mean?" he might say, "Well, it's what you do when you are parting company," so obviously he clearly understands at a conceptual level what's expected. Furthermore, his hands are not paralyzed or clumsy: He can move individual fingers as gracefully and independently as any of us. What's missing is the ability to conjure up a vibrant, dynamic internal picture of the required action which can be used to guide the orchestration of muscle twitches to mime the action. Not surprisingly, putting the actual hammer in his hand may (as it does in some patients) lead to accurate performance since it doesn't require him to rely on an internal image of the hammer.

Three additional points about these patients. First, they cannot judge whether someone else is performing the requested action correctly or not, reminding us that their problem lies in neither motor ability nor perception but in *linking* the two. Second, some patients with apraxia have difficulty imitating novel gestures produced by the examining physician. Third and most surprisingly, they are completely unaware that they themselves are miming incorrectly; there is no sign of frustration. All of these missing abilities sound compellingly reminiscent of the abilities traditionally attributed to mirror neurons. Surely it can't be a coincidence that the IPL in monkeys is rich in mirror neurons. Based on this reasoning my postdoctoral colleague Paul McGeoch and I suggested in 2007 that apraxia is fundamentally a disorder of mirror-neuron function. Intriguingly, many autistic children also have apraxia, an unexpected link that lends support to our idea that a mirror-neuron deficit might underlie both disorders. Paul and I opened a bottle to celebrate having clinched the diagnosis.

But what caused the accelerated evolution of the IPL—and the angular gyrus part of it—in the first place? Did the selection pressure come from the need for higher forms of abstraction? Probably not. The most likely cause of its explosive development in primates was the need to achieve an exquisitely refined, fine-grained interaction between vision and muscle and joint position sense while negotiating branches on tree-tops. This resulted in the capacity of cross-modal abstraction, for example, when a branch is signaled as being horizontal both by the image falling on the retina and the dynamic stimulation of touch, joint, and muscle receptors in the hands.

The next step was critical: The lower part of the IPL split accidentally, possibly as a result of gene duplication, a frequent occurrence in evolution. The upper part, the supramarginal gyrus, retained the old function of its ancestral lobule—hand-eye coordination—elaborating it to the new levels of sophistication required for skilled tool use and imitation in humans. In the angular gyrus the very same computational ability set the stage (became an exaptation) for other types of abstraction as well: the ability to extract the common denominator among superficially dissimilar entities. A weeping willow looks sad because you project sadness on to it. Juliet is the sun because you can abstract certain things they have in common. Five donkeys and five apples have "fiveness" in common.

A tangential piece of evidence for this idea comes from my examination of patients who have damage to the IPL of the left hemisphere. These patients usually have anomia. (difficulty finding words), but I found that some of them failed the bouba-kiki test and were also abysmal at interpreting proverbs, often interpreting them literally instead of metaphorically. One patient I saw in India recently got 14 out of 15 proverbs wrong even though he was perfectly intelligent in other respects. Obviously this study needs to be repeated on additional patients but it promises to be a fruitful line of enquiry.

The angular gyrus is also involved in naming objects, even common objects such as comb or pig. This reminds us that a word, too, is a form of abstraction from multiple instances (for example, multiple views of a comb seen in different contexts but always serving the *function* of hairdressing). Sometimes they will substitute a related word ("cow" for "pig") or try to define the word in absurdly comical ways. (One patient said "eye medicine" when I pointed to my glasses.) Even more intriguing was an observation I made in India on a fifty-year-old physician with anomia. Every Indian child learns about many gods in Indian mythology, but two great favorites are Ganesha (the elephant-headed god) and Hanuman (the monkey god) and each has an elaborate family history.

When I showed him a sculpture of Hanuman, he picked it up, scrutinized it, and misidentified it as Ganesha, which belongs to the same category, namely god. But when I asked him to tell me more about the sculpture, which he continued to inspect, he said it was the son of Shiva and Parvati—a statement that is true for Ganesha, not Hanuman. It's as if the mere act of mislabeling the sculpture vetoed its visual appearance, causing him to give incorrect attributes to Hanuman! Thus the name of an object, far from being just any other attribute of the object, seems to be a magic key that opens a whole treasury of meanings associated with the object. I can't think of a simpler explanation for this phenomenon, but the existence of such unsolved mysteries fuels my interest in neurology just as much as the explanations for which we can generate and test specific hypotheses.

LET US TURN now to the aspect of language that is most unequivocally human: syntax. The so-called syntactic structure, which I mentioned earlier, gives human language its enormous range and flexibility. It seems to have evolved rules that are intrinsic to this system, rules that no ape has been able to master but every human language has. How did this particular aspect of language evolve? The answer comes, once again, from the exaptation principle—the notion that adaptation to one specific function becomes assimilated into another, entirely different function. One intriguing possibility is that the hierarchical tree structure of syntax may have evolved from a more primitive neural circuit that was already in place for tool use in the brains of our early hominin ancestors.

Let's take this a step further. Even the simplest type of opportunistic tool use, such as using a stone to crack open a coconut, involves an action—in this case, cracking (the verb)—performed by the right hand of the tool user (the subject) on the object held passively by the left hand (the object). If this basic sequence were already embedded in the neural circuitry for manual actions, it's easy to see how it might have set the stage for the subject-verb-object sequence that is an important aspect of natural language.

In the next stage of hominin evolution, two amazing new abilities emerged that were destined to transform the course of human evolution. First was the ability to find, shape, and store a tool for future use, leading to our sense of planning and anticipation. Second—and especially important for subsequent language origin—was use of the subassembly technique in tool manufacture. Taking an axe head and hafting (tying) it to a long wooden handle to create a composite tool is one example. Another is hafting a small knife at an angle to a small pole and then tying this assembly to another pole to lengthen it so that fruits can be reached and yanked off trees. The wielding of a composite structure bears a tantalizing resemblance to the embedding of, say, a noun phrase within a longer sentence. I suggest that this isn't just a superficial analogy. It's entirely possible that the brain mechanism that implemented the hierarchical subassembly strategy in tool use became coopted for a totally novel function, the syntactic tree structure.

But if the tool-use subassembly mechanism were borrowed for aspects of syntax, then wouldn't the tool-use skills deteriorate correspondingly as syntax evolved, given limited neural space in the brain? Not necessarily. A frequent occurrence in evolution is the duplication of preexisting body parts brought about by actual gene duplication. Just think of multisegmented worms, whose bodies are composed of repeating, semi-independent body sections, a bit like a chain of railroad cars. When such duplicated structures are harmless and not metabolically costly, they can endure many generations. And they can, under the right circumstances, provide the perfect opportunity for that duplicate structure to become specialized for a different function. This sort of thing has happened repeatedly in the evolution of the rest of the body, but its role in the evolution of brain mechanisms is not widely appreciated by psychologists. I suggest that an area very close to what we now call Broca's area originally evolved in tandem with the IPL (especially the supramarginal portion) for the multimodal and hierarchical subassembly routines of tool use. There was a subsequent duplication of this ancestral area, and one of the two new subareas became further specialized for syntactic structure that is divorced from actual manipulation of physical objects in the world—in other words, it became Broca's area. Add to this cocktail the influence of semantics, imported from Wernicke's area, and aspects of abstraction from the angular gyrus, and you have a potent mix ready for the explosive development of full-fledged language. Not coincidentally, perhaps, these are the very areas in which mirror neurons abound.

Bear in mind that my argument thus far focuses on evolution and exaptation. Another question remains. Are the concepts of subassembly tool use, hierarchical tree structure of syntax (including recursion), and conceptual recursion mediated by separate modules in the brains of modern humans? How autonomous, really, are these modules in our brains? Would a patient with apraxia (the inability to mime the use of tools) caused by damage to the supramarginal gyrus also have problems with subassembly in tool use? We know that patients with Wernicke's aphasia produce syntactically normal gibberish—the basis for suggesting that, at least in modern brains, syntax doesn't depend on the recursiveness of semantics or indeed of high-level embedding of concepts within concepts.[3]

But how syntactically normal is their gibberish? Does their speech—mediated entirely by Broca's area on autopilot—really have the kinds of syntactic tree structure and recursion that characterize normal speech? If not, are we really justified in calling Broca's area a "syntax box"? Can a Broca's aphasic do algebra, given that algebra also requires recursion to some extent? In other words, does algebra piggyback on preexisting neural circuits that evolved for natural syntax? Earlier in this chapter I gave the example of a single patient with Broca's aphasia who could do algebra, but there are precious few studies on these topics, each of which could generate a PhD thesis.

SO FAR I have taken you on an evolutionary journey that culminated in the emergence of two key human abilities: language and abstraction. But there is another feature of human uniqueness that has puzzled philosophers for centuries, namely, the link between language and sequential thinking, or reasoning in logical steps. Can we think without silent internal speech? We have already discussed language, but we need to be clear about what is meant by thinking before we try grappling with this question. Thinking involves, among other things, the ability to engage in open-ended symbol manipulation in your brain following certain rules. How closely are these rules related to those of syntax? The key phrase here is "open-ended."

To understand this, think of a spider spinning a web and ask yourself, Does the spider have knowledge about Hooke's law regarding the tension of stretched strings? The spider must "know" about this in some sense, otherwise the web would fall apart. Would it be more accurate to say that the spider's brain has tacit, rather than explicit, knowledge of Hooke's law? Although the spider behaves as though it knows this law—the very existence of the web attests to this—the spider's brain (yes, it has one) has no explicit representation of it. It cannot use the law for any purpose other than weaving webs and, in fact, it can only weave webs according to a fixed motor sequence. This isn't true of a human engineer who consciously deploys Hooke's law, which she learned and understood from physics textbooks. The human's deployment of the law is open-ended and flexible, available for an infinite number of applications. Unlike the spider he has an explicit representation of it in his mind—what we call understanding. Most of the knowledge of the world that we have falls in between these two extremes: the mindless knowledge of a spider and the abstract knowledge of the physicist.

What do we mean by "knowledge" or "understanding"? And how do billions of neurons achieve them? These are complete mysteries. Admittedly, cognitive neuroscientists are still very vague about the exact meaning of words like "understand," "think," and indeed the word "meaning" itself. But it is the business of science to find answers step by step through speculation and experiment. Can we approach some of these mysteries experimentally? For instance, what about the link between language and thinking? How might you experimentally explore the elusive interface between language and thought?

Common sense suggests that some of the activities regarded as thinking don't require language. For example, I can ask you to fix a lightbulb on a ceiling and show you three wooden boxes lying on the floor. You would have the internal sense of juggling the visual images of the boxes—stacking them up in your mind's eye to reach the bulb socket—before actually doing so. It certainly doesn't feel like you are engaging in silent internal speech—"Let me stack box A on box B," and so on. It feels as if we do this kind of thinking visually and not by using language. But we have to be careful with this deduction because introspection about what's going in one's head (stacking the three boxes) is not a reliable guide to what's actually going on. It's not inconceivable that what feels like the internal juggling of visual symbols actually taps into the same circuitry in the brain that mediates language, even though the task feels purely geometric or spatial. However much this seems to violate common sense, the activation of visual image–like representations may be incidental rather than causal.

Let's leave visual imagery aside for the moment and ask the same question about the formal operations underlying logical thinking. We say, "If Joe is bigger than Sue, and if Sue is bigger than Rick, then Joe must be bigger than Rick." You don't have to conjure up mental images to realize that the deduction ("then Joe must be . . .") follows from the two premises ("If Joe is . . . and if Sue is . . ."). It's even easier to appreciate this if you substitute their names with abstract tokens like A, B, and C: If A > B and B > C, then it must be true that A > C. We also can intuit that if A > C and B > C, it doesn't necessarily follow that A > B.

But where do these obvious deductions, based on the rules of transitivity, come from? Is it hardwired into your brain and present at birth? Was it learned from induction because every time in the past, when any entity A was bigger than B and B was bigger than C, it was always the case that A was bigger than C as well? Or was it learned initially through language? Whether this ability is innate or learned, does it depend on some kind of silent internal language that mirrors and partially taps into the same neural machinery used for spoken language? Does language precede propositional logic, or vice versa? Or perhaps neither is necessary for the other, even though they mutually enrich each other.

These are intriguing theoretical questions, but can we translate them into experiments and find some answers? Doing so has proved to be notoriously difficult in the past, but I'll propose what philosophers would call a thought experiment (although, unlike philosophers' thought experiments, this one can actually be done). Imagine I show you three boxes of three different sizes on the floor and a desirable object dangling from a high ceiling. You will instantly stack the three boxes, with the largest one at the bottom and the smallest at the top, and then climb up to retrieve the reward. A chimp can also solve this problem but presumably requires physical trial-and-error exploration of the boxes (unless you pick an Einstein among chimps).

But now I modify the experiment: I put a colored luminous spot on each of the boxes—red (on the big box), blue (intermediate box), and green (small box)—and have the boxes lying separately on the floor. I bring you into the room for the first time and expose you to the boxes long enough for you to realize which box has which spot. Then I switch the room lights off so that only the luminous colored dots are visible. Finally, I bring a luminous reward into the dark room and dangle it from the ceiling.

If you have a normal brain you will, without hesitation, put the red-dotted box at the bottom, the blue-dotted box in the middle, and the green-dotted box on top, and then climb to the top of the pile to retrieve the dangling reward. (Let's assume the boxes have handles sticking out that you use to pick them up with, and that the boxes have been made equal weight so that you can't use tactile cues to distinguish them.) In other words, as a human being you can create arbitrary symbols (loosely analogous to words) and then juggle them entirely in your brain, doing a virtual-reality simulation to discover a solution. You could even do this if during the first phase you were shown only the red- and green-dotted boxes, and then separately shown the green- and blue-dotted boxes, followed finally in the test phase by seeing the red- and green-dotted boxes alone. (Assume that stacking even two boxes gives you better access to the reward.) Even though the relative sizes of the boxes were not currently visible during these three viewing stages, I bet you could now juggle the symbols entirely in your head to establish the transitivity using conditional (if-then) statements—"If red is bigger than blue and blue is bigger than green, then red must be bigger than green"—and then proceed to stack the green box on the red box in the dark to reach the reward. An ape would almost certainly fail at this task, which requires off-line (out of sight) manipulation of arbitrary signs, the basis of language.

But to what extent is language an actual requirement for conditional statements mentally processed off-line, especially in novel situations? Perhaps one could find out by carrying out the same experiment on a patient who has Wernicke's aphasia. Given the claim that the patient can produce sentences like "If Blaka is bigger than Guli, then Lika tuk," the question is whether she understands the transitivity implied in the sentence. If so, would she pass the three-boxes test we designed for chimps? Conversely, what about a patient with Broca's aphasia, who purportedly has a broken syntax box? He no longer uses "ifs," "buts," and "thens" in his sentences and doesn't comprehend these words when he hears or reads them. Would such a patient nevertheless be able to pass the three-boxes test, implying he doesn't need the syntax module to understand and deploy the rules of deductive if-then inferences in a versatile manner? One could ask the same question of a number of other rules of logic as well. Without such experiments the interface between language and thought will forever remain a nebulous topic reserved for philosophers.

I have used the three-boxes idea to illustrate that one can, in principle, experimentally disentangle language and thought. But if the experiment proves impractical to carry out, one could conceivably confront the patient with cleverly designed video games that embody the same logic but do not require explicit verbal instructions. How good would the patient be at such games? And indeed, can the games themselves be used to slowly coax language comprehension back into action?

Another point to consider is that the ability to deploy transitivity in abstract logic may have evolved initially in a social context. Ape A sees ape B bullying and subduing ape C, who has on previous occasions successfully subdued A. Would A then spontaneously retreat from B, implying the ability to employ transitivity? (As a control, one would have to show that A doesn't retreat from B if B is only seen subduing some other random ape C.)

The three-boxes test given to Wernicke's aphasics might help us to disentangle the internal logic of our thought processes and the extent to which they interact with language. But there is also a curious emotional aspect to this syndrome that has received scant attention, namely, aphasics' complete indifference—indeed, ignorance—of the fact that they are producing gibberish and their failure to register the expression of incomprehension on the faces of people they are talking to. Conversely, I once wandered into a clinic and started saying "Sawadee Khrap. Chua alai? Kin Krao la yang?" to an American patient and he smiled and nodded acknowledgment. Without his language comprehension module he couldn't tell nonsense speech and normal speech apart, whether the speech emerged from his own mouth or from mine. My postdoctoral colleague Eric Altschuler and I have often toyed with the idea of introducing two Wernicke's aphasics to each other. Would they talk incessantly to each other all day, and without getting bored? We joked about the possibility that Wernicke's aphasics are *not* talking gibberish; maybe they have a private language comprehensible only to each other.

———

WE HAVE BEEN speculating on the evolution of language and thought, but still haven't resolved it. (The three-boxes experiment or its video-game analog hasn't been tried yet.) Nor have we considered the modularity of language itself: the distinction between semantics and syntax (including what we defined earlier in the chapter as recursive embedding, for example, "The girl who killed the cat that ate the rat started to sing"). Presently, the strongest evidence for the modularity of syntax comes from neurology, from the observation that patients with a damaged Wernicke's area produce elaborate, grammatically correct sentences that are devoid of meaning. Conversely, in patients who have a damaged Broca's area but an intact Wernicke's area, like Dr. Hamdi, meaning is preserved, but there is no syntactic deep structure. If semantics ("thought") and syntax were mediated by the same brain region or by diffuse neural networks, such an "uncoupling" or dissociation of the two functions couldn't occur. This is the standard view presented by psycholinguists, but is it really true? The fact that the deep structure of language is deranged in Broca's aphasia is beyond question, but does it follow that this brain region is specialized exclusively for key aspects of language such as recursion and hierarchical embedding? If I lop off your hand you can't write, but your writing center is in the angular gyrus, not in your hand. To counter this argument psycholinguists usually point out that the converse of this syndrome occurs when Wernicke's area is damaged: Deep structure underlying grammar is preserved but meaning is abolished.

My postdoctoral colleagues Paul McGeoch and David Brang and I decided to take a closer look. In an influential and brilliant paper written in 2001 in the journal *Science*, the linguist Noam Chomsky and cognitive neuroscientist Marc Hauser surveyed the whole field of psycholinguistics and the conventional wisdom that language is unique to humans (and probably modular). They found that almost every aspect of language could be seen in other species, after adequate training, such as in chimps, but the one aspect that makes the deep grammatical structure in humans unique is recursive embedding. When people say that deep structure and syntactic organization are normal in Wernicke's aphasia, they are usually referring to the more obvious aspects, such as the ability to generate a fully formed sentence employing nouns, prepositions, and conjunctions but carrying no meaningful content ("John and Mary went to the joyful bank and paid hat"). But clinicians have long known that, contrary to popular wisdom, the speech output of Wernicke's aphasics isn't entirely normal even in its syntactic structure. It's usually somewhat impoverished. However, these clinical observations were largely ignored because they were made long before recursion was recognized as the sine qua non of human language. Their true importance was missed.

When we carefully examined the speech output of many Wernicke's aphasics, we found that, in addition to the absence of meaning, the most striking and obvious loss was in recursive embedding. Patients spoke in loosely strung together phrases using conjunctions: "Susan came and hit John and took the bus and Charles fell down," and so forth. But they could almost never construct recursive sentences such as "John who loved Julie used a spoon." (Even without setting "who loved Julie" off with commas, we know instantly that John used the spoon, not Julie.) This observation demolishes the long-standing claim that Broca's area is a syntax box that is autonomous from Wernicke's area. Recursion may turn out to be a property of Wernicke's area, and indeed may be a general property common to many brain functions. Furthermore, we mustn't confuse the issue of functional autonomy and modularity in the modern human brain with the question of evolution: Did one module provide a substrate for the other or even evolve into another, or did they evolve completely independently in response to different selection pressures?

Linguists are mainly interested in the former question—the autonomy of rules intrinsic to the module—whereas the evolutionary question usually elicits a yawn (just as any talk of evolution or brain modules would seem pointless to a number theorist interested in rules intrinsic to the number system). Biologists and developmental psychologists, on the other hand, are interested not only in the rules that govern language but also in the evolution, development, and neural substrates of language, including (but not confined to) syntax. A failure to make this distinction has bedeviled the whole language evolution debate for nearly a century. The key difference, of course, is that language capacity evolved through natural selection over two hundred thousand years, whereas number theory is barely two thousand years old. So for what it is worth, my own (entirely unbiased) view is that on this particular issue the biologists are right. As an analogy, I'll invoke again my favorite example, the relationship between the chewing and hearing. All mammals have three tiny bones—malleus, stapes, and incus—inside the middle ear. These bones transmit and amplify sounds from the eardrum to the inner ear. Their sudden emergence in vertebrate evolution (mammals have them but their reptilian ancestors don't) was a complete mystery and often used as ammunition by creationists until comparative anatomists, embryologists, and paleontologists discovered that they actually evolved from the back of the jawbone of the reptile. (Recall that the back of your jaw articulates very close to your ear.) The sequence of steps makes a fascinating story.

The mammalian jaw has a single bone, the mandible, whereas our reptilian ancestors had three. The reason is that reptiles, unlike mammals, frequently consume enormous prey rather than frequent small meals. The jaw is used exclusively for swallowing, not chewing, and due to reptiles' slow metabolic rate, the unchewed food in the stomach can take weeks to break down and digest. This kind of eating requires a large, flexible, multihinged jaw. But as reptiles evolved into metabolically active mammals, the survival strategy switched to consumption of frequent small meals to maintain a high metabolic rate.

Remember also that reptiles lie low on the ground with their limbs sprawled outward, thereby swinging the neck and head close to the ground while they sniff for prey. The three bones of the jaw lying on the ground allowed reptiles to also transmit sounds made by other animals' nearby footsteps to the vicinity of the ear. This is called bone conduction, as opposed air conduction which is used by mammals.

As they evolved into mammals, reptiles raised themselves up from the sprawling position to stand higher up off the ground on vertical legs. This allowed two of the three jaw bones to become progressively assimilated into the middle ear, being taken over entirely for hearing airborne sounds and giving up their chewing function altogether. But this change in function was only possible because they were already strategically located—in the right place at the right time—and were already beginning to be used for hearing terrestrially transmitted sound vibrations. This radical shift in function also served the additional purpose of transforming the jaw into a single, rigid nonhinged bone—the mandible—which was much stronger and more useful for chewing.

The analogy with language evolution should be obvious. If I were to ask you whether chewing and hearing are modular and independent of each other, both structurally and functionally, the answer would obviously be yes. And yet we know that the latter evolved from the former, and we can even specify the steps involved. Likewise, there is clear evidence that language functions such as syntax and semantics are modular and autonomous and furthermore are also distinct from thinking, perhaps as distinct as hearing is from chewing. Yet it is entirely possible that one of these functions, such as syntax, evolved from other, earlier functions such as tool use and/or thinking. Unfortunately, since language doesn't fossilize like jaws or ear bones, we can only construct plausible scenarios. We may have to live with not knowing what the exact sequence of events was. But hopefully I have given you a glimpse of the kind of theory that we need to come up with, and the kinds of experiments we need to do, to account for the emergence of full-fledged language, the most glorious of all our mental attributes.

CHAPTER 7

Beauty and the Brain:
The Emergence of Aesthetics

———

Art is a lie that makes us realize the truth.

—PABLO PICASSO

Aₙ OLD INDIAN MYTH SAYS THAT BRAHMA CREATED THE UNI-
verse and all the beautiful snow-clad mountains, rivers, flowers, birds,
and trees—even humans. Yet soon afterward, he was sitting on a chair,
his head in his hands. His consort, Saraswati, asked him, "My lord—you
created the whole beautiful Universe, populated with men of great valor
and intellect who worship you—why are you so despondent?" Brahma
replied, "Yes, all this is true, but the men whom I have created have no
appreciation of the beauty of my creations and, without this, all their
intellect means nothing." Whereupon Saraswati reassured Brahma, "I
will give mankind a gift called art." From that moment on people developed an aesthetic sense, started responding to beauty, and saw the divine
spark in all things. Saraswati is therefore worshipped throughout India
as the goddess of art and music—as humankind's muse.

This chapter and the next are concerned with a deeply fascinating question: How does the human brain respond to beauty? How are
we special in terms of how we respond to and create art? How does
Saraswati work her magic? There are probably as many answers to this
question as there are artists. At one end of the spectrum is the lofty idea
that art is the ultimate antidote to the absurdity of the human predicament—the only "escape from this vale of tears," as the British surrealist
and poet Roland Penrose once said. At the other extreme is the school
of Dada, the notion that "anything goes," which says that what we call

art is largely contextual or even entirely in the mind of the beholder. (The most famous example is Marcel Duchamp putting a urinal bowl in a gallery and saying, in effect, "I call it art; therefore it's art.") But is Dada really art? Or is it merely art mocking itself? How often have you walked into a gallery of contemporary art and felt like the little boy who knew instantly that the emperor had no clothes?

Art endures in a staggering diversity of styles: Classical Greek art, Tibetan art, African Art, Khmer art, Chola bronzes, Renaissance art, impressionism, expressionism, cubism, fauvism, abstract art—the list is endless. But beneath all this variety, might there some general principles or artistic universals that cut across cultural boundaries? Can we come up with a science of art? Science and art seem fundamentally antithetical. One is a quest for general principles and tidy explanations while the other is a celebration of the individual imagination and spirit, so that the very notion of a science of art seems like an oxymoron. Yet that is my goal for this chapter and the next: to convince you that our knowledge of human vision and of the brain is now sophisticated enough that we can speculate intelligently on the neural basis of art and maybe begin to construct a scientific theory of artistic experience. Saying this does not in any way detract from the originality of the individual artist, for the manner in which she deploys these universal principles is entirely hers.

First, I want to make a distinction between art as defined by historians and the broad topic of aesthetics. Because both art and aesthetics require the brain to respond to beauty, there is bound to be a great deal of overlap. But art includes such things as Dada (whose aesthetic value is dubious), whereas aesthetics includes such things as fashion design, which is not typically regarded as high art. Maybe there can never be a science of high art, but I suggest there can be of the principles of aesthetics that underlie it.

Many principles of aesthetics are common to both humans and other creatures and therefore cannot be the result of culture. Can it be a coincidence that we find flowers to be beautiful even though they evolved to be beautiful to bees rather than to us? This is not because our brains evolved from bee brains (they didn't), but because both groups independently converged on some of the same universal principles of aesthetics. The same is true for why we find male birds of

FIGURE 7.1   The elaborately constructed "nest," or bower, of the male bowerbird, designed to attract females. Such "artistic" principles as grouping by color, contrast, and symmetry are in evidence.

paradise such a feast for the eyes—to the point of using them as head-dresses—even though they evolved for females of their own species and not for *Homo sapiens*.

Some creatures, such as bowerbirds from Australia and New Guinea, possess what we humans perceive as artistic talent. The males of the genus are drab little fellows but, perhaps as a Freudian compensation, they build enormous gorgeously decorated bowers—bachelor pads—to attract mates (Figure 7.1). One species builds a bower that is eight feet tall with elaborately constructed entrances, archways, and even lawns in front of the entryway. On different parts of the bower, he arranges clusters of flowers into bouquets, sorts berries of various types by color, and forms gleaming white hillocks out of bits of bone and eggshell. Smooth shiny pebbles arranged into elaborate designs are often part of the display. If the bowers are near human habitation, the bird will borrow bits of cigarette foil or shiny shards of glass (the avian equivalent of jewelry) to provide accent.

The male bowerbird takes great pride in the overall appearance and even fine details of his structure. Displace one berry, and he will hop over to put it back, showing the kind of fastidiousness seen in many a human artist. Different species of bowerbirds build discernibly different nests, and most remarkable of all, individuals within a species have different styles. In short, the bird shows artistic originality which serves to impress and attract individual females. If one of these bowers were displayed in a Manhattan art gallery without revealing that it was created by a bird brain, I'd wager it would elicit favorable comments.

Returning to humans, one problem concerning aesthetics has always puzzled me. What, if anything, is the key difference between kitsch art and real art? Some would argue that one person's kitsch might be another person's high art. In other words, the judgment is entirely subjective. But if a theory of art cannot objectively distinguish kitsch from the real, how complete is that theory, and in what sense can we claim to have really understood the meaning of art? One reason for thinking that there's a genuine difference is that you can learn to like real art after enjoying kitsch, but it's virtually impossible to slide back into kitsch after knowing the delights of high art. Yet the difference between the two remains tantalizingly elusive. In fact, I will lay out a challenge that no theory of aesthetics can be said to be complete unless it confronts this problem and can objectively spell out the distinction.

In this chapter, I'll speculate on the possibility that real art—or indeed aesthetics—involves the proper and effective deployment of certain artistic universals, whereas kitsch merely goes through the motions, as if to make a mockery of the principles without a genuine understanding of them. This isn't a full theory, but it's a start.

FOR A LONG time I had no real interest in art. Well, that isn't entirely true, because any time I'd attend a scientific meeting in a big city I would visit the local galleries, if only to prove to myself that I was cultured. But it's fair to say I had no deep passion for art. But all that changed in 1994 when I went on a sabbatical to India and began what was to become a lasting love affair with aesthetics. During a three-month visit to Chennai (also known as Madras), the city in southern India where I was born, I found myself with extra time on my hands. I was there as a visiting professor at the Institute of Neurology to work on patients with stroke, phantom limbs following amputation, or a sensory loss caused by leprosy. The clinic was undergoing a dry spell, so there weren't many patients to see. This gave me ample opportunity for leisurely walks through the Shiva temple in my neighborhood in Mylapore, which dates back to the first millennium B.C.E.

A strange thought occurred to me as I looked at the stone and bronze sculptures (or "idols," as the English used to call them) in the temple. In the West, these are now found mostly in museums and galleries and are referred to as Indian art. Yet I grew up praying to these as a child and never thought of them as art. They are so well integrated into the fabric of life in India—the daily worship, music, and dance—that it's hard to know where art ends and where ordinary life begins. Such sculptures are not separate strands of existence the way they are here in the West.

Until that particular visit to Chennai, I had a rather colonial view of Indian sculptures thanks to my Western education. I thought of them largely as religious iconography or mythology rather than fine art. Yet on this visit, these images had a profound impact on me as beautiful works of art, not as religious artifacts.

When the English arrived in India during Victorian times, they regarded the study of Indian art mainly as ethnography and anthropology. (This would be equivalent to putting Picasso in the anthropology section of the national museum in Delhi.) They were appalled by the nudity and often described the sculptures as primitive or not realistic. For example, the bronze sculpture of Parvati (Figure 7.2a), which dates back to the zenith of southern Indian art during the Chola period (A.D. twelfth century), is regarded in India as the very epitome of feminine sensuality, grace, poise, dignity, and charm—indeed, of all that is feminine. Yet when the Englishmen looked at this and other similar sculptures (Figure 7.2b), they complained that it wasn't art because the sculptures didn't resemble real women. The breasts and hips were too big, the waist too narrow. Similarly, they pointed out that the miniature paintings of the Mogul or Rajasthani school often lacked the perspective found in natural scenes.

In making these criticisms they were, of course, unconsciously

FIGURE 7.2   (a) A bronze sculpture of the goddess Parvati created during the Chola period (tenth to thirteenth century) in southern India. (b) Replica of a sandstone sculpture of a stone nymph standing below an arched bough, from Khajuraho, India, in the twelfth century, demonstrating "peak shift" of feminine form. The ripe mangos on the branch are a visual echo of her ripe, young breasts and (like the breasts) a metaphor of the fertility and fecundity of nature.

comparing ancient Indian art with the ideals of Western art, especially classical Greek and Renaissance art in which realism is emphasized. But if art is about realism, why even create the images? Why not just walk around looking at things around you? Most people recognize that the purpose of art is not to create a realistic replica of something but the exact opposite: It is to deliberately distort, exaggerate—even transcend—realism in order to achieve certain pleasing (and sometimes disturbing) effects in the viewer. And the more effectively you do this, the bigger the aesthetic jolt.

Picasso's Cubist pictures were anything but realistic. His women—with two eyes on one side of the face, hunchbacks, misplaced limbs, and so on—were considerably more distorted than any Chola bronze or Mogul miniature. Yet the Western response to Picasso was that he was a genius who liberated us from the tyranny of realism by showing us that art doesn't have to even try to be realistic. I do not mean to detract from Picasso's brilliance, but he was doing what Indian artists had done a millennium earlier. Even his trick of depicting multiple views of an object in a single plane was used by Mogul artists. (I might add that I am not a great fan of Picasso's art.)

Thus the metaphorical nuances of Indian art were lost on Western art historians. One eminent bard, the nineteenth-century naturalist and writer Sir George Christopher Molesworth Birdwood, considered Indian art to be mere "crafts" and was repulsed by the fact that many of the gods had multiple arms (often allegorically signifying their many divine attributes). He referred to Indian art's greatest icon, *The Dancing Shiva*, or *Nataraja*, which appears in the next chapter, as a multiarmed monstrosity. Oddly enough, he didn't have the same opinion of angels depicted in Renaissance art—human children with wings sprouting on their scapulae—which were probably just as monstrous to some Indian eyes. As a medical man, I might add that multiple arms in humans do occasionally crop up—a staple of freak shows in the old days—but a human being sprouting wings is impossible. (However, a recent survey revealed that about one-third of all Americans claim they have seen angels, a frequency that's higher than even Elvis sightings!)

So works of art are not photocopies; they involve deliberate hyperbole and distortion of reality. But you can't just randomly distort an image and call it art (although, here in La Jolla, many do). The question is, what types of distortion are effective? Are there any rules that the artist deploys, either consciously or unconsciously, to change the image in a systematic way? And if so, how universal are these rules?

While I was struggling with this question and poring over ancient Indian manuals on art and aesthetics, I often noticed the word *rasa*. This Sanskrit word is difficult to translate, but roughly it means "capturing the very essence, the very spirit of something, in order to evoke a specific mood or emotion in the viewer's brain." I realized that, if you want to understand art, you have to understand *rasa* and how it is represented in the neural circuitry in the brain. One afternoon, in a whimsical mood, I sat at the entrance of the temple and jotted down what I thought might be the "eight universal laws of aesthetics," analogous to the Buddha's eightfold path to wisdom and enlightenment. (I later came up with an additional ninth law—so there, Buddha!) These are rules of thumb that the artist or even fashion designer deploys to create visually pleasing images that more optimally titillate the visual areas in the brain compared with what he could accomplish using realistic images or real objects.

In the pages that follow I will elaborate on these laws. Some I believe are genuinely new, or at least haven't been stated explicitly in the context of visual art. Others are well known to artists, art historians, and philosophers. My goal is not to provide a complete account of the neurology of aesthetics (even assuming such a thing were possible) but to tie strands together from many different disciplines and to provide a coherent framework. Semir Zeki, a neuroscientist at the University College of London, has embarked on a similar venture which he calls "neuroesthetics." Please be assured that this type of analysis doesn't in any way detract from the more lofty spiritual dimensions of art any more than describing the physiology of sexuality in the brain detracts from the magic of romantic love. We are dealing with different levels of descriptions that complement rather than contradict each other. (No one would deny that sexuality is a strong component of romantic love.)

In addition to identifying and cataloging these laws, we also need to understand what their function might be, if any, and why they evolved. This is an important difference between the laws of biology and the laws of physics. The latter exist simply because they exist, even though the physicist may wonder why they always seem so simple and elegant to the human mind. Biological laws, on the other hand, must have evolved because they helped the organism deal with the world reliably, enabling it to survive and transmit its genes more efficiently. (This isn't always true, but it's true often enough to make it worthwhile for a biologist to constantly keep it in mind.) So the quest for biological laws shouldn't be driven by a quest for simplicity or elegance. No woman who has been through labor would say that it's an elegant solution to giving birth to a baby.

Moreover, to assert there might be universal laws of aesthetics and art does not in any way diminish the important role of culture in the creation and appreciation of art. Without cultures, there wouldn't be distinct styles of art such as Indian and Western. My interest is not in the differences between various artistic styles but in principles that cut across cultural barriers, even if those principles account for only, say 20 percent of the variance seen in art. Of course, cultural variations in art are fascinating, but I would argue that certain systematic principles lie behind these variations.

Here are the names of my nine laws of aesthetics:

1. Grouping
2. Peak shift
3. Contrast
4. Isolation
5. Peekaboo, or perceptual problem solving
6. Abhorrence of coincidences
7. Orderliness
8. Symmetry
9. Metaphor

It isn't enough to just list these laws and describe them; we need a coherent biological perspective. In particular, when exploring any universal human trait such as humor, music, art, or language, we need to keep in mind three basic questions: roughly speaking, What? Why? and How? First, what is the internal logical structure of the particular trait you are looking at (corresponding roughly to what I call laws)? For example, the law of grouping simply means that the visual system tends to group similar elements or features in the image into clusters. Second, why does the particular trait have the logical structure that it does? In other words, what is the biological function it evolved for? And third, how is the trait or law mediated by the neural machinery in the brain?[1] All three of these questions need to be answered before we can genuinely claim to have understood any aspect of human nature.

In my view, most older approaches to aesthetics have either failed or remained frustratingly incomplete with regard to these questions. For example, the Gestalt psychologists were good at pointing out laws of perception but didn't correctly answer why such laws may have evolved or how they came to be enshrined in the neural architecture of the brain. (Gestalt psychologists regarded the laws as byproducts of some undiscovered physical principles such as electrical fields in the brain.) Evolutionary psychologists are often good at pointing out what function a law might serve but are typically not concerned with specifying in clear logical terms what the law actually is, with exploring its underlying neural mechanisms, or even with establishing whether the law exists or not! (For instance, is there a law of cooking in the brain because most cultures cook?) And last, the worst offenders are neurophysiologists (except the very best ones), who seem interested in neither the functional logic nor the evolutionary rationale of the neural circuits they explore so diligently. This is amazing, given that as Theodosius Dobzhansky famously said, "Nothing in biology makes any sense except in the light of evolution."

A useful analogy comes from Horace Barlow, a British visual neuroscientist whose work is central to understanding the statistics of natural scenes. Imagine that a Martian biologist arrives on Earth. The Martian is asexual and reproduces by duplication, like an amoeba, so it doesn't know anything about sex. The Martian dissects a man's testicles, studies its microstructure in excruciating detail, and finds innumerable sperm swimming around. Unless the Martian knew about sex (which it doesn't), it wouldn't have the foggiest understanding of the structure and function of the testes despite all its meticulous dissections. The Martian would be mystified by these spherical balls dangling in half the human population and might even conclude that the wriggling sperm were parasites. The plight of many of my colleagues in physiology is not unlike that of the Martian. Knowing the minute detail doesn't necessarily mean you comprehend the function of the whole from its parts.

So with the three overarching principles of internal logic, evolutionary function, and neural mechanics in mind, let's see the role each of my individual laws plays in constructing a neurobiological view of aesthetics. Let's begin with a concrete example: grouping.

The Law of Grouping

The law of grouping was discovered by Gestalt psychologists around the turn of the century. Take a moment to look again at Figure 2.7, the Dalmatian dog in Chapter 2. All you see at first is a set of random splotches, but after several seconds you start grouping some of the splotches together. You see a Dalmatian dog sniffing the ground. Your brain glues the "dog" splotches together to form a single object that is clearly delineated from the shadows of leaves around it. This is well known, but vision scientists frequently overlook the fact that successful grouping

FIGURE 7.3   In this Renaissance painting, very similar colors (blues, dark brown, and beige) are scattered spatially throughout the painting. The grouping of similar colors is pleasing to the eye even if they are on different objects.

feels good. You get an internal "Aha!" sensation as if you have just solved a problem.

Grouping is used by both artists and fashion designers. In some well-known classic Renaissance paintings (Figure 7.3), the same azure blue color repeats all over the canvas as part of various unrelated objects. Likewise the same beige and brown are used in halos, clothes, and hair throughout the scene. The artist uses a limited set of colors rather than an enormous range of colors. Again, your brain enjoys grouping similar-colored splotches. It feels good, just as it felt good to group the "dog" splotches, and the artist exploits this. He doesn't do this because he is stingy with paint or has only a limited palette. Think of the last time you selected a mat to frame a painting. If there are bits of blue in the painting you pick a matte that's tinted blue. If there are mainly green earth tones in the painting, then a brown mat looks most pleasing to the eye.

The same holds for fashion. When you go to Nordstrom's department store to buy a red skirt, the salesperson will advise you to buy a red scarf and a red belt to go with it. Or if you are a guy buying a blue suit, the salesperson may recommend a tie with some identical blue flecks to go with the suit.

But what's all this really about? Is there a logical reason for grouping colors? Is it just marketing and hype, or is this telling you something fundamental about the brain? This is the "why" question. The answer is that grouping evolved, to a surprisingly large extent, to defeat camouflage and to detect objects in cluttered scenes. This seems counterintuitive because when you look around, objects are clearly visible— certainly not camouflaged. In a modern urban environment, objects are so commonplace that we don't realize vision is mainly about detecting objects so that you can avoid them, dodge them, chase them, eat them, or mate with them. We take the familiar for granted, but just think of one of your arboreal ancestors trying to spot a lion hidden behind a screen of green splotches (a tree branch, say). Only visible are several yellow splotches of lion fragments (Figure 7.4). But your brain says (in effect), "What's the likelihood that all these fragments are exactly the same color by coincidence? Zero. So they probably belong to one object. So let me glue them together to see what it is. Aha! *Oops!* It's a lion—run!" This seemingly esoteric ability to group splotches may have made all the difference between life and death.

FIGURE 7.4   A lion seen through foliage. The fragments are grouped by the prey's visual system before the overall outline of the lion becomes evident.

Little does the salesperson at Nordstrom's realize that when she picks the matching red scarf for your red skirt, she is tapping into a deep principle underlying brain organization, and that she's taking advantage of the fact that your brain evolved to detect predators seen behind foliage. Again, grouping feels good. Of course the red scarf and red skirt are not one object, so logically they shouldn't be grouped, but that doesn't stop her from exploiting the grouping law anyway, to create an attractive combination. The point is, the rule worked in the treetops in which our brains evolved. It was valid often enough that incorporating it as a law into visual brain centers helped our ancestors leave behind more babies, and that's all that matters in evolution. The fact that an artist can misapply the rule in an individual painting, making you group splotches from different objects, is irrelevant because your brain is fooled and enjoys the grouping anyway.

Another principle of perceptual grouping, known as good continuation, states that graphic elements suggesting a continued visual contour will tend to be grouped together. I recently tried constructing a version of it that might be especially relevant to aesthetics (Figure 7.5). Figure 7.5b is unattractive, even though it is made of components whose shapes

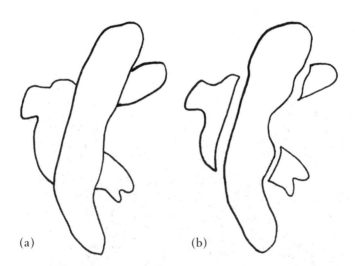

(a)   (b)

FIGURE 7.5   (a) Viewing the diagram on the left gives you a pleasing sensation of completion: The brain enjoys grouping. (b) In the right-hand diagram, the smaller blobs flanking the central vertical blob are not grouped by the visual system, creating a sort of perceptual tension.

and arrangement are similar to Figure 7.5a, which is pleasing to the eye. This is because of the "Aha!" jolt you get from completion (grouping) of object boundaries behind occluders (7.5a, whereas in 7.5b there is irresolvable tension).

And now we need to answer the "how" question, the neural mediation of the law. When you see a large lion through foliage, the different yellow lion fragments occupy separate regions of the visual field, yet your brain glues them together. How? Each fragment excites a separate cell (or small cluster of cells) in widely separated portions of the visual cortex and color areas of the brain. Each cell signals the presence of the feature by means of a volley of nerve impulses, a train of what are called spikes. The exact sequence of spikes is random; if you show the same feature to the same cell it will fire again just as vigorously, but there's a new random sequence of impulses that isn't identical to the first. What seems to matter for recognition is not the exact pattern of nerve impulses but which neurons fire and how much they fire—a principle known as Müller's law of specific nerve energies. Proposed in 1826, the law states that the different perceptual qualities evoked in the brain by sound, light, and pinprick—namely, hearing, seeing, and pain—are not caused by differences in patterns of activation but by different locations of nervous structures excited by those stimuli.

That's the standard story, but an astonishing new discovery by two neuroscientists, Wolf Singer of the Max Planck Institute for Brain Research in Frankfurt, Germany, and Charles Gray from Montana State University, adds a novel twist to it. They found that if a monkey looks at a big object of which only fragments are visible, then many cells fire in parallel to signal the different fragments. That's what you would expect. But surprisingly, as soon as the features are grouped into a whole object (in this case, a lion), all the spike trains become perfectly synchronized. And so the exact spike trains *do* matter. We don't yet know how this occurs, but Singer and Gray suggest that this synchrony tells higher brain centers that the fragments belong to a single object. I would take this argument a step further and suggest that this synchrony allows the spike trains to be encoded in such a way that a coherent output emerges which is relayed to the emotional core of the brain, creating an "Aha! Look here, it's an object!" jolt in you. This jolt arouses you and makes you swivel your eyeballs and head toward the object, so you can pay attention to it, identify it, and take action. It's this "Aha!" signal that the artist or designer exploits when she uses grouping. This isn't as far-fetched as it sounds; there are known back projections from the amygdala and other limbic structures (such as the nucleus accumbens) to almost every visual area in the hierarchy of visual processing discussed in Chapter 2. Surely these projections play a role in mediating the visual "Aha!"

The remaining universal laws of aesthetics are less well understood, but that hasn't stopped me from speculating on their evolution. (This isn't easy; some laws may not themselves have a function but may be byproducts of other laws that do.) In fact, some of the laws actually seem to contradict each other, which may actually turn out to be a blessing. Science often progresses by resolving apparent contradictions.

The Law of Peak Shift

My second universal law, the peak-shift effect, relates to how your brain responds to exaggerated stimuli. (I should point out that the phrase "peak shift" has a purportedly precise meaning in the animal learning literature, whereas I am using it more loosely.) It explains why caricatures are so appealing. And as I mentioned earlier, ancient Sanskrit manuals on aesthetics often use the word *rasa*, which translates roughly to "capturing the very essence of something." But how exactly does the artist extract the very essence of something and portray it in a painting or a sculpture? And how does your brain respond to *rasa*?

A clue, oddly enough, comes from studies in animal behavior, especially the behavior of rats and pigeons that are taught to respond to certain visual images. Imagine a hypothetical experiment in which a rat is being taught to discriminate a rectangle from a square (Figure 7.6). Every time the animal approaches the rectangle, you give it a piece of cheese, but if it goes to the square you don't. After a few dozen trials, the rat learns that "rectangle = food," it begins to ignore the square and go toward the rectangle alone. In other words, it now likes the rectangle. But amazingly, if you now show the rat a longer and skinnier rectangle than the one you showed it originally, it actually prefers that rectangle to the original! You may be tempted to say, "Well, that's a bit silly. Why would the rat actually choose the new rectangle rather than the one you trained it with?" The answer is the rat isn't being silly at all. It has learned a rule—"rectangularity"—rather than a particular prototype rectangle, so from its point of view, the more rectangular, the better. (By that, one means "the higher the ratio of a longer side to a shorter side, the better.") The more you emphasize the contrast between the rectangle and the square, the more attractive it is, so when shown the long skinny one the rat thinks, "Wow! What a rectangle."

This effect is called peak shift because ordinarily when you teach an animal something, its peak response is to the stimulus you trained it with. But if you train the animal to discriminate something (in this case, a rectangle) from something else (the square), the peak response is to a totally new rectangle that is shifted away even further from the square in its rectangularity.

What has peak shift got to do with art? Think of caricatures. As I mentioned in Chapter 2, if you want to draw a caricature of Nixon's face, you take all those features of Nixon that make his face special and different from the average face, such as his big nose and shaggy eyebrows, and you amplify them. Or to put it differently, you take the mathematical average of all male faces and subtract this average from Nixon's face, and then amplify the difference. By doing this you have created a picture that's even more Nixon-like than the original Nixon! In short, you have captured the very essence—the *rasa*—of Nixon. If you overdo it, you get

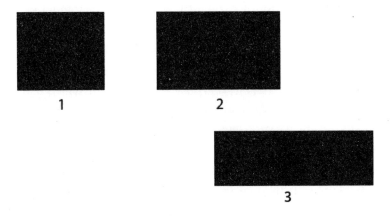

FIGURE 7.6   Demonstration of the peak shift principle: The rat is taught to prefer the rectangle (2) over the square (1) but then spontaneously prefers the longer, skinnier rectangle (3).

a humorous effect—a caricature—because it doesn't look even human; but if you do it right, you get great portraiture.

Caricatures and portraits aside, how does this principle apply to other art forms? Take a second look at the goddess Parvati (Figure 7.2a), which conveys the essence of feminine sensuality, poise, charm, and dignity. How does the artist achieve this? A first-pass answer is that he has subtracted the average male form from the average female form and amplified the difference. The net result is a woman with exaggerated breasts and hips and an attenuated hourglass waist: slender yet voluptuous. The fact that she doesn't look like your average real woman is irrelevant; you like the sculpture just as the rat liked the skinnier rectangle more than the original prototype, saying, in effect, "Wow! What a woman!" But there's surely more to it than that, otherwise any *Playboy* pinup would be a work of art (although, to be sure, I've never seen a pinup whose waist is as narrow as the goddess's).

Parvati is not merely a sexy babe; she is the very embodiment of feminine perfection—of grace and poise. How does the artist achieve this? He does so by accentuating not merely her breasts and hips but also her feminine posture (formally known as *tribhanga*, or "triple flexion," in Sanskrit). There are certain postures that a woman can adopt effortlessly but are impossible (or highly improbable) in a man because of anatomical differences such as the width of the pelvis, the angle between the neck and shaft of the femur, the curvature of the lumbar spine. Instead of subtracting male form from female form, the artist goes into a more abstract posture space, subtracting the average male posture from the average female posture, and then amplifies the difference. The result is an exquisitely feminine posture, conveying poise and grace.

Now take a look at the dancing nymph in Figure 7.7 whose twisting torso is almost anatomically absurd but who nevertheless conveys an incredibly beautiful sense of movement and dance. This is probably achieved, once again, by the deliberate exaggeration of posture that may activate—indeed hyperactivate—mirror neurons in the superior temporal sulcus. These cells respond powerfully when a person is viewing changing postures and movements of the body as well as changing facial expressions. (Remember pathway 3, the "so what" stream in vision processing discussed in Chapter 2?) Perhaps sculptures such as the dancing nymph are producing an especially powerful stimulation of certain

FIGURE 7.7   Dancing stone nymph from Rajasthan, India, eleventh century. Does it stimulate mirror neurons?

classes of mirror neurons, resulting in a correspondingly heightened reading of the body language of dynamic postures. It's hardly surprising, then, that even most types of dance—Indian or Western—involve clever ritualized exaggerations of movements and postures that convey specific emotions. (Remember Michael Jackson?)

The relevance of the peak-shift law to caricatures and to the human body is obvious, but how about other kinds of art?[2] Can we even begin to approach Van Gogh, Rodin, Gustav Klimt, Henry Moore, or Picasso? What can neuroscience tell us about abstract and semiabstract art? This is where most theories of art either fail or start invoking culture, but I'd like to suggest that we don't really need to. The important clue to understanding these so-called higher art forms comes from a very unexpected source: ethology, the science of animal behavior, in particular, from the work of the Nobel Prize–winning biologist Nikolaas Tinbergen, who did his pioneering work on seagulls in the 1950s.

Tinbergen studied herring gulls, common on both the English and American coasts. The mother gull has a prominent red spot on her long yellow beak. The gull chick, soon after it hatches from the egg, begs for food by pecking vigorously on the red spot on the mother's beak. The mother then regurgitates half-digested food into her chick's gaping mouth. Tinbergen asked himself a very simple question: How does the chick recognize its mom? Why doesn't it beg for food from any animal that's passing by?

Tinbergen found that to elicit this begging behavior in the chick you don't really need a mother seagull. When he waved a disembodied beak in front of the chick, it pecked at the red spot just as vigorously, begging the beak-wielding human for food. The chick's behavior—confusing a human adult for a mother seagull—might seem silly, but it isn't. Remember, vision evolved to discover and respond to objects (recognize them, dodge them, eat them, catch them, or mate with them) quickly and reliably by doing as little work as needed for the job at hand—taking short-cuts where necessary to minimize computational load. Through millions of years of accumulated evolutionary wisdom, the gull chick's brain has learned that the only time it will see a long yellow thing with a red spot on the end is when there's a mom attached to it at the other end. After all, in nature the chick is never likely to encounter a mutant pig with a beak or a malicious ethologist waving around a fake beak. So the chick's brain can take advantage of this statistical redundancy in nature and the equation "long thing with red spot = mom" gets hardwired into its brain.

In fact Tinbergen found that you don't even need a beak; you can just have a rectangular strip of cardboard with a red dot on the end, and the chick will beg for food equally vigorously. This happens because the chick brain's visual machinery isn't perfect; it's wired up in such a way that it has a high enough hit rate in detecting mom to survive and leave offspring. So you can readily fool these neurons by providing a visual stimulus that approximates the original (just as a key doesn't have to be absolutely perfect to fit a cheap lock; it can be rusty or slightly corroded.)

But the best was yet to come. To his amazement, Tinbergen found that if he had a very long thick stick with three red stripes on the end, the chick goes berserk, pecking at it much more intensely than at a real beak. It actually prefers this strange pattern, which bears almost no resemblance to the original! Tinbergen doesn't tell us why this happens, but it's almost as though the chick had stumbled on a superbeak (Figure 7.8).

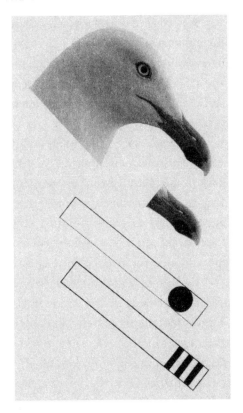

FIGURE 7.8   The gull chick pecks at a disembodied beak or, a stick with a spot that is a reasonable approximation of the beak given the limits of sophistication of visual processing. Paradoxically, a stick with three red stripes is even more effective than a real beak; it is an ultranormal stimulus.

Why could such a thing happen? We really don't know the "alphabet" of visual perception, whether in gulls or humans. Obviously, neurons in the visual centers of the gull's brain (which have fancy Latin names like nucleus rotundum, hyperstriatum, and ectostriatum) are not optimally functioning machines; they are merely wired up in such a way that they can detect beaks, and therefore mothers, reliably enough. Survival is the only thing evolution cares about. The neuron may have a rule like "the more red outline the better," so if you show it a long skinny stick with three stripes, the cell actually likes it even more! This is related to the peak-shift effect on rats mentioned earlier, except for one key difference: in the case of the rat responding to the skinnier rectangle, it's perfectly obvious what rule the animal has learned and what you are amplifying. But in the case of the seagull, the stick with three stripes is hardly an exaggerated version of a real beak; it isn't clear at all what rule you are tapping into or amplifying. The heightened response to the striped beak may be an inadvertent consequence of the way the cells are wired up rather than the deployment of a rule with an obvious function.

We need a new name for this type of stimulus, so I'll call it an "ultranormal" stimulus (to distinguish it from "supernormal," a phrase that already exists). The response to an ultranormal stimulus pattern (such as the three-striped beak) cannot be predicted from looking at the original (the single-spot beak). You could predict the response—at least in theory—if you knew in detail the functional logic of the circuitry in the chick's brain that allows the rapid, efficient detection of beaks. You could then devise patterns that actually excite these neurons even more effectively than the original stimulus, so the chick's brain goes "Wow! What a sexy beak!" Or you might be able to discover the ultranormal stimulus by trial and error, stumbling on it as Tinbergen did.

This brings me to my punch line about semiabstract or even abstract art for which no adequate theory has been proposed so far. Imagine that seagulls had an art gallery. They would hang this long thin stick with three stripes on the wall. They would call it a Picasso, worship it, fetishize it, and pay millions of dollars for it, while all the time wondering why they are turned on by it so much, even though (and this is the key point) it doesn't resemble anything in their world. I suggest this is exactly what human art connoisseurs are doing when they look at or purchase abstract works of art; they are behaving exactly like the gull chicks.

By trial and error, intuition or genius, human artists like Picasso or Henry Moore have discovered the human brain's equivalent of the seagull brain's stick with three stripes. They are tapping into the figural primitives of our perceptual grammar and creating ultranormal stimuli that more powerfully excite certain visual neurons in our brains as opposed to realistic-looking images. This is the essence of abstract art. It may sound like a highly reductionist, oversimplified view of art, but bear in mind that I'm not saying that's *all* there is to art, only that it's an important component.

The same principle may apply to impressionist art—a Van Gogh or a Monet canvas. In Chapter 2, I noted that visual space is organized in the brain so that spatially adjacent points are mapped one-to-one onto adjacent points on the cortex. Moreover, out of the thirty or so areas in the human brain, a few—especially V4—are devoted primarily to color. But in the color area, wavelengths adjacent in an abstract "color space" are mapped onto adjacent points in the brain even when they are not near each other in external space. Perhaps Monet and Van Gogh were introducing peak shifts in abstract color space rather than "form space," even deliberately smudging form when required. A black-and-white Monet is an oxymoron.

This principle of ultranormal stimuli may be relevant not just to art but to other quirks of aesthetic preference as well, like whom you are attracted to. Each of us carries templates for members of the opposite sex (such as your mother or father, or your first really sizzling amorous encounter), and maybe those whom you find inexplicably and disproportionately attractive later in life are ultranormal versions of these early prototypes. So the next time you are unaccountably—even perversely— attracted to someone who is not beautiful in any obvious sense, don't jump to the conclusion that it's just pheromones or "the right chemistry." Consider the possibility that she (or he) is an ultranormal version of the gender you're attracted to buried deep in your unconscious. It's a strange thought that human life is built on such quicksand, governed largely by vagaries and accidental encounters from the past, even though we take such great pride in our aesthetic sensibilities and freedom of choice. On this one point I am in complete agreement with Freud.

There is a potential objection to the notion that our brains are at least partially hardwired to appreciate art. If this were really true, then why doesn't everyone like Henry Moore or a Chola bronze? This is an important question. The surprising answer might be that everyone does "like" a Henry Moore or Parvati, but not everyone knows it. The key to understanding this quandary is to recognize that the human brain has many quasi-independent modules that can at times signal inconsistent information. It may be that all of us have basic neural circuits in our visual areas which show a heightened response to a Henry Moore sculpture, given that it is constructed out of certain form primitives that hyperactivate cells that are tuned to respond to these primitives. But perhaps in many of us, other higher cognitive systems (such as the mechanisms of language and thought in the left hemisphere) kick in and censor or veto the output of the face neurons by saying, in effect, "There is something wrong with this sculpture; it looks like a funny twisted blob. So ignore that strong signal from cells at an earlier stage in your visual processing." In short, I am saying all of us do like Henry Moore but many of us are in denial about it! The idea that people who claim not to like Henry Moore are closet Henry Moore enthusiasts could in principle be tested with brain imaging. (And the same holds for the Victorian Englishman's response to the Chola bronze Parvati.)

An even more striking example of quirky aesthetic preference is the manner in which certain guppies prefer decoys of the opposite sex that are painted blue, even though there's nothing in the guppy that's blue. (If a chance mutation were to occur making one guppy blue, I predict the emergence of a future race of guppies in the next few millennia that evolve to become uselessly, intensely blue.) Could the appeal of silver foil to bowerbirds and the universal appeal of shiny metallic jewelry and precious stones to people also be based on some idiosyncratic quirk of brain wiring? (Maybe evolved for detecting water?) It's a sobering thought when you consider how many wars have been fought, loves lost, and lives ruined for the sake of precious stones.

SO FAR I have discussed only two of my nine laws. The remaining seven are the subject of the next chapter. But before we continue, I want to take up one final challenge. The ideas I have considered so far on abstract and semiabstract art and portraiture sound plausible, but how do we know they actually are true? The only way to find out would be to do experiments. This may seem obvious, but the whole concept of an experiment—the need to test your idea by manipulating one variable alone while keeping everything else constant—is new and surprisingly alien to the human mind. It's a relatively recent cultural invention that began with Galileo's experiments. Before him, people "knew" that if a heavy stone and a peanut were dropped simultaneously from the top of a tower, the heavier one would obviously fall faster. All it took was a five-minute experiment by Galileo to topple two thousand years of wisdom. This experiment, moreover, that can be repeated by any ten-year-old schoolgirl.

A common fallacy is that science begins with naïve unprejudiced observations about the world while in fact the opposite is true. When exploring new terrain, you always begin with a tacit hypothesis of what might be true—a preconceived notion or prejudice. As the British zoologist and philosopher of science Peter Medawar once said, we are not "cows grazing on the pasture of knowledge." Every act of discovery involves two critical steps: first, unambiguously stating your conjecture of what might be true, and second, devising a crucial experiment to test your conjecture. Most theoretical approaches to aesthetics in the past have been concerned mainly with step 1 but not step 2. Indeed, the theories are usually not stated in a manner that permits either confirmation or refutation. (One notable exception is Brent Berlin's pioneering work on the use of the galvanic skin response.)

Can we experimentally test our ideas about peak shift, supernormal stimuli, and other laws of aesthetics? There are at least three ways of doing so. The first one is based on the galvanic skin response (GSR); the second is based on recording nerve impulses from single nerve cells in the visual area in the brain; and the third is based on the idea that if there is anything to these laws, we should be able to use them to devise new pictures that are more attractive than what you might have predicted from common sense (what I refer to as the "grandmother test": If an elaborate theory cannot predict what your grandmother knows using common sense, then it isn't worth much).

You already know about GSR from previous chapters. This test provides an excellent, highly reliable index of your emotional arousal when you look at anything. If you look at something scary, violent, or sexy (or, as it turns out, a familiar face like your mother or Angelina Jolie), there is a big jolt in GSR, but nothing happens if you look at a shoe or furniture. This is a better test of someone's raw, gut-level emotional reactions to the world than asking what she feels. A person's verbal response is likely to be inauthentic. It may be contaminated by the "opinions" of other areas of the brain.

So GSR gives us a handy experimental probe for understanding art. If my conjectures about the appeal of Henry Moore sculptures are correct, then the Renaissance scholar who denies an interest in such abstract works (or, for that matter, the English art historian who feigns indifference to Chola bronzes) should nevertheless register a whopping GSR to the very images whose aesthetic appeal he denies. His skin can't lie. Similarly, we know that you will show a higher GSR to a photo of your mother than to a photo of a stranger, and I predict that the difference will be even greater if you look at a caricature or evocative sketch of your mother rather than at a realistic photo. This would be interesting because it's counterintuitive. As a control for comparison, you could use a countercaricature, by which I mean a sketch that deviates from the prototype toward the average face rather than away from it (or indeed, a face outline that deviates in a random direction). This would ensure that any enhanced GSR you observed with the caricature wasn't simply because of the surprise caused by the distortion. It would be genuinely due to its appeal as a caricature.

But GSR can only take us so far; it is a relatively coarse measure because it pools several types of arousal and it can't discriminate positive from negative responses. But even though it's a crude measure, it's not a bad place to start because it can tell the experimenter when you are indifferent to a work of art and when you are feigning indifference. The criticism that the test can't discriminate negative arousal from positive arousal (at least not yet!) isn't as damaging as it sounds because who is to say that negative arousal isn't also part of art? Indeed, attention grabbing—whether initially positive or negative—is often a prelude to attraction. (After all, slaughtered cows pickled in formaldehyde were displayed in the venerable MOMA [Museum of Modern Art] in New York, sending shock waves throughout the art world). There are many layers of reaction to art, which contribute to its richness and appeal.

A second approach is to use eye movements, in particular, a technique pioneered by the Russian psychologist Alfred Yarbus. You can use an electronic optical device to see where a person is fixating and how she is moving her eyes from one region to another in a painting. The fixations tend to be clustered around eyes and lips. One could therefore show a normally proportioned cartoon of a person on one side of the image and a hyperbolic version on the other side. I would predict that even though the normal cartoon looks more natural, the eye fixations will cluster more around the caricature. (A randomly distorted cartoon could be included to control for novelty.) These findings could be used to complement the GSR results.

The third experimental approach to aesthetics would be to record from cells along the visual pathways in primates and compare their responses to art versus any old picture. The advantage of recording from single cells is that it may eventually allow a more fine-grained analysis of the neurology of aesthetics than what could be achieved with GSR alone. We know that there are cells in a region called the fusiform gyrus that respond mainly to specific familiar faces. You have brain cells that fire in response to a picture of your mother, your boss, Bill Clinton, or Madonna. I predict that a "boss cell" in this face recognition region should show an even bigger response to a caricature of your boss than to an authentic, undistorted face of your boss (and perhaps an even smaller response to a plain-looking countercaricature). I first suggested this in a paper I wrote with Bill Hirstein in the mid-1990s. The experiment has now been done on monkeys by researchers at Harvard and MIT, and sure enough the caricatures hyperactivate the face cells as expected. Their results provide grounds for optimism that some of the other laws of aesthetics I have proposed may also turn out to be true.

THERE IS A widespread fear among scholars in the humanities and arts that science may someday take over their discipline and deprive them of employment, a syndrome I have dubbed "neuron envy." Nothing could be further from the truth. Our appreciation of Shakespeare is not diminished by the existence of a universal grammar or Chomskian deep structure underlying all languages. Nor should the diamond you are about to give your lover lose its radiance or romance if you tell her that it is made of carbon and was forged in the bowels of Earth when the solar system was born. In fact, the diamond's appeal should be enhanced! Similarly, our conviction that great art can be divinely inspired and may have spiritual significance, or that it transcends not only realism but reality itself, should not stop us from looking for those elemental forces in the brain that govern our aesthetic impulses.

CHAPTER 8

The Artful Brain: Universal Laws

———

Art is the accomplishment of our desire to find ourselves among the phenomena of the external world.

—RICHARD WAGNER

BEFORE MOVING ON TO THE NEXT SEVEN LAWS, I WANT TO CLARify what I mean by "universal." To say that the wiring in your visual centers embodies universal laws does not negate the critical role of culture and experience in shaping your brain and mind. Many cognitive faculties that are fundamental to your human way of life are only partly specified by your genes. Nature and nurture interact. Genes wire up your brain's emotional and cortical circuits to a certain extent and then leave it to faith that the environment will shape your brain the rest of the way, producing you, the individual. In this respect the human brain is absolutely unique—as symbiotic with culture as a hermit crab is with its shell. While the laws are hardwired, the content is learned.

Consider face recognition. While your ability to learn faces is innate, you are not born knowing your mother's face or the mail carrier's face. Your specialized face cells learn to recognize faces through exposure to the people you encounter.

Once face knowledge is acquired, the circuitry may spontaneously respond more effectively to caricatures or Cubist portraits Once your brain learns about other classes of objects or shapes—bodies, animals, automobiles, and such—your innate circuitry may spontaneously display the peak-shift principle or respond to bizarre ultranormal stimuli analogous to the stick with stripes. Because this ability emerges in all human brains that develop normally, we are safe in calling it universal.

Contrast

It is hard to imagine a painting or sketch without contrast. Even the simplest doodle requires contrasting brightness between the black line and white background. White paint on a white canvas could hardly be called art (although in the 1990s the purchase of an all-white painting figured in Yasmina Reza's hilarious award-winning play *"Art,"* poking fun at how easily people are influenced by art critics).

In scientific parlance, contrast is a relatively sudden change in luminance, color, or some other property between two spatially contiguous homogeneous regions. We can speak of luminance contrast, color contrast, texture contrast, or even depth contrast. The bigger the difference between the two regions, the higher the contrast.

Contrast is important in art or design; in a sense it's a minimum requirement. It creates edges and boundaries as well as figures against background. With zero contrast you see nothing at all. Too little contrast and a design can be bland. And too much contrast can be confusing.

Some contrast combinations are more pleasing to the eye than others. For example, high-contrast colors such as a blue splotch on a yellow background are more attention grabbing than low-contrast pairings like a yellow splotch on an orange background. It's puzzling at first glance. After all, you can easily see a yellow object against an orange background but that combination does not draw your attention the same way as blue on yellow.

The reason a boundary of high color contrast is more attention getting can be traced to our primate origins, to when we swung arm over arm like Spiderman in the unruly treetops, in dim twilight or across great distances. Many fruits are red on green so our primate eyes will see them. The plants advertise themselves so animals and birds can spot them from a great distance, knowing they are ripe and ready to eat and be dispersed through defecation of the seeds. If trees on Mars were mainly yellow, we would expect to see blue fruits.

The law of contrast—juxtaposing dissimilar colors and/or luminances—might seem to contradict the law of grouping, which involves connecting similar or identical colors. And yet the evolutionary function of both principles is, broadly speaking, the same: to delineate and direct attention to object boundaries. In nature, both laws help species survive. Their main difference lies in the area over which the comparison or integration of colors occurs. Contrast detection involves comparing regions of color that lie right next to each other in visual space. This makes evolutionary sense because object boundaries usually coincide with contrasting luminance or color. Grouping, on the other hand, performs comparisons over wider distances. Its goal is to detect an object that is partially obscured, like a lion hiding behind a bush. Glue those yellow patches together perceptually, and it turns out to be one big lump shaped like a lion.

In modern times we harness contrast and grouping to serve novel purposes unrelated to their original survival function. For example, a good fashion designer will emphasize the salience of an edge by using dissimilar, highly contrasting colors (contrast), but will use similar colors for far-flung regions (grouping). As I mentioned in Chapter 7, red shoes go with a red shirt (conducive to grouping). It's true, of course, that the red shoes aren't an innate part of the red shirt, but the designer is tapping into the principle that, in your evolutionary past, they would have belonged to a single object. But vermilion scarf on a ruby-red shirt is hideous. Too much low contrast. Yet a high-contrast blue scarf on a red shirt will work fine, and it's even better if the blue is flecked with red polka dots or floral prints.

Similarly, an abstract artist will use a more abstract form of the law of contrast to capture your attention. The San Diego Museum of Contemporary Art has in its contemporary art collection a large cube about three feet in diameter, densely covered with tiny metal needles pointing in random directions (by Tara Donovan). The sculpture resembles fur made of shining metal. Several violations of expectations are at work here. Large metal cubes usually have smooth surfaces but this one is furry. Cubes are inorganic while fur is organic. Fur is usually a natural brown or white, and is soft to touch, not metallic and prickly. These shocking conceptual contrasts endlessly titillate your attention.

Indian artists use a similar trick in their sculptures of voluptuous nymphs. The nymph is naked except for a few strings of very ornate coarsely textured jewelry draped on her (or flying off her chest if she is dancing). The baroque jewelry contrasts sharply with her body, making her bare skin look even more smooth and sensuous.

Isolation

Earlier I suggested that art involves creating images that produce heightened activation of visual areas in your brain and emotions associated with visual images. Yet any artist will tell you that a simple outline or doodle—say, Picasso's doves or Rodin's sketches of nudes—can be much more effective than a full color photo of the same object. The artist emphasizes a single source of information—such as color, form, or motion—and deliberately plays down or deletes other sources. I call this the "law of isolation."

Again we have an apparent contradiction. Earlier I emphasized peak shift—hyperbole and exaggeration in art—but now I am emphasizing understatement. Aren't the two ideas polar opposites? How can less be more? The answer: They aim to achieve different goals.

If you look in standard physiology and psychology textbooks, you will learn that a sketch is effective because cells in your primary visual cortex, where the earliest stage of visual processing occurs, only care about lines. These cells respond to the boundaries and edges of things but are insensitive to the feature-poor fill regions of an image. This fact about the circuitry of the primary visual area is true, but does it explain why a mere outline sketch can convey an extra vivid impression of what's being depicted? Surely not. It only predicts that an outline sketch should be adequate, that it should be as effective as a halftone (the reproduction of a black-and-white photo). It doesn't tell you why it's more effective.

A sketch can be more effective because there is an attentional bottleneck in your brain. You can pay attention to only one aspect of an image or one entity at a time (although what we mean by "aspect" or "entity" is far from clear). Even though your brain has 100 billion nerve cells, only a small subset of them can be active at any given instant. In the dynamics of perception, one stable percept (perceived image) automatically excludes others. Overlapping patterns of neural activity and the neural networks in your brain constantly compete for limited attentional resources. Thus when you look at a full-color picture, your attention is distracted by the clutter of texture and other details in the image. But a sketch of the same object allows you to allocate all your attentional resources to the outline, where the action is.

(a)   (b)   (c)

FIGURE 8.1   Comparison between (a) Nadia's drawing of a horse, (b) da Vinci's drawing, and (c) the drawing of a normal eight-year-old.

Conversely, if an artist wants to evoke the *rasa* of color by introducing peak shifts and ultranormal stimuli in color space, then she would be better off playing down the outlines. She might deemphasize boundaries, deliberately smudging the outlines or leaving them out entirely. This reduces the competitive bid from outlines on your attentional resources, freeing up your brain to focus on color space. As mentioned in Chapter 7, that is what Van Gogh and Monet do. It's called impressionism.

Great artists intuitively tap into the law of isolation, but evidence for it also comes from neurology—cases in which many areas in the brain are dysfunctional—and the "isolation" of a single brain module allows the brain to gain effortless access to its limited attentional resources, without the patient even trying.

One striking example comes from an unexpected source: autistic children. Compare the three illustrations of horses in Figure 8.1. The one on the right (Figure 8.1c) is by a normal eight-year-old child. Pardon me for saying so, but it's quite hideous—completely lifeless, like a cardboard cutout. The one on the left (Figure 8.1a), amazingly, is by a seven-year-old mentally retarded autistic child named Nadia. Nadia can't converse with people and can barely tie a shoelace, yet her drawing brilliantly conveys the *rasa* of a horse; the beast seems to almost leap out of the canvas. Finally, in the middle (Figure 8.1b) is a horse drawn by Leonardo da Vinci. When giving lectures, I often conduct informal polls by asking the audience to rank-order the three horses by how well they are drawn without telling them in advance who drew them. Surprisingly, more people prefer Nadia's horse to da Vinci's. Here again we have a paradox. How is it possible that a retarded autistic child who can barely talk can draw better than one of the greatest geniuses of the Renaissance?

The answer comes from the law of isolation as well as the brain's modular organization. (Modularity is a fancy term for the notion that different brain structures are specialized for different functions.) Nadia's social awkwardness, emotional immaturity, language deficits, and retardation all stem from the fact that many areas in her brain are damaged and function abnormally. But maybe—as I suggested in my book *Phantoms in the Brain*—there is a spared island of cortical tissue in her right parietal lobe, a region known to be involved in many spatial skills, including our sense of artistic proportion. If the right parietal lobe is damaged by a stroke or tumor, a patient often loses the ability to draw even a simple sketch. The pictures they manage to draw are usually detailed but lack fluidity of line and vividness. Conversely, I have noticed that when a patient's left parietal lobe is damaged, his drawings sometimes actually improve. He starts leaving out irrelevant details. You might wonder if the right parietal lobe is the brain's *rasa* module for artistic expression.

I suggest that poor functioning in many of Nadia's brain areas results in freeing her spared right parietal—her *rasa* module—to get the lion's share of her attentional resources. You and I could achieve such a thing only through years of training and effort. This hypothesis would explain why her art is so much more evocative than Leonardo's. It may turn out that a similar explanation holds for autistic calculating prodigies: profoundly retarded children who can nonetheless perform astonishing feats of arithmetic like multiplying two 13-digit numbers in a matter of seconds. (Notice I said, "calculating," not math. True mathematical talent may require not just calculation but a combination of several skills, including spatial visualization.) We know that the left parietal lobe is involved in numerical computation, since a stroke there will typically knock out a patient's ability to subtract or divide. In calculating savants, the left parietal may be spared relative to the right. If all of the autistic child's attention is allocated to this number module in the left parietal, the result would be a calculating prodigy rather than a drawing prodigy.

In an ironic twist, once Nadia reached adolescence, she became less autistic. She also completely lost her ability to draw. This observation lends credibility to the isolation idea. Once Nadia matured and gained some higher abilities, she could no longer allocate the bulk of her attention to the *rasa* module in her right parietal (implying, perhaps, that formal education can actually stifle some aspects of creativity).

In addition to reallocating attention, there may be actual anatomical changes in the brains of autistics that explain their creativity. Perhaps spared areas grow larger, attaining enhanced efficacy. So Nadia may have had an enlarged right parietal, especially the right angular gyrus, which would explain her profound artistic skills. Autistic children with savant skills are often referred to me by their parents, and one of these days I will get around to having their brains scanned to see if there are indeed spared islands of supergrown tissue. Unfortunately, this isn't as easy as it sounds, as autistic children often find it very difficult to sit still in the scanner. Incidentally, Albert Einstein had huge angular gyri, and I once made the whimsical suggestion that this allowed him to combine numerical (left parietal) and spatial (right parietal) skills in extraordinary ways that we lesser mortals cannot even begin to imagine.

Evidence for the isolation principle in art can also be found in clinical neurology. For example, not long ago a physician wrote to me about epileptic seizures originating in his temporal lobes. (Seizures are uncontrolled volleys of nerve impulses that course through the brain the way feedback amplifies through a speaker and microphone.) Until his seizures began quite unexpectedly at the age of sixty, the physician had no interest whatsoever in poetry. Yet all of a sudden, voluminous rhyme poured out. It was a revelation, a sudden enrichment of his mental life, just when he was starting to get jaded.

A second example, from the elegant work of Bruce Miller, a neurologist at the University of California, San Francisco, concerns patients who late in life develop a form of rapidly progressive dementia and blunting of intellect. Called frontotemporal dementia, the disorder selectively affects the frontal lobes—the seat of judgment and of crucial aspects of attention and reasoning—and the temporal lobes, but it spares islands of parietal cortex. As their mental faculties deteriorate, some of these patients suddenly, much to their surprise and to the surprise of those around them, develop an extraordinary ability to paint and draw. This is consistent with my speculations about Nadia—that her artistic skills were the result of her spared, hyperfunctioning right parietal lobe.

These speculations on autistic savants and patients with epilepsy and frontotemporal dementia raise a fascinating question. Is it possible that we less-gifted, normal people also have latent artistic or mathematical talents waiting to be liberated by brain disease? If so, would it be possible to unleash these talents without actually damaging our brains or paying the price of destroying other skills? This seems like science fiction, but as the Australian physicist Allan Snyder has pointed out, it could be true. Maybe the idea could be tested.

I was mulling over this possibility during a recent visit to India when I received what must surely be the strangest phone call of my life (and that's saying a lot). It was long distance, from a reporter at an Australian newspaper.

"Dr. Ramachandran, I'm sorry to bother you at home," he said. "An amazing new discovery has been made. Can I ask you some questions about it?"

"Sure, go ahead."

"You know Dr. Snyder's idea about autistic savants?" he asked.

"Yes," I said. "He suggests that in a normal child's brain, lower visual areas create sophisticated three-dimensional representations of a horse or any other object. After all, that's what vision evolved for. But as the child gradually learns more about the world, higher cortical areas generate more abstract, conceptual descriptions of a horse; for example, 'it's an animal with a long snout and four legs and a whisklike trail, etc.' With time, the child's view of the horse becomes dominated by these higher abstractions. He becomes more concept driven and has less access to the earlier, more visual representations that capture art. In an autistic child these higher areas fail to develop, so he is able to access these earlier representations in a manner that you and I can't. Hence the child's amazing talent in art. Snyder presents a similar argument for math savants that I find hard to follow."

"What do you think of his idea?" the reporter asked.

"I agree with it and have made many of the same arguments," I said. "But the scientific community has been highly skeptical, arguing that Snyder's idea is too vague to be useful or testable. I disagree. Every neurologist has at least one story up her sleeve about a patient who suddenly developed a quirky new talent following a stroke or brain trauma. But the best part of his theory," I continued, "is a prediction he made that now seems obvious in hindsight. He suggested that if you were to somehow temporarily inactivate 'higher' centers in a normal person's brain, that person might suddenly be able to access the so-called lower representations and create beautiful drawings or start generating prime numbers.

"Now, what I like about this prediction is that it's not just a thought experiment. We can use a device called a transcranial magnetic stimulator, or TMS, to harmlessly and temporarily inactivate portions of a normal adult's brain. Would you then see a sudden efflorescence of artistic or mathematical talent while the inactivation lasted? And would this teach that person to transcend his usual conceptual blocks? If so, would he pay the penalty of losing his conceptual skills? And once the stimulation has caused him to overcome a block (if it does), can he then do it on his own without the magnet?"

"Well, Dr. Ramachandran," said the reporter, "I have news for you. Two researchers, here in Australia, who were inspired in part by Dr. Snyder's suggestion, actually tried the experiment. They recruited normal student volunteers and tried it out."

"Really?" I said, fascinated. "What happened?"

"Well, they zapped the student's brains with a magnet, and suddenly these students could effortlessly produce beautiful sketches. And in one case the student could generate prime numbers the same way some idiot savants do."

The reporter must have sensed my bewilderment, because I remained silent.

"Dr. Ramachandran, are you still there? Can you still hear me?"

It took a whole minute for the impact to sink in. I have heard many strange things in my career as a behavioral neurologist, but this was without doubt the strangest.

I must confess I had (and still have) two very different reactions to this discovery. The first is sheer incredulity and skepticism. The observation doesn't contradict anything we know in neurology (partly because we know so little), but it sounds outlandish. The very notion of some skill being enhanced by knocking out parts of the brain is bizarre—the sort of thing you would expect to see on *The X-Files*. It also smacks of the kind of pep talk you hear from motivational gurus who are forever telling you about all your hidden talents waiting to be awakened by purchasing their tapes. Or drug peddlers claiming their magic potions will elevate your mind to whole new dimensions of creativity and imagination. Or that absurd but tenaciously popular factoid about how people only use 10 percent of their brains—whatever that's supposed to mean. (When reporters ask me about the validity of this claim, I usually tell them, "Well, that's certainly true here in California.")

My second reaction was, Why not? After all, we know that astonishing new talent can emerge relatively suddenly in frontotemporal dementia patients. That is, we know such unmasking by brain reorganization can happen. Given this existence proof, why should I be so shocked by the Australian discovery? Why should their observation with TMS be any less likely than Bruce Miller's observations of patients with profound dementia?

The surprising aspect is the timescale. Brain disease takes years to develop and the magnet works in seconds. Does that matter? According to Allan Snyder, the answer is no. But I'm not so sure.

Perhaps we can test the idea of isolated brain regions more directly. One approach would be to use functional brain imaging such as fMRI, which you may recall measures magnetic fields in the brain produced by changes in blood flow while the subject is doing something or looking at something. My ideas about isolation, along with Allan Snyder's ideas, predict that, when you look at cartoon sketches or doodles of faces, you should get a higher activation of the face area than of areas dealing with color, topography, or depth. Alternatively, when you look at a color photo of a face, you should see the opposite: a decrement in the relative response to the face. This experiment has not been done.

Peekaboo, or Perceptual Problem Solving

The next aesthetic law superficially resembles isolation but is really quite different. It's the fact that you can sometimes make something more attractive by making it less visible. I call it the "peekaboo principle." For example, a picture of a nude woman seen behind a shower curtain or wearing diaphanous, skimpy clothes—an image that men would say approvingly "leaves something to the imagination"—can be much more alluring than a pinup of the same nude woman. Similarly, disheveled tresses that conceal half a face can be enchanting. But why is this so?

After all, if I am correct in saying that art involves hyperactivation of visual and emotional areas, a fully visible naked woman should be more attractive. If you are a heterosexual man, you would expect an unimpeded view of her breasts and genitalia to excite your visual centers more effectively than her partially concealed private parts. Yet often the opposite is true. Similarly, many women will find images of hot and sexy but partially clad men to be more attractive than fully naked men.

We prefer this sort of concealment because we are hardwired to love solving puzzles, and perception is more like puzzle solving than most people realize. Remember the Dalmation dog? Whenever we successfully solve a puzzle, we get rewarded with a zap of pleasure that is not all that different from the "Aha!" of solving a crossword puzzle or scientific problem. The act of searching for a solution to a problem—whether purely intellectual, like a crossword or logic puzzle, or purely visual, like "Where's Waldo?"—is pleasing even before the solution is found. It's fortunate that your brain's visual centers are wired up to your limbic reward mechanisms. Otherwise, when you try to figure out how to convince the girl you like to sneak off into the bushes with you (working out a social puzzle) or chase that elusive prey or mate through the underbrush in dense fog (solving a fast-changing series of sensorimotor puzzles), you might give up too easily!

So, you like partial concealment and you like solving puzzles. To understand the peekaboo law you need to know more about vision. When you look at a simple visual scene, your brain is constantly resolving ambiguities, testing hypotheses, searching for patterns, and comparing current information with memories and expectations.

One naïve view of vision, perpetuated mainly by computer scientists, is that it involves a serial hierarchical processing of the image. Raw data comes in as picture elements, or pixels, in the retina and gets handed up through a succession of visual areas, like a bucket brigade, undergoing more and more sophisticated analysis at each stage, culminating in the eventual recognition of the object. This model of vision ignores the massive feedback projections that each higher visual area sends back to lower areas. These back projections are so massive that it's misleading to speak of a hierarchy. My hunch is that at each stage in processing, a partial hypothesis, or best-fit guess, is generated about the incoming data and then sent back to lower areas to impose a small bias on subsequent processing. Several such best fits may compete for dominance, but eventually, through such bootstrapping, or successive iterations, the final perceptual solution emerges. It's as though vision works top down rather than bottom up.

Indeed, the line between perceiving and hallucinating is not as crisp as we like to think. In a sense, when we look at the world, we are hallucinating all the time. One could almost regard perception as the act of choosing the one hallucination that best fits the incoming data, which is often fragmentary and fleeting. Both hallucinations and real perceptions emerge from the same set of processes. The crucial difference is that when we are perceiving, the stability of external objects and events helps anchor them. When we hallucinate, as when we dream or float in a sensory deprivation tank, objects and events wander off in any direction.

To this model I'd add the notion that each time a partial fit is discovered, a small "Aha!" is generated in your brain. This signal is sent to limbic reward structures, which in turn prompt the search for additional, bigger "Ahas!," until the final object or scene crystallizes. In this view, the goal of art is to create images that generate as many mutually consistent mini-"Aha!" signals as possible (or at least a judicious saturation of them) to titillate the visual areas in your brain. Art in this view is a form of visual foreplay for the grand climax of object recognition.

The law of perceptual problem solving, or peekaboo, should now make more sense. It may have evolved to ensure that the search for visual solutions is inherently pleasurable rather than frustrating, so that you don't give up too easily. Hence the appeal of a nude behind semitransparent clothes or the smudged water lilies of Monet.[1]

The analogy between aesthetic joy and the "Aha!" of problem solving is compelling, but analogies can only get us so far in science. Ultimately, we need to ask, What is the actual neural mechanism in the brain that generates the aesthetic "Aha!"?

One possibility is that when certain aesthetic laws are deployed, a signal is sent from your visual areas directly to your limbic structures. As I noted, such signals may be sent from other brain areas at every stage in the perceptual process (by grouping, boundary recognition, and so on) in what I call visual foreplay, and not just from the final stage of object recognition ("Wow! It's Mary!"). How exactly this happens is unclear, but there are known anatomical connections that go back and forth between limbic structures, such as the amygdala, and other brain areas at almost every stage in the visual hierarchy. It's not hard to imagine these being involved in producing mini-"Ahas!" The phrase "back and forth" is critical here; it allows artists to simultaneously tap into multiple laws to evoke multiple layers of aesthetic experience.

Back to grouping: There may be a powerful synchronization of nerve impulses from widely separated neurons signaling the features that are grouped. Perhaps this synchrony itself is what subsequently activates limbic neurons. Some such process may be involved in creating the pleasing and harmonious resonance between different aspects of what appears on the surface to be a single great work of art.

We know there are neural pathways directly linking many visual areas with the limbic structures. Remember David, the patient with Capgras syndrome from Chapter 2? His mother looks like an imposter to him because the connections from his visual centers and his limbic structures were severed by an accident, so he doesn't get the expected emotional jolt when seeing his mom. If such a disconnection between vision and emotion is the basis of the syndrome, then Capgras patients should not be able to enjoy visual art. (Although they should still enjoy music, since hearing centers in their cortices are not disconnected from their limbic systems.) Given the rarity of the syndrome this isn't easy to test, but there are, in fact, cases of Capgras patients in the older literature who claimed that landscapes and flowers were suddenly no longer beautiful.

Furthermore, if my reasoning about multiple "Ahas!" is correct—in that the reward signal is generated at every stage in the visual process, not just in the final stage of recognition—then people with Capgras syndrome should not only have problems enjoying a Monet but also take much longer to find the Dalmatian dog. They should also have problems solving simple jigsaw puzzles. These are predictions that, to my knowledge, have not been directly tested.

Until we have a clearer understanding of the connections between the brain's reward systems and visual neurons, it's also best to postpone discussing certain questions like these: What's the difference between mere visual pleasure (as when seeing a pinup) and a visual aesthetic response to beauty? Does the latter merely produce a heightened pleasure response in your limbic system (as the stick with three stripes does for the gull chick, described in Chapter 7), or is it, as I suspect, an altogether richer

and more multidimensional experience? And how about the difference between the "Aha!" of mere arousal versus the "Aha!" of aesthetic
arousal? Isn't the "Aha!" signal just as big with any old arousal—such
as being surprised, scared, or sexually stimulated—and if so, how does
the brain distinguish these other types of arousal from a true aesthetic
response? It may turn out that these distinctions aren't as watertight as
they seem; who would deny that eros is a vital part of art? Or that an artist's creative spirit often derives its sustenance from a muse?

I'm not saying these questions are unimportant; in fact, it's best to be
aware of them right up front. But we have to be careful not to give up the
whole enterprise just because we cannot yet provide complete answers to
every quandary. On the contrary, we should be pleased that the process
of trying to discover aesthetic universals has thrown up these questions
we are forced to confront.

Abhorrence of Coincidences

When I was a ten-year-old schoolboy in Bangkok, Thailand, I had a
wonderful art teacher named Mrs. Vanit. During a class assignment, we
were asked to produce landscapes, and I produced a painting that looked
a bit like Figure 8.2a—a palm tree growing between two hills.

Mrs. Vanit frowned as she looked at the picture and said, "Rama, you
should put the palm tree a bit off to one side, not exactly between the
hills."

I protested, "But Mrs. Vanit, surely there's nothing logically impossible about this scene. Maybe the tree is growing in such a way that its
trunk coincides exactly with the V between the hills. So why do you say
the picture is wrong?"

(a)   (b)

FIGURE 8.2   Two hills with a tree in the middle. (a) The brain dislikes unique
vantage points and (b) prefers generic ones.

"Rama, you can't have coincidences in pictures," said Mrs. Vanit.

The truth was neither Mrs. Vanit nor I knew the answer to my question at that time. I now realize that my drawing illustrates one of the most important laws in aesthetic perception: the abhorrence of coincidences.

Imagine that Figure 8.2a depicts a real visual scene. Look carefully and you'll realize that in real life, you could only see the scene in Figure 8.2a from one vantage point, whereas you could see the one in Figure 8.2b from any number of vantage points. One viewpoint is unique and one is generic. As a class, images like the one in Figure 8.2b are much more common. So Figure 8.2a is—to use a phrase introduced by Horace Barlow—"a suspicious coincidence." And your brain always tries to find a plausible alternate, generic interpretation to avoid the coincidence. In this case it doesn't find one and so the image isn't pleasing.

Now let's look at a case where a coincidence does have an interpretation. Figure 8.3 shows the famous illusory triangle described by Italian psychologist Gaetano Kanizsa. There really isn't a triangle. It's just three black Pac-Man-like figures facing one another. But you perceive an opaque white triangle whose three corners partially occlude three black circular discs. Your brain says (in effect), "What's the likelihood that these three Pac-Men are lined up exactly like this simply by chance? It's too

FIGURE 8.3   Three black discs with pie-shaped wedges removed from them: The brain prefers to see this arrangement as an opaque white triangle whose corners partially occlude circular discs.

much of a suspicious coincidence. A more plausible explanation is that it depicts an opaque white triangle occluding three black discs." Indeed, you can almost hallucinate the edges of the triangle. So in this case your visual system has found a way of explaining the coincidence (eliminating it, you might say) by coming up with an interpretation that feels good. But in the case of the tree centered in the valley, your brain struggles to find an interpretation of the coincidence and is frustrated because there isn't one.

Orderliness

The law of what I loosely call "orderliness," or regularity, is clearly important in art and design, especially the latter. Again, this principle is so obvious that it's hard to talk about it without sounding banal, but a discussion of visual aesthetics is not complete without it. I will lump a number of principles under this category which have in common an abhorrence for deviation from expectations (for instance, the preference for rectilinearity and parallel edges and for the use of repetitive motifs in carpets). I will touch on these only briefly because many art historians, like Ernst Gombrich and Rudolf Arnheim, have already discussed them extensively.

Consider a picture frame hanging on the wall, slightly tilted. It elicits an immediate negative reaction that is wildly out of proportion to the deviation. The same holds for a drawer that doesn't close completely because there's a piece of crumpled paper wedged in it and sticking out. Or an envelope with a single tiny hair accidentally caught under the sealed portion. Or a tiny piece of lint on an otherwise flawless suit. Why we react this way is far from clear. Some of it seems to be simple hygiene, which has both learned and instinctive components. Disgust with dirty feet is surely a cultural development, while picking a piece of lint out of your child's hair might derive from the primate grooming instinct.

The other examples, such as the tilted frame or slightly disarrayed pile of books, seem to imply that our brains have a built-in need to impose regularity or predictability, although this doesn't explain much.

It's unlikely that all examples of regularity or predictability embody the same law. A closely related law, for example, is our love of visual repetition or rhythm, such as floral motifs used in Indian art and Persian carpets. But it's hard to imagine that this exemplifies the same law as our fondness for a straightly hung picture frame. The only thing the two have in common, at a very abstract level, is that both involve predictability. In each case the need for regularity or order may reflect a deeper need your visual system has for economy of processing.

Sometimes deviations from predictability and order are used by designers and artists to create pleasing effects. So why should some deviations, like a tilted frame, be ugly while others—say, a beauty spot placed asymmetrically near the angle of the mouth of Cindy Crawford, rather than being in the middle of her chin or nose—be attractive? The artist seems to strike a balance between extreme regularity, which is boring, and complete chaos. For example, if she uses a motif of repeating small flowers framing a sculpture of a goddess, she may try to break the monotony of the repetition by adding some more widely spaced large flowers to create two overlapping rhythms of different periodicity. Whether there has to be a certain mathematical relationship between the two scales of repetition and what kind of phase shifts between the two are permissible are good questions—yet to be answered.

Symmetry

Any child who has played with a kaleidoscope and any lover who has seen the Taj Mahal has been under the spell of symmetry. Yet even though designers recognize its allure and poets use it to flatter, the question of why symmetrical objects should be pretty is rarely raised.

Two evolutionary forces might explain the allure of symmetry. The first explanation is based on the fact that vision evolved mainly for discovering objects, whether for grabbing, dodging, mating, eating, or catching. But your visual field is always crammed full of objects: trees, fallen logs, splotches of color on the ground, rushing brooks, clouds, outcroppings of rocks, and on and on. Given that your brain has limited attentional capacity, what rules of thumb might it employ to ensure attention gets allocated to where it's most needed? How does your brain come up with a hierarchy of precedence rules? In nature, "important" translates into "biological objects" such as prey, predator, member of the same species, or mate, and all such objects have one thing in common: symmetry. This would explain why symmetry grabs your attention and arouses you, and by extension, why the artist or architect can exploit this trait to good use. It would explain why a newborn baby prefers looking at symmetrical inkblots over asymmetrical ones. The preference likely taps a rule of thumb in the baby's brain that says, in effect, "Hey, something symmetrical. That feels important. I should keep looking."

The second evolutionary force is more subtle. By presenting a random sequence of faces with varying degrees of symmetry to college undergraduates (the usual guinea pigs in such experiments), psychologists have found that the most symmetrical faces are generally judged to be the most attractive. This in itself is hardly surprising; no one expects the twisted visage of Quasimodo to be attractive. But intriguingly, even minor deviations are not tolerated. Why?

The surprising answer comes from parasites. Parasitic infestation can profoundly reduce the fertility and fecundity of a potential mate, so evolution places a very high premium on being able to detect whether your mate is infected. If the infestation occurred in early fetal life or infancy, one of the most obvious externally visible signs is a subtle loss of symmetry. Therefore, symmetry is a marker, or flag, for good health, which in turn is an indicator of desirability. This argument explains why your visual system finds symmetry appealing and asymmetry disturbing. It's an odd thought that so many aspects of evolution—even our aesthetic preferences—are driven by the need to avoid parasites. (I once wrote a satirical essay that "gentlemen prefer blondes" for the same reason. It's much easier to detect anemia and jaundice caused by parasites in a light-skinned blonde than in a swarthy brunette.)

Of course, this preference for symmetrical mates is largely unconscious. You are completely unaware that you are doing it. What a fitting bit of symmetry that the same evolutionary quirk in the great Mogul emperor Shah Jahan's brain that caused him to select the perfectly symmetrical, parasite-free face of his beloved Mumtaz, also caused him to construct the exquisitely symmetrical Taj Mahal itself, a universal symbol of eternal love!

But we must now deal with the apparent exceptions. Why is a *lack* of symmetry appealing at times? Imagine you are arranging furniture, pictures, and other accessories in a room. You don't need a professional designer to tell you that total symmetry won't work (although within the room you can have islands of symmetry, such as a rectangular table with symmetrically placed chairs). On the contrary, you need carefully chosen asymmetry to create the most dramatic effects. The clue to resolving this paradox comes from the observation that the symmetry rule applies only to objects, not to large-scale scenes. This makes perfect evolutionary sense because a predator, a prey, a friend, or a mate is always an isolated, independent object.

Your preference for symmetrical objects and asymmetrical scenes is also reflected in the "what" and "how" (sometimes called "where") streams in your brain's visual processing stream. The "what" stream (one of two subpathways in the new pathway) flows from your primary visual areas toward your temporal lobes, and concerns itself with discrete objects and the spatial relationships of features within objects, such as the internal proportions of a face. The "how" stream flows from your primary visual area toward your parietal lobes and concerns itself more with your general surroundings and the relationships between objects (such as the distance between you, the gazelle you're chasing, and the tree it's about to dodge behind). It's no surprise that a preference for symmetry is rooted in the "what" stream, where it is needed. So the detection and enjoyment of symmetry is based on object-centered algorithms in your brain, not scene-centered ones. Indeed, objects placed symmetrically in a room would look downright silly because, as we have seen, the brain dislikes coincidences it can't explain.

Metaphor

The use of metaphor in language is well known, but it's not widely appreciated that it's also used extensively in visual art. In Figure 8.4 you see a sandstone sculpture from Kajuraho in Northern India, circa A.D. 1100. The sculpture depicts a voluptuous celestial nymph who arches her back to gaze upward as if aspiring to God or heaven. She probably occupied a niche at the base of a temple. Like most Indian nymphs she has a narrow waist weighed down heavily by big hips and breasts. The arch of the bough over her head closely follows the curvature of her arm (a postural example of a grouping principle called closure). Notice the plump, ripe mangoes dangling from the branch which, like the nymph herself, are a metaphor of the fertility and fecundity of nature. In addition, the plumpness of the mangoes provides a sort of visual echo of the plumpness and ripeness of her breasts. So there are multiple layers of metaphor and meaning in the sculpture, and the result is incredibly beautiful. It's almost as though the multiple metaphors amplify each other, although why this internal resonance and harmony should be especially pleasing is anybody's guess.

I find it intriguing that the visual metaphor is probably understood by the right hemisphere long before the more literal-minded left hemisphere can spell out the reasons. (Unlike a lot of flaky pop psychology lore about hemispheric specialization, this particular distinction probably does have a grain of truth.) I am tempted to suggest that there is ordinarily a translation barrier between the left hemisphere's language-based, propositional logic and the more oneiric (dream like), intuitive "thinking" (if that's the right word) of the right, and great art sometimes succeeds by dissolving this barrier. How often have you listened to a strain of music that evokes a richness of meaning that is far more subtle than what can be articulated by the philistine left hemisphere?

A more mundane example is the use of certain attention-drawing tricks used by designers. The word "tilt" printed in visually tilted letters produces a comical yet pleasing effect. This tempts me

FIGURE 8.4   A stone nymph below an arching bough, looking heavenward for divine inspiration. Khajuraho, India, eleventh century.

to posit a separate law of aesthetics, which we might call "visual resonance," or "echo" (although I am wary of falling into the trap that some Gestaltists fell into of calling every observation a law). Here the resonance is between the concept of the word "tilt" with its actual literal tilt, blurring the boundary between conception and perception.

In comics, words like "scared," "fear," or "shiver" are often printed in wiggly lines as if the letters themselves were trembling. Why is this so effective? I'd say it is because the wiggly line is a spatial echo of your own shiver, which in turn resonates with the concept of fear. It may be that watching someone tremble (or tremble as depicted metaphorically by a wiggly letters) makes you echo the tremble ever so slightly because it prepares you to run away, anticipating the predator that may have caused the other person to tremble. If so, your reaction time for detecting the word "fear" depicted in wiggly letters might be much shorter than if the word were depicted in straight lines (smooth letters), an idea that can be tested in the laboratory.[2]

I will conclude my comments on the aesthetic law of metaphor with Indian art's greatest icon: *The Dancing Shiva*, or *Nataraja*. In Chennai (Madras), there is bronze gallery in the state museum that houses a magnificent collection of southern Indian bronzes. One of its prize works is a twelfth-century *Nataraja* (Figure 8.5). One day around the turn of the twentieth century, an elderly *firangi* ("foreigner" or "white" in Hindi) gentleman was observed gazing at the *Nataraja* in awe. To the amazement of the museum guards and patrons, he went into a sort of trance and proceeded to mimic the dance postures. A crowd gathered around, but the gentleman seemed oblivious until the curator finally showed up to see what was going on. He almost had the poor man arrested until he realized the European was none other than the world-famous sculptor Auguste Rodin. Rodin was moved to tears by *The Dancing Shiva*. In his writings he referred to it as one of the greatest works of art ever created by the human mind.

You don't have to be religious or Indian or Rodin to appreciate the grandeur of this bronze. At a very literal level, it depicts the cosmic dance of Shiva, who creates, sustains, and destroys the Universe. But the sculpture is much more than that; it is a metaphor of the dance of the Universe itself, of the movement and energy of the cosmos. The artist depicts this sensation through the skillful use of many devices. For example, the centrifugal motion of Shiva's arms and legs flailing in different directions and the wavy tresses flying off his head symbolize the agitation and frenzy of the cosmos. Yet right in the midst of all this turbulence—this fitful fever of life—is the calm spirit of Shiva himself. He gazes at his own creation with supreme tranquility and poise. How skillfully the artist has combined these seemingly antithetical elements of movement and

FIGURE 8.5 *Nataraja* depicting the cosmic dance of Shiva. Southern India, Chola period, twelfth century.

energy, on the one hand, and eternal peace and stability on the other. This sense of something eternal and stable (God, if you like) is conveyed partly by Shiva's slightly bent left leg, which gives him balance and poise even in the midst of his frenzy, and partly by his serene, tranquil expression, which conveys a sense of timelessness. In some Nataraja sculptures this peaceful expression is replaced by an enigmatic half-smile, as though the great god were laughing at life and death alike.

This sculpture has many layers of meaning, and indologists like

Heinrich Zimmer and Ananda Coomaraswamy wax lyrically about them. While most Western sculptors try to capture a moment or snapshot in time, the Indian artist tries to convey the very nature of time itself. The ring of fire symbolizes the eternal cyclical nature of creation and destruction of the Universe, a common theme in Eastern philosophy, which is also occasionally hit upon by thinkers in the West. (I am reminded in particular of Fred Hoyle's theory of the oscillating universe.) One of Shiva's right hands holds a tambour, which beats the Universe into creation and also represents perhaps the pulse beat of animate matter. But one of his left hands holds the fire that not only heats up and energizes the universe but also consumes it, allowing destruction to perfectly balance out creation in the eternal cycle. And so it is that the *Nataraja* conveys the abstract, paradoxical nature of time, all devouring yet ever creative.

Below Shiva's right foot is a hideous demonic creature called Apasmara, or "the illusion of ignorance," which Shiva is crushing. What is this illusion? It's the illusion that all of us scientific types suffer from, that there is nothing more to the Universe than the mindless gyrations of atoms and molecules, that there is no deeper reality behind appearances. It is also the delusion of some religions that each of us has a private soul who is watching the phenomena of life from his or her own special vantage point. It is the logical delusion that after death there is nothing but a timeless void. Shiva is telling us that if you destroy this illusion and seek solace under his raised left foot (which he points to with one of his left hands), you will realize that behind external appearances (Maya), there is a deeper truth. And once you realize this, you see that, far from being an aloof spectator, here to briefly watch the show until you die, you are in fact part of the ebb and flow of the cosmos—part of the cosmic dance of Shiva himself. And with this realization comes immortality, or *moksha*: liberation from the spell of illusion and union with the supreme truth of Shiva himself. There is, in my mind, no greater instantiation of the abstract idea of god—as opposed to a personal God—than the *Shiva/ Nataraja*. As the art critic Coomaraswamy says, "This is poetry, but it is science nonetheless."

I am afraid I have strayed too far afield. This is a book about neurology, not Indian art. I showed you the *Shiva/Nataraja* only to underscore that the reductionist approach to aesthetics presented in this chapter is in no way meant to diminish great works of art. On the contrary, it may actually enhance our appreciation of their intrinsic value.

I OFFER THESE nine laws as a way to explain why artists create art and why people enjoy viewing it.[3] Just as we consume gourmet food to generate complex, multidimensional taste and texture experiences that titillate our palate, we appreciate art as gourmet food for the visual centers in the brain (as opposed to junk food, which is analogous to kitsch). Even though the rules that artists exploit originally evolved because of their survival value, the production of art itself doesn't have survival value. We do it because it's fun and that's all the justification it needs.

But is that the whole story? Apart from its role in pure enjoyment, I wonder if there might be other, less obvious reasons why humans engage in art so passionately. I can think of four candidate theories. They are about the value of art itself, not merely of aesthetic enjoyment.

First, there is the very clever, if somewhat cheeky and cynical, suggestion favored by Steven Pinker that acquiring or owning unique, one-of-a-kind works may have been a status symbol to advertise superior access to resources (a psychological rule of thumb evolved for assessing superior genes). This is especially true today as the increasing availability of mass copying methods places an ever higher premium (from the art buyer's perspective) on owning an original—or at least (from the art seller's perspective) on fooling the buyer into the mock status conferred by purchasing limited-edition prints. No one who has been to an art show cocktail reception in Boston or La Jolla can fail to see that there is some truth to this view.

Second, an ingenious idea has been proposed by Geoffrey Miller, the evolutionary psychologist at the University of New Mexico, and by others that art evolved to advertise to potential mates the artist's manual dexterity and hand-eye coordination. This was promptly dubbed the "come up and see my etchings" theory of art. Like the male bowerbird, the male artist is in effect telling his muse, "Look at my pictures. They show I have excellent hand-eye coordination and a complex, well-integrated brain—genes I'll pass on to your babies." There is an irritating grain of truth to Miller's idea, but personally I don't find it very convincing. The main problem is that it doesn't explain why the advertisement should take the form of art. It seems like overkill. Why not directly advertise this ability to potential mates by showing off your skills in archery or athletic prowess in soccer? If Miller is right, women should find the ability to knit and embroider to be very attractive in potential husbands, given that it requires superb manual dexterity—even though most women, not even feminists, don't value such skills in a man. Miller might argue that women value not the dexterity and skill per se but the creativity that underlies the finished product. But despite its supreme cultural importance to humans, the biological survival value of art as an index of creativity is dubious given that it doesn't necessarily spill over into other domains. (Just look at the number of starving artists!)

Notice that Pinker's theory predicts that the women should hover around the buyers, whereas Miller's theory predicts they should hover around the starving artists themselves.

To these ideas I'll add two more. To understand them you need to consider thirty-thousand-year-old cave art from Lascaux, France. These cave-wall images are hauntingly beautiful even to the modern eye. To achieve them, the artists must have used some of the same aesthetic laws used by modern artists. For example, the bisons are mostly depicted as outline drawings (isolation), and bison-like characteristics such as small head and large hump are grossly exaggerated. Basically, it's a caricature (peak shift) of a bison created by unconsciously subtracting the average generic hoofed quadruped from a bison and amplifying the differences. But apart from just saying, "They made these images just to enjoy them," can we say anything more?

Humans excel at visual imagery. Our brains evolved this ability to create an internal mental picture or model of the world in which we can rehearse forthcoming actions, without the risks or the penalties of doing them in the real world. There are even hints from brain-imaging studies by Harvard University psychologist Steve Kosslyn showing that your brain uses the same regions to imagine a scene as when you actually view one.

But evolution has seen to it that such internally generated representations are never as authentic as the real thing. This is a wise bit of self-restraint on your genes' part. If your internal model of the world were a perfect substitute, then anytime you felt hungry you could simply imagine yourself at a banquet, consuming a feast. You would have no incentive to find real food and would soon starve to death. As the Bard said, "You cannot cloy the hungry edge of appetite by bare imagination of a feast."

Likewise, a creature that developed a mutation that allowed it to imagine orgasms would fail to pass on its genes and would quickly become extinct. (Our brains evolved long before porn videos, *Playboy* magazine, and sperm banks.) No "imagine orgasm" gene is likely to make a big splash in the gene pool.

Now what if our hominin ancestors were worse than us at mental imagery? Imagine they wanted to rehearse a forthcoming bison or lion hunt. Perhaps it was easier to engage in realistic rehearsal if they had actual props, and perhaps these props are what we today call cave art. They may have used these painted scenes in much the way that a child enacts imaginary fights between his toy soldiers, as a form of play to educate his internal imagery. Cave art could also have been used for teaching hunting skills to novices. Over several millennia these skills would become assimilated into culture and acquired religious significance. Art, in short, may be nature's own virtual reality.

Finally, a fourth, less prosaic reason for art's timeless appeal may be that it speaks an oneiric, right-hemisphere-based language that is unintelligible—alien, even—to the more literal-minded left hemisphere. Art conveys nuances of meaning and subtleties of mood that can only be dimly apprehended or conveyed through spoken language. The neural codes used by the two hemispheres for representing higher cognitive functions may be utterly different. Perhaps art facilitates communion between these two modes of thinking that would otherwise remain mutually unintelligible and walled off. Perhaps emotions also need a virtual reality rehearsal to increase their range and subtlety for future use, just as we engage in athletics for motor rehearsal and frown over crossword puzzles or ponder over Gödel's theorem for intellectual invigoration. Art, in this view, is the right hemisphere's aerobics. It's a pity that it isn't emphasized more in our schools.

SO FAR, WE have said very little about the creation—as opposed to the perception—of art. Steve Kosslyn and Martha Farah of Harvard have used brain-imaging techniques to show that creatively conjuring up a visual image probably involves the inner (ventromedial cortex) portion of the frontal lobes. This portion of the brain has back-and-forth connections with parts of the temporal lobes concerned with visual memories. A crude template of the desired image is initially evoked through these connections. Back-and-forth interactions between this template and what's being painted or sculpted lead to progressive embellishments and refinements of the painting, resulting in the multiple, stage-by-stage mini-"Ahas!" we spoke of earlier. When the self-amplifying echoes between these layers of visual processing reach a critical volume, they get delivered as a final, kick-ass "Aha!" to reward centers such as the septal nuclei and the nucleus accumbens. The artist can then relax with her cigarette, cognac, and muse.

Thus the creative production of art and the appreciation of art may be tapping into the same pathways (except for the frontal involvement in the former). We have seen that faces and objects enhanced through peak shifts (caricatures, in other words) hyperactivate cells in the fusiform gyrus. Overall scene layout—as in landscape paintings—probably requires the right inferior parietal lobule, whereas "metaphorical," or conceptual aspects of art might require both the left and right angular gyri. A more thorough study of artists with damage to different portions of either the right or left hemisphere might be worthwhile—especially bearing in mind our laws of aesthetics.

Clearly we have a long way to go. Meanwhile, it's fun to speculate. As Charles Darwin said in his *Descent of Man,*

> false facts are highly injurious to the progress of science, for they often endure long; but false views, if supported by some evidence, do little harm, for everyone takes a salutary pleasure in proving their falseness; and when this is done, one path toward errors is closed and the road to truth is often at the same time opened.

CHAPTER 9

An Ape with a Soul:
How Introspection Evolved

———

Hang up philosophy! Unless philosophy can make a Juliet . . .

—WILLIAM SHAKESPEARE

JASON MURDOCH WAS AN INPATIENT AT A REHABILITATION CEN-
ter in San Diego. After a serious head injury in a car accident near the
Mexican border, he had been in a semiconscious state of vigilant coma
(also called akinetic mutism) for nearly three months before my colleague, Dr. Subramaniam Sriram, examined him. Because of damage
to the anterior cingulate cortex in the front of his brain, Jason couldn't
walk, talk, or initiate actions. His sleep-wake cycle was normal but he
was bedridden. When awake he seemed alert and conscious (if that's
the right word—words lose their resolving power when dealing with
such states). He sometimes had slight "ouch" withdrawal in response to
pain, but not consistently. He could move his eyes, often swiveling them
around to follow people. Yet he couldn't recognize anyone—not even his
parents or siblings. He could not talk or comprehend speech, nor could
he interact with people meaningfully.

But if his father, Mr. Murdoch, phoned him from next door, Jason
suddenly became alert and talkative, recognizing his dad and engaging him in conversation. That is until Mr. Murdoch went back into the
room. Then Jason lapsed back into his semiconscious "zombie" state.
Jason's cluster of symptoms has a name: telephone syndrome. He could
be made to flip back and forth between the two states, depending on
whether his father was directly in his presence or not.

Think of what this means. It is almost as if there are two Jasons trapped inside one body: the Jason on the phone, who is fully alert and conscious, and the Jason in person, who is a barely conscious zombie. How can this be? The answer has to do with how the accident affected the visual and auditory pathways in Jason's brain. To a surprising extent, the activity of each pathway—vision and hearing—must be segregated all the way up to the critically important anterior cingulate. This collar of tissue, as we shall see, is where your sense of free will partly originates.

If the anterior cingulate is extensively damaged, the result is the full picture of akinetic mutism; unlike Jason, the patient is in a permanent twilight state, not interacting with anyone under any circumstances. But what if the damage to the anterior cingulate is more subtle—say, the visual pathway to the anterior cingulate is damaged selectively at some stage, but the auditory pathway is fine. The result is telephone syndrome: Jason springs to action (speaking metaphorically!) when chatting on the phone but lapses into akinetic mutism when his father walks into the room. Except when he is on the telephone, Jason is no longer a person.

I am not making this distinction arbitrarily. Although Jason's visuo-motor system can still track and automatically attend to objects in space, he cannot recognize or attribute meaning to what he sees. Except when he is on the phone with his father, Jason lacks the ability to form rich, meaningful metarepresentations, which are essential to not only our uniqueness as a species but also our uniqueness as individuals and our sense of self.

Why is Jason a person when he is on the phone but not otherwise? Very early in evolution the brain developed the ability to create first-order sensory representations of external objects that could elicit only a very limited number of reactions. For example a rat's brain has only a first-order representation of a cat—specifically, as a furry, moving thing to avoid reflexively. But as the human brain evolved further, there emerged a second brain—a set of nerve connections, to be exact—that was in a sense parasitic on the old one. This second brain creates meta-representations (representations of representations—a higher order of abstraction) by processing the information from the first brain into man-ageable chunks that can be used for a wider repertoire of more sophisti-cated responses, including language and symbolic thought. This is why, instead of just "the furry enemy" that it is for the rat, the cat appears to you as a mammal, a predator, a pet, an enemy of dogs and rats, a thing that has ears, whiskers, a long tail, and a meow; it even reminds you of Halle Berry in a latex suit. It also has a name, "cat," symbolizing the whole cloud of associations. In short, the second brain imbues an object with meaning, creating a metarepresentation that allows you to be consciously aware of a cat in a way that the rat isn't.

Metarepresentations are also a prerequisite for our values, beliefs, and priorities. For example, a first-order representation of disgust is a visceral "avoid it" reaction, while a metarepresentation would include, among other things, the social disgust you feel toward something you consider morally wrong or ethically inappropriate. Such higher-order representations can be juggled around in your mind in a manner that is unique to humans. They are linked to our sense of self and enable us to find meaning in the outside world—both material and social—and allow us to define ourselves in relation to it. For example, I can say, "I find her attitude toward emptying the cat litter box disgusting."

The visual Jason is essentially dead and gone as a person, because his ability to have metarepresentations of what he sees is compromised.[1] But the auditory Jason lives on; his metarepresentations of his father, his self, and their life together are largely intact as activated via the auditory channels of his brain. Intriguingly, the hearing Jason is temporarily switched off when Mr. Murdoch appears in person to talk to his son. Perhaps because the human brain emphasizes visual processing, the visual Jason stifles his auditory twin.

Jason presents a striking case of a fragmented self. Some of the "pieces" of Jason have been destroyed, yet others have been preserved and retain a surprising degree of functionality. Is Jason still Jason if he can be broken into fragments? As we shall see, a variety of neurological conditions show us that the self is not the monolithic entity it believes itself to be. This conclusion flies directly in the face of some of our most deep-seated intuitions about ourselves—but data are data. What the neurology tells us is that the self consists of many components, and the notion of one unitary self may well be an illusion.

SOMETIME IN THE twenty-first century, science will confront one of its last great mysteries: the nature of the self. That lump of flesh in your cranial vault not only generates an "objective" account of the outside world but also directly experiences an internal world—a rich mental life of sensations, meanings, and feelings. Most mysteriously, your brain also turns its view back on itself to generate your sense of self-awareness.

The search for the self—and the solutions to its many mysteries—is hardly a new pursuit. This area of study has traditionally been the preserve of philosophers, and it is fair to say that on the whole they haven't made a lot of progress (though not for want of effort; they have been at it for two thousand years). Nonetheless, philosophy has been extremely useful in maintaining semantic hygiene and emphasizing the need for clarity in terminology.[2] For example, people often use the word "consciousness" loosely to refer to two different things. One is qualia—the immediate experiential qualities of sensation, such as the redness of red or the pungency of curry—and the second is the self who experiences these sensations. Qualia are vexing to philosophers and scientists alike because even though they are palpably real and seem to lie at the very core of mental experience, physical and computational theories about brain function are utterly silent on the question of how they might arise or why they might exist.

Let me illustrate the problem with a thought experiment. Imagine an intellectually highly advanced but color-blind Martian scientist who sets out to understand what humans mean when they talk about color. With his *Star Trek*–level technology he studies your brain and completely figures out down to every last detail what happens when you have mental experiences involving the color red. At the end of his study he can account for every physicochemical and neurocomputational event that occurs when you see red, think of red, or say "red." Now ask yourself: Does this account encompass everything there is to the ability to see and think about redness? Can the color-blind Martian now rest assured that he understands your alien mode of visual experience even though his brain is not wired to respond to that particular wavelength of electromagnetic radiation? Most people would say no. Most would say that no matter how detailed and accurate this outside-objective description of color cognition might be, it has a gaping hole at its center because it leaves out the quale of redness. ("Quale," pronounced "kwah-lee," is the singular form of "qualia.") Indeed, there is no way you can convey the ineffable quality of redness to someone else short of hooking up your brain directly to that person's brain.

Perhaps science will eventually stumble on some unexpected method or framework for dealing with qualia empirically and rationally, but such advances could easily be as remote from our present-day grasp as molecular genetics was to those living in the Middle Ages. Unless there is a potential Einstein of neurology lurking around somewhere.

I suggested that qualia and self are different. Yet you can't solve the former without the latter. The notion of qualia without a self experiencing/introspecting on them is an oxymoron. In similar vein Freud had argued that we cannot equate the self with consciousness. Our mental life, he said, is governed by the unconscious, a roiling cauldron of memories, associations, reflexes, motives, and drives. Your "conscious life" is an elaborate after-the-fact rationalization of things you really do for other reasons. Because technology had not yet advanced sufficiently to allow observation of the brain, Freud lacked the tools to take his ideas beyond the couch, and so his theories were caught in the doldrums between true science and untethered rhetoric.[3]

Might Freud have been right? Could most of what constitutes our "self" be unconscious, uncontrollable, and unknowable?[4] Despite Freud's current unpopularity (to put it mildly), modern neuroscience has in fact revealed that he was right in arguing that only a limited part of the brain is conscious. The conscious self is not some sort of "kernel" or concentrated essence that inhabits a special throne at the center the neural labyrinth, but neither is it a property of the whole brain. Instead, the self seems to emerge from a relatively small cluster of brain areas that are linked into an amazingly powerful network. Identifying these regions is important since it helps narrow the search. We know, after all, that the liver and the spleen are not conscious; only the brain is. We are simply taking a step further and saying that only some parts of the brain are conscious. Knowing which parts are and what they are doing is the first step toward understanding consciousness.

The phenomenon of blindsight is a particularly clear indicator that there may be a grain of truth in Freud's theory of the unconscious. Recall from Chapter 2 that someone with blindsight has damage to the V1 area in the visual cortex, and as a result cannot see anything. She is blind. She experiences none of the qualia associated with vision. If you project a spot of light on the wall in front of her, she will tell you categorically that she does not see anything. Yet if asked to reach out to touch the spot, she can do so with uncanny accuracy even though to her it feels like a wild guess. She is able to do this, as we saw earlier, because the old pathway between her retina and her parietal lobe is intact. So even though she can't see the spot, she can still reach out and touch it. Indeed, a blindsight patient can often even guess the color and orientation of a line (vertical or horizontal) using this pathway even though she cannot perceive it consciously.

This is astonishing. It implies that only the information streaming through your visual cortex is associated with consciousness and linked to your sense of self. The other parallel pathway can go about its business performing the complex computations required for hand guidance (or even correctly guessing color) without consciousness ever coming into the picture. Why? These two paths for visual information are made up of identical-looking neurons, after all, and they seem to be performing equally complex computations, yet only the new pathway casts the light of consciousness on visual information. What's so special about these circuits that they "require" or "generate" consciousness? In other words, why aren't all aspects of vision and vision-guided behavior similar to blindsight, chugging along with competence and accuracy but without conscious awareness and qualia? Might the answer to this question give clues to solving the riddle of consciousness?

The example of blindsight is suggestive not only because it supports the idea of the unconscious mind (or several unconscious minds). It also demonstrates how neuroscience can marshal evidence about the innermost workings of the brain in order to make its way through the cold-case file, so to speak, addressing some of the unanswered questions about the self that have plagued philosophers and scientists for millennia. By studying patients who have disturbances in self-representation and observing how specific brain areas malfunction, we can better understand how a sense of self arises in the normal human brain. Each disorder becomes a window on a specific aspect of the self.

First, let's define these aspects of the self, or at the very least, our intuitions about them.

1. *Unity:* Despite the teeming diversity of sensory experiences that you are deluged with moment to moment, you feel like one person. Moreover, all of your various (and sometimes

contradictory) goals, memories, emotions, actions, beliefs, and present awareness seem to cohere to form a single individual.

2. *Continuity:* Despite the enormous number of distinct events punctuating your life, you feel a sense of continuity of identity through time—moment to moment, decade to decade. And as Endel Tulving has noted, you can engage in mental "time travel," starting from early childhood and projecting yourself into the future, sliding to and fro effortlessly. This Proustian virtuosity is unique to humans.

3. *Embodiment:* You feel anchored and at home in your body. It never occurs to you that the hand you just used to pick up your car keys might not belong to you. Nor would you think you're in any danger of believing the arm of a waiter or a cashier is in fact your own arm. However, scratch the surface and it turns out your sense of embodiment is surprisingly fallible and flexible. Believe it or not, you can be optically tricked into temporarily leaving your body and experiencing yourself in another location. (This happens to some extent when you view a live, real-time video of yourself or stand in a carnival hall of mirrors.) By wearing heavy makeup to disguise yourself and looking at your own video image (which doesn't have to do a left-right reversal like a mirror), you can get an inkling of an out-of-body experience, especially if you move various body parts and change your expression. Furthermore, as we saw in Chapter 1, your body image is highly malleable; it can be altered in position and size using mirrors. And as we will see later in this chapter, it can be profoundly disturbed in disease.

4. *Privacy:* Your qualia and mental life are your own, unobservable by others. You can empathize with your neighbor's pain thanks to mirror neurons, but you can't literally experience his pain. Yet, as we noted in Chapter 4, there are circumstances under which your brain generates touch sensations that precisely simulate the sensations being experienced by another individual. For instance, if I anesthetize your arm and have you watch me touch my own arm, you begin to feel my touch sensations. So much for the privacy of self.

5. *Social embedding:* The self maintains an arrogant sense of privacy and autonomy that belies how closely it is linked to other

brains. Can it be coincidental that almost all of our emotions make sense only in relation to other people? Pride, arrogance, vanity, ambition, love, fear, mercy, jealousy, anger, hubris, humility, pity, even self-pity—none of these would have any meaning in a social vacuum. It makes perfect evolutionary sense to feel grudges, gratitude, or bonhomie, for example, toward other people based on your shared interpersonal histories. You take intent into account and attribute the faculty of choice, or free will, to fellow social beings and apply your rich palette of social emotions to their actions on that basis. But we are so deeply hardwired for imputing things such as motive, intent, and culpability to the actions of others that we often overextend our social emotions to nonhuman, nonsocial objects, or situations. You can get "angry" with the tree branch that fell on you, or even with the freeways or the stock market. It is worth noting that this is one of the major roots of religion: We tend to imbue nature itself with human-like motives, desire, and will, and hence we feel compelled to supplicate, pray to, bargain with, and look for reasons why God or karma or what have you has seen fit to punish us (individually or collectively) with natural disasters or other hardships. This persistent drive reveals just how much the self needs to feel part of a social environment that it can interact with and understand on its own terms.

6. *Free will:* You have a sense of being able to consciously choose between alternative courses of action with the full knowledge that you could have chosen otherwise. You normally don't feel like an automaton or as though your mind is a passive thing buffeted by chance and circumstance—although in some "diseases" such as romantic love, you come close. We don't yet know how free will works, but, as we shall see later in the chapter, at least two brain regions are crucially involved. The first is the supramarginal gyrus on the left side of the brain, which allows you to conjure up and envisage different potential courses of action. The second is the anterior cingulate, which makes you desire (and helps you choose) one action based on a hierarchy of values dictated by the prefrontal cortex.

7. *Self-awareness:* This aspect of the self is almost axiomatic; a

self that is not aware of itself is an oxymoron. Later in this chapter I will argue that your self-awareness might partly depend on your brain using mirror neurons recursively, allowing you to see yourself from another person's (allocentric) viewpoint. Hence the use of terms like "self-conscious" (embarrassed), when what you really mean is being conscious of someone else being conscious of you.

These seven aspects, like the legs of a table, work together to hold up what we call the self. However, as you can already see, they are vulnerable to illusions, delusions, and disorders. The table of the self can continue to stand without one of these legs, but if too many are lost then its stability becomes severely compromised.

How did these multiple attributes of self emerge in evolution? What parts of the brain are involved, and what are the underlying neural mechanisms? There are no simple answers to these questions—certainly nothing to rival the simplicity of a statement like "because that is how God made us"—but just because the answers are complicated and counterintuitive is no reason to give up the quest. By exploring several syndromes that straddle the boundary between psychiatry and neurology, I believe we can glean invaluable clues to how the self is created and sustained in normal brains. In this regard my approach is similar to that used elsewhere in the book: considering odd cases to illuminate normal function.[5] I do not claim to have "solved" the problem of self (I wish!), but I believe these cases provide very promising ways it can be approached. Overall, I think this is not a bad start for tackling a problem that is not even considered legitimate by many scientists.

Several points are worth noting before we examine particular cases. One is that despite the bizarreness of symptoms, each patient is relatively normal in other respects. A second is that each patient is completely sincere and confident in his belief and this belief is immune from intellectual correction (just like persistent superstitions in otherwise rational people). A patient with panic attacks might agree with you intellectually that his forebodings of doom are not "real," but during the attack itself, nothing will convince him that he isn't dying.

One last caveat: We need to be careful when drawing insights from psychiatric syndromes because some of them (none, I hope, that I am examining here) are bogus. Take for example de Clérambault syndrome, which is defined as a young woman developing an obsessive delusion that a much older and famous man is madly in love with her but he is in denial about it. Google it if you don't believe me. (Ironically there's no name for the very real and common delusion in which an older gentleman believes that a young hottie is in love with him but doesn't know it! One reason for this might be that the psychiatrists who "discover" and name syndromes have historically been men.)

Then there is Koro, the alleged disorder said to afflict Asian gentlemen who claim that their penis is shrinking and will eventually wither away. (Again the converse does exist in some elderly Caucasian men— the delusion that the penis is expanding—when it actually isn't. This was pointed out to me by my colleague Stuart Anstis.) Koro is likely to have been fabricated by Western psychiatrists, though it is not inconceivable that it might arise from a reduced representation of the penis in the body-image center, the right superior parietal lobule.

And let's not forget another notable invention, "oppositional defiant disorder." This diagnosis is sometimes given to smart, spirited youngsters who dare to question the authority of older establishment figures, such as psychiatrists. (Believe it or not, this is a diagnosis for which a psychologist can actually bill the patient's insurance company.) The person who concocted this syndrome, whoever he or she is, is brilliant, for any attempt by the patient to challenge or protest the diagnosis can itself be construed as evidence for its validity! Irrefutability is built into its very definition. Another pseudomalady, again officially recognized, is "chronic underachievement syndrome"—what used to be called stupidity.

With these caveats in mind let us try to tackle the syndromes themselves and explore their relevance to the self and to human uniqueness.

Embodiment

We will begin with three disorders that allow us to examine the mechanisms involved in creating a sense of embodiment. These conditions reveal that the brain has an innate body image, and when that body image doesn't match up with the sensory input from the body—whether visual or somatic—the ensuing disharmony can disrupt the self's sense of unity as well.

APOTEMNOPHILIA: DOCTOR, REMOVE MY ARM PLEASE

Vital to the human sense of self is a person's feeling of inhabiting his own body and owning his body parts. Although a cat has an implicit body image of sorts (it doesn't try to squeeze into a rat hole), it can't go on a diet seeing that it is obese or contemplate its paw and wish it weren't there. Yet the latter is precisely what happens in some patients who develop apotemnophilia, a curious disorder in which a completely normal individual has an intense and ever-present desire to amputate an arm or a leg. ("Apotemnophila" derives from the Greek: *apo*, "away from"; *temnein*, "to cut"; and *philia,* "emotional attachment to.") He may describe his body as being "overcomplete" or his arm as being "intrusive." You get the feeling that the subject is trying to convey something ineffable. For instance he might say, "It's not as if I feel it doesn't belong to me, Doctor. On the contrary, it feels like it's too present." More than half the patients go on to actually have the limb removed.

Apotemnophilia is often viewed as being "psychological." It has even been suggested that it arises from a Freudian wish-fulfillment fantasy, the stump resembling a large penis. Others have regarded the condition as attention-seeking behavior, although why the desire for attention should take this strange form and why so many of these people keep their desires secret for much of their lives is never explained.

Frankly, I find these psychological explanations unconvincing. The condition usually begins early in life, and it is unlikely that a ten-year-old would desire a giant penis (although an orthodox Freudian wouldn't rule it out). Moreover, the subject can point to the specific line—say, two centimeters above the elbow—along which she desires amputation. It isn't simply a vague desire to eliminate a limb, as one would expect from a psychodynamic account. Nor can it be a desire to attract attention, for if that were the case, why be so particular about where the cut should be made? Finally, the subject usually has no other psychological issues of any consequence.

There are also two other observations I made of these patients that strongly suggest a neurological origin for the condition. First, in more than two-thirds of cases the left limb is involved. This disproportionate involvement of the left arm reminds me of the decidedly neurological disorder of somatoparaphrenia (described later), in which the patient, who has a right-hemisphere stroke, not only denies the paralysis of his left arm but also insists that the arm doesn't belong to him. This is rarely seen in those with left-hemisphere strokes. Second, my students Paul McGeoch and David Brang and I have found that touching the limb below the line of the desired amputation produces a big jolt in the patient's GSR (galvanic skin response), but touching above the line or touching the other limb does not. The patient's alarm bells really and truly go off when the affected limb is touched below the line. Since it's hard to fake a GSR, we can be fairly sure of a neurological basis for the disorder.

How does one explain this strange disorder in terms of the known anatomy? As we saw in Chapter 1, nerves for touch, muscle, tendon, and joint sensation project to your primary (S1) and secondary (S2) somatosensory cortices in and just behind the postcentral gyrus. Each of these areas of the cortex contains a systematic, topographically organized map of bodily sensations. From there, somatosensory information gets sent to your superior parietal lobule (SPL), where it gets combined with balance information from your inner ear and visual feedback about the limbs' positions. Together these inputs construct your body image: a unified, real-time representation of your physical self. This representation of the body in the SPL (and probably its connections with the posterior insula) is partly innate. We know this because some patients with arms missing from birth experience vivid phantom arms, implying the existence of scaffolding that is hardwired by genes.[6] It doesn't require a leap of faith to suggest that this multisensory body image is organized topographically in the SPL the same way it is in S1 and S2.

If a particular body part such as an arm or a leg failed to be represented in this hardwired scaffolding of your body image, the result could conceivably be a sense of strangeness or possibly revulsion toward it. But why? Why is the patient not merely indifferent to the limb? After all, patients with nerve damage to the arm resulting in a complete loss of sensation don't say they want their arm removed.

The answer to this question lies in the key concept of mismatch aversion, which as you will see plays a crucial role in many forms of mental illness. The general idea is that lack of coherence, or mismatch, between the outputs of brain modules can create alienation, discomfort, delusion, or paranoia. The brain abhors internal anomalies—such as the mismatch between emotion and identification in Capgras syndrome—and will often go to absurd lengths to deny them or explain them away. (I emphasize "internal" because generally speaking, the brain is more tolerant of anomalies in the external world. It may even enjoy them: Some people love the thrill of solving baffling mysteries.) It isn't clear where the internal mismatch is detected to create unpleasantness. I suggest it's done by the insula (especially the insula in the right hemisphere), a small patch of tissue which receives signals from S2 and sends outputs to the amygdala, which in turn sends sympathetic arousal signals down to the rest of the body.

In the case of nerve damage, the input to S1 and S2 itself is lost, so there is no mismatch or discrepancy between S2 and the multisensory body image in the SPL. In apotemnophilia, by contrast, there is normal sensory input from the limb to the body maps in S1 and S2, but there is no "place" for the limb signals to output to in the SPL body image maintained by the SPL.[7] The brain does not tolerate this mismatch well, and so this discrepancy is crucial for creating the feelings of "overpresence" and mild aversiveness of the limb, and the accompanying desire for amputation. This explanation of apotemnophilia would account for the heightened GSR and also the essentially ineffable and paradoxical nature of the experience: part of the body and not part of the body at the same time.

Consistent with this overall framework I have noticed that merely having the patient look at his affected limb through a minifying lens to optically shrink it makes the limb feel far less unpleasant, presumably by reducing the mismatch. Placebo-controlled experiments are needed to confirm this.

Finally, my lab conducted a brain-scanning study on four patients with apotemnophilia and compared the results with four normal control subjects. In the controls, touching any part of the body activated right SPL. In all four patients, touching the part of the limb each one wanted removed evoked no activity in the SPL—the brain's map of the body didn't light up, so to speak, on the scans. But touching the unaffected limb did. If we can replicate this finding with a larger number of patients, our theory will be well supported.

One curious aspect of apotemnophilia that is unexplained by our model is the associated sexual inclinations in some subjects: desire for intimacy with another amputee. These sexual overtones are probably what misled people to propose a Freudian view of the disorder.

Let me suggest something different. Perhaps one's sexual "aesthetic preference" for certain body morphology is dictated in part by the shape of the body image as represented—and hardwired—in the right SPL and possibly insular cortex. This would explain why ostriches prefer ostriches as mates (presumably even when smell cues are eliminated) and why pigs prefer porcine shapes over humans.

Expanding on this, I suggest that there is a genetically specified mechanism that allows a template of one's body image (in the SPL) to become transcribed into limbic circuitry, thereby determining aesthetic visual preference. If this idea is right, then someone whose body image was congenitally armless or legless would be attracted to people missing the same limb. Consistent with this view, people who wish to have their leg amputated are almost always attracted to leg amputees, not arm amputees.

SOMATOPARAPHRENIA: DOCTOR, THIS IS MY MOTHER'S ARM

Distortion of body-part ownership also occurs in one of the strangest syndromes in neurology, which has the tongue-twisting name "somatoparaphrenia." Patients with a left-hemisphere stroke have damage to the band of fibers issuing from the cortex down into the spinal cord. Because the left side of the brain controls the right side of the body (and vice versa), this leaves the right side of their bodies paralyzed. They complain about their paralysis, asking the doctor whether the arm will ever recover, and not surprisingly they are often depressed.

When the stroke is in the right hemisphere, the paralysis is on the left. The majority of such patients are troubled by the paralysis as expected, but a small minority deny the paralysis (anosognosia), and an even smaller subset actually deny ownership of the left arm, ascribing it to the examining physician or to a spouse, sibling, or parent. (Why a particular person is chosen isn't clear, but it reminds me of the manner in which the Capgras delusion often also involves a specific individual.)

In this subset of patients there is usually damage to the body maps in S1 and S2. In addition to this, the stroke has destroyed the corresponding body-image representation in the right SPL, which would ordinarily receive input from S1 and S2. Sometimes there is also additional dam-age to the right insula—which receives input the directly from S2 and also contributes to the construction of the person's body image. The net result of this combination of lesions—S1, S2, SPL, and insula—is a com-plete sense of *dis*ownership of the arm. The ensuing tendency to ascribe it to someone else may be a desperate, unconscious attempt to explain the alienation of the arm (shades of Freudian "projection" here).

Why is somatoparaphrenia only seen when the right parietal is dam-aged but not when the left one is? To understand this we have to invoke the idea of division of labor between the two hemispheres (hemispheric specialization), a topic I will consider in some detail later in this chapter. Rudiments of such specialization probably exist even in the great apes, but in humans it is much more pronounced and may be yet another fac-tor contributing to our uniqueness.

TRANSSEXUALITY: DOCTOR, I'M TRAPPED IN THE WRONG KIND OF BODY!

The self also has a sex: You think of yourself as male or female and expect others to treat you as such. It is such an ingrained aspect of your self-identity that you hardly ever pause to think about it—until things go awry, at least by the standards of a conservative, conformist society. The result is the "disorder" called transsexuality.

As with somatoparaphrenia, distortions or mismatches in the SPL can also explain the symptoms of transsexuals. Many male-to-female trans-sexuals report feeling that their penis seems to be redundant or, again, overpresent and intrusive. Many female-to-male transsexuals report feel-ing like a man in a woman's body, and a majority of them have had a phantom penis since early childhood. Many of these women also report having phantom erections.[8] In both kinds of transsexuals the discrep-ancy between internally specified sexual body image—which, surpris-ingly, includes details of sexual anatomy—and external anatomy leads to an intense discomfort and, again, a yearning to reduce the mismatch.

Scientists have shown that during fetal development, different aspects of sexuality are set in motion in parallel: sexual morphology (external anatomy), sexual identity (what you see yourself as), sexual orientation (what sex you are attracted to), and sexual body image (your brain's internal representation of your body parts). Normally these harmonize during physical and social development to culminate in normal sexuality, but they can become uncoupled, leading to deviations that shift the individual toward one or the other end of the spectrum of normal distribution.

I am using the words "normal" and "deviation" here only in the statistical sense relative to the overall human population. I do not mean to imply that these ways of being are undesirable or perverse. Many transsexuals have told me that they would rather have surgery than be "cured" of their desire. If this seems strange, think of intense but unrequited romantic love. Would you request that your desire be removed? There is no simple answer.

Privacy

In Chapter 4, I explained the role of the mirror-neuron system in viewing the world from another person's point of view, both spatially and (perhaps) metaphorically. In humans this system may have turned inward, enabling a representation of one's *own* mind. With the mirror-neuron system thus "bent back" on itself full-circle, self-awareness was born. There is a subsidiary evolutionary question of which came first—other-awareness or self-awareness—but that's tangential. My point is that the two coevolved, enriching each other enormously and culminating in the kind of reciprocity between self-awareness and other-awareness seen only in humans.

Although mirror neurons allow you to tentatively adopt another person's vantage point, they don't result in an out-of-body experience. You don't literally float out to where that other vantage point is, nor do you lose your identity as a person. Similarly, when you watch another person being touched, your "touch" neurons fire, but even though you empathize, you don't actually feel the touch. It turns out that in both cases, your frontal lobes inhibit the activated mirror neurons at least enough to stop all this from happening so you remain anchored in your own body. Additionally, "touch" neurons in your skin send a null signal to your mirror neurons, saying, "Hey, you are not being touched" to ensure that you don't literally feel the other guy being touched. Thus in the normal brain a dynamic interplay of three sets of signals (mirror neurons, frontal lobes, and sensory receptors) is responsible for preserving both the individuality of your own mind and body, and your mind's reciprocity with others—a paradoxical state of affairs unique to humans. Disturbances in this system, we shall see, would lead to a dissolution of interpersonal boundaries, personal identity, and body image—allowing us to explain a wide spectrum of seemingly incomprehensible symptoms seen in psychiatry. For example, derangements in frontal inhibition of mirror-neuron system may lead to a disturbing out-of-body experience—as though you were really watching yourself from above. Such syndromes reveal how blurred the boundary between reality and illusion can become under certain circumstances.

MIRROR NEURONS AND "EXOTIC" SYNDROMES

Mirror-neuron activity can go awry in many ways, sometimes in full-blown neurological disorders but also, I suspect, in numerous, more subtle ways as well. For instance, I wonder whether a dissolution of interpersonal boundaries may also explain more exotic syndromes such as folie à deux, in which two people, such as Bush and Cheney, share each other's madness. Romantic love is a minor form of folie à deux, a mutual delusional fantasy that often afflicts otherwise normal people. Another example is Munchausen syndrome by proxy, in which hypochondriasis (where every trifling symptom is experienced as a harbinger of fatal illness) is unconsciously projected onto another (the "proxy")— often by a parent onto his or her child—instead of onto oneself.

Much more bizarre is the Couvade syndrome, in which men in Lamaze classes start developing pseudocyesis, or false signs of pregnancy. (Perhaps mirror-neuron activity results in the release of empathy hormones such as prolactin, which act on the brain and body to generate a phantom pregnancy.)

Even Freudian phenomena such as projection begin to make sense: You wish to deny your unpleasant emotions, but they are too salient to deny completely so you ascribe them to others; it's the I-you confusion again. As we will see, this is not unlike a patient with somatoparaphrenia "projecting" her paralyzed arm to her mother. Lastly, there is Freudian countertransference, in which the psychoanalyst's self starts fusing with the patient's, which can sometimes land the psychoanalist in legal trouble if the patient is of the opposite sex.

Obviously, I am not claiming to have "explained" these syndromes; I am merely pointing out how they might fit into our overall scheme and how they may give us hints about the manner in which the normal brain constructs a sense of self.

AUTISM

In Chapter 5, I presented evidence that a paucity of mirror neurons, or the circuits they project to, may underlie autism. If mirror neurons do indeed play a role in self-representation, then one would predict that an autistic person, even a high-functioning one, could probably not introspect, could never feel self-esteem or self-deprecation—let alone experience self-pity or self-aggrandizement—or even know what these words mean. Nor could the child experience the embarrassment—and the blush—that accompanies the state of being self-conscious. Casual observations of autistic people suggest that all this might be true, but there have been no systematic experiments to determine the limits of their introspective abilities. For example, if I were to ask you what's the difference between need and desire (you need toothpaste; you desire a woman or man), or between pride and arrogance, hubris and humility, or sadness and sorrow, you would typically think for a bit before being able to spell out the distinction. An autistic child may be incapable of these distinctions while still being capable of other abstract distinctions (such as "What's the difference between a Democrat and a Republican, other than IQ?").

Another subtle test might be to see whether a high-functioning autistic child (or adult) can understand a conspiratorial wink, which usually involves a three-way social interaction between you, the person you are winking at, and a third person—real or imaginary—in the vicinity. This requires representing one's own as well as the other two people's minds. If I give you a sly wink when telling a lie to someone else (who can't see the wink), then I have an implied social contract with you: "I am letting you in on this—see how I am tricking that person?" A wink is also used when flirting with someone, unbeknownst to others in the vicinity, although I don't know if this is universal to all cultures. (And, lastly, you wink to someone to whom you are saying something in jest as if to say, "You realize I am only are joking, right?") I once asked the famous high-functioning autist and writer Temple Grandin whether she knew what winking meant. She told me that she understands winking intellectually but doesn't ever do it and has no intuitive feel for it.

More directly relevant to the framework of the present chapter is the observation made by Leo Kanner (who first described autism) that autistic children often confuse the pronouns "me" and "you" in conversation. This shows a poor differentiation of ego boundaries and a failure of the self-other distinction which, as we have seen, depends partially on mirror neurons and associated frontal inhibitory circuitry.

THE FRONTAL LOBES AND THE INSULA

Earlier in this chapter, I suggested that apotemnophilia results from a mismatch between somatosensory cortices S1 and S2, on the one hand, and on the other the superior (and inferior) parietal lobules, the region where you normally construct a dynamic image of your body in space. But where exactly is the mismatch detected? Probably in the insula, which is buried in the temporal lobes. The posterior (back) half of this structure combines multiple sensory inputs—including pain—from internal organs, muscles, joints, and vestibular (sense of balance) organs in the ear to generate an unconscious sense of embodiment. Discrepancies between different inputs here produce vaguely articulated discomfort, as when your vestibular and visual senses are put in conflict on a ship and you feel queasy.

The posterior insula then relays to the front (anterior) part of the insula. The eminent neuroanatomist, Arthur D. (Bud) Craig, from the Barrow Neurological Institute in Phoenix, has suggested that the posterior insula registers only rudimentary unconscious sensations, which need to be "re-represented" in more sophisticated form in the anterior insula before your body image can be consciously experienced.

Craig's "re-representations" are loosely similar to what I called "metarepresentations" in *Phantoms in the Brain*. But in my scheme, further back-and-forth interactions with the anterior cingulate and other frontal structures are required for constructing your full sense of being a person reflecting on your sensations and making choices. Without these interactions it makes little sense to speak of a conscious self, whether embodied or not.

So far in this book, I have said very little about the frontal lobes, which became especially well developed in hominins and must play an important role in our uniqueness. Technically the frontal lobes are comprised of the motor cortex as well as the bulk of the cortex in front of it—the prefrontal cortex. Each prefrontal lobe has three subdivisions: the ventromedial prefrontal (VMF), or bottom inner part; the dorsolateral (DLF), or upper outer part; and the dorsomedial (DMF), or upper inner part (see Figure Int.2, in the Introduction). (Because the colloquial term "frontal lobes" includes the prefrontal cortex as well, I use "F" in these abbreviations, not "P.") Let's consider some of the functions of these three prefrontal regions.

I invoked the VMF in Chapter 8 when discussing pleasurable aesthetic responses to beauty. The VMF also receives signals from the anterior insula to generate your conscious sense of being embodied. In conjunction with parts of the anterior cingulate cortex (ACC), it motivates "desire" to take action. For instance, the discrepancy in body image in apotemnophilia, picked up in the right anterior insula, would be relayed to the VMF and the anterior cingulate to motivate a conscious plan of action: "Go to Mexico and get the arm removed!" In parallel, the insula projects directly to the amygdala, which activates the autonomic fight-or-flight response via the hypothalamus. That would explain the heightened skin sweating (galvanic skin response, or GSR) that we saw in our patients with apotemnophilia.

Of course, all this is pure speculation; at this point we don't even know whether my explanation of apotemnophilia is correct. Nonetheless, my hypothesis illustrates the style of reasoning needed to explain many brain disorders. Just brushing such disorders aside as being "mental" or "psychological" problems serves no purpose; such labeling neither illuminates normal function nor helps the patient.

Given their extensive connections with limbic structures, it is hardly surprising that the medial frontal lobes—the VMF and possibly the DMF—are also involved in setting up the hierarchy of values that govern your ethics and morality, traits that are especially well developed in humans. Unless you are a sociopath (who has disturbances in these circuits, as shown by Antonio Damasio), you don't usually lie or cheat, even when 100 percent sure you could get away with it if you tried. Indeed, your sense of morality and your concern for what others think of you are so powerful that you even act to extend them beyond your death. Imagine you have been diagnosed with terminal cancer and have old letters in your drawers that could be dredged up after your death, incriminating you in a sex scandal. If you are like most people, you will promptly destroy the evidence, even though logically, why should your posthumous reputation matter to you once you are gone?

I have already hinted at the role of mirror neurons in empathy. Apes almost certainly have empathy of sorts, but humans have both empathy and "free will," the two necessary ingredients for moral choice. This trait requires a more sophisticated deployment of mirror neurons—acting in conjunction with the anterior cingulate—than any ape before us has achieved.

Let's turn now to the dorsomedial prefrontal area (DMF). The DMF has been found in brain-imaging studies to be involved in conceptual aspects of the self. If you are asked to describe your own attributes and personality traits (rather than someone else's), this area lights up in brain-imaging studies. On the other hand, if you were to describe the raw feel of your embodiment, one would expect your VMF to light up, but this hasn't been tested yet.

Lastly, there is the dorsolateral prefrontal area. The DLF is required for holding things in your current, ongoing mental landscape, so you can use your ACC to direct attention to different aspects of the information and act according to your desires. (The technical name for this function is working memory.) The DLF is also required for logical reasoning, which involves paying attention to different facets of a problem and juggling abstractions—such as words and numbers—synthesized in the inferior parietal lobules (see Chapter 4). How and where the precise rules for this juggling arise is anybody's guess.

The DLF also interacts with the parietal lobe. The two act jointly to construct a consciously experienced, animated body moving in space and time (which complements the insula-VMF pathway's creation of a more viscerally felt anchoring of your self in your body). The subjective boundary between these two types of body image is somewhat blurred, reminding us of the sheer complexity of connections needed for even something as "simple" as your body image. This point will be driven home later; we will encounter a patient with a phantom twin next to him. Vestibular stimulation caused the twin to shrink and move. This implies powerful interactions between (a) vestibular input to the insula, which produces a visceral anchoring of the body, and (b) vestibular input to the right parietal lobe, which—along with muscle, joint sense, and vision—constructs a vivid sense of a consciously experienced, moving body.

Unity

What if the self is produced not by a single entity but by the push and pull of multiple forces of which we are largely unconscious? Now I'll use the lenses of anosognosia and out-of-body experiences to examine the unity—and disunity—of the self.

HEMISPHERIC SPECIALIZATION: DOCTOR, I AM IN TWO MINDS

A great deal of pop psychology deals with the question of how the two hemispheres might be specialized for different roles. For example, the right hemisphere is thought to be more intuitive, creative, and emotional than the left, which is said to be more linear, rational, and Spock-like in its mentality. Many a New Age guru has used the idea to promote ways of unleashing the hidden potential of the right hemisphere.

As with most pop ideas, there is a kernel of truth to all this. In *Phantoms in the Brain*, I postulated that the two hemispheres have different, but complementary, coping styles in dealing with the world. Here I will consider the relevance of this to understanding anosognosia, the denial of paralysis seen in some stroke patients. Speaking more generally, it can help us understand why even most normal people—including you and me—engage in minor denials and rationalizations to cope with the stresses of our daily lives. What is the evolutionary function of these hemispheric differences, if any?

Information arriving through the senses is ordinarily merged with preexisting memories to create a belief system about yourself and the world. This internally consistent belief system, I suggest, is constructed mainly by the left hemisphere. If there is a small piece of anomalous information that doesn't fit your "big picture" belief system, the left hemisphere tries to smooth over the discrepancies and anomalies in order to preserve the coherence of self and the stability of behavior. In a process called confabulation, the left hemisphere sometimes even fabricates information to preserve its harmony and overall view of itself. A Freudian might say that the left hemisphere does this to avoid shattering the ego, or to reduce what psychologists refer to as cognitive dissonance, a disharmony between different internal aspects of self. Such disconnects give rise to the confabulations, denials, and delusions that one sees in psychiatry. In other words, Freudian defenses originate mainly in the left hemisphere. In my account, however, unlike in orthodox Freudianism, they evolved not to "protect the ego" but to stabilize behavior and impose a sense of coherence and narrative to your life.

But there has to be a limit. If left unchecked, the left hemisphere would likely render a person delusional or manic. It is one thing to play down some of your weaknesses to yourself (an unrealistic "optimism" may be useful temporarily for forging ahead), but another thing to delude yourself into thinking you are rich enough to buy a Ferrari (or that your arm is not paralyzed) when neither is true. So it seems reasonable to postulate a "devil's advocate" in the right hemisphere that allows "you" to adopt a detached, objective (allocentric) view of yourself.[9] This right-brain system would often be able to detect major discrepancies that your egocentric left hemisphere has ignored or suppressed but shouldn't have. You are then alerted to this, and the left hemisphere is jolted into revising its narrative.

The notion that many aspects of the human psyche might arise from a push-pull antagonism between complementary regions of the two hemispheres might seem like a gross oversimplification; indeed, the theory itself might be the result of "dichotomania," the brain's tendency to simplify the world by dividing things into polarized opposites (night and day, yin and yang, male and female, and so on). But it makes perfect sense from a systems engineering point of view. Control mechanisms that stabilize a system and help avoid oscillations are the rule rather than the exception in biology.

I will now explain how the difference between coping styles of the two hemispheres accounts for anosognosia—the denial of disability, in this case paralysis. As we saw earlier, when either hemisphere is damaged by stroke the result is hemiplegia, a complete paralysis of one side of the body. If the stroke is in the left hemisphere, then the right side of the body is paralyzed, and as expected the patient will complain about the paralysis and request treatment. The same is true for a majority of right-hemisphere strokes, but a significant minority of patients remain indifferent. They play down the extent of the paralysis and stubbornly deny that they cannot move—or even deny ownership of a paralyzed limb! Such denial usually happens as a result of additional damage to the postulated "devil's advocate" in the right hemisphere's frontoparietal regions, which allows the left hemisphere to go into an "open loop," taking its denials to absurd limits.

I recently examined an intelligent, sixty-year-old patient named Nora, who had an especially striking version of this syndrome.

"Nora, how are you today?" I asked.

"Fine, Sir, except the hospital food. It's terrible."

"Well, let's take a look at you. Can you walk?"

"Yes." (Actually, she hadn't taken a single step in the last week.)

"Nora, can you use your hands, can you move them?"

"Yes."

"Both hands?"

"Yes." (Nora had not used a fork in a week.)

"Can you move your left hand?"

"Yes, of course."

"Touch my nose with your left hand."

Nora's hand remains motionless.

"Are you touching my nose?"

"Yes."

"Can you see your hand touching my nose?"

"Yes, it's now almost touching your nose."

A few minutes later I grabbed Nora's lifeless left arm, raised it toward her face, and asked, "Whose hand is this, Nora?"

"That's my mother's hand, Doctor."

"Where is your mother?"

At this point Nora looked puzzled and glanced around for her mother. "She is hiding under the table."

"Nora, you said you can move your left hand?"

"Yes."

"Show me. Touch your own nose with your left hand."

Without the slightest hesitation Nora moved her right hand toward her flaccid left hand, grabbed it and used it like a tool to touch her nose. The amazing implication is that even though she was denying that her left arm was paralyzed, she must have known at some level that it was, for if not, why would she spontaneously reach out to grab it? And why does she use "her mother's" left hand as a tool to touch her own nose? It would appear that there are many Noras within Nora.

Nora's case is an extreme manifestation of anosognosia. More commonly the patient tries to play down the paralysis, rather than engaging in outright denial or confabulation. "No problem, Doc. It's getting better every day!" Over the years I have seen many such patients and been struck by the fact that many of their comments bear a striking resemblance to the kinds of everyday denials and rationalizations that we all engage in to tide over the discrepancies in our daily lives. Sigmund (and more especially his daughter Anna) Freud referred to these as "defense mechanisms," suggesting that their function is to "protect the ego"—whatever that means. Examples of such Freudian defenses would include denial, rationalization, confabulation, reaction formation, projection, intellectualization, and repression. These curious phenomena have only a tangential relevance to the problem of Consciousness (with a big *C*), but—as Freud urged—they represent the dynamic interplay of between the conscious and unconscious, so studying them may indirectly illuminate our understanding of consciousness and other related aspects of human nature. So I'll list them.

1. *Outright denial*—"My arm isn't paralyzed."

2. *Rationalization*—The tendency we all have to ascribe some unpleasant fact about ourselves to an external cause: For example, we might say, "The exam was too hard" rather than "I didn't study hard enough," or "The professor is sadistic" rather than "I am not smart." This tendency is amplified in patients.

For example, when I asked a patient, Mr. Dobbs, "Why are you not moving your left hand like I asked you to?" his replies varied:

"I am an army officer, Doctor. I don't take orders."

"The medical students have been testing me all day. I am tired."

"I have severe arthritis in my arm; it's too painful to move."

3. *Confabulation*—The tendency to make things up to protect your self image: This is done unconsciously; there is no deliberate intention to deceive. "I can see my hand moving, Doctor. It's an inch from your nose."

4. *Reaction formation*—The tendency to assert the opposite of what you unconsciously know to be true about yourself, or, to paraphrase Hamlet, the tendency to protest too much. An example of this is closeted homosexuals engaging in vehement disapproval of same-sex marriages.

Another example: I remember pointing to a heavy table in a stroke clinic and asking a patient whose left arm was paralyzed, "Can you lift that table with your right hand?"

"Yes."

"How high can you lift it?"

"By about an inch."

"Can you lift the table with your left hand?"

"Yes, by two inches."

Clearly "someone" in there knew she was paralyzed for, if not, why would she exaggerate the arm's ability?

5. *Projection*—Ascribing your own deficiencies to another person. In the clinic: "The [paralyzed] arm belongs to my mother." In ordinary life: "He is a racist."

6. *Intellectualization*—Transforming an emotionally threatening fact into an intellectual problem, thereby deflecting attention from and blunting its emotional impact. Many a person with a terminally ill spouse or family member, unable to face the potential loss, starts treating the illness as a purely intellectual challenge. This could be regarded as a combination of denial and intellectualization, though the terminology is unimportant.

7. *Repression*—The tendency to block the retrieval of painful memories, which if dredged up would be "painful to the ego." Although the word has made it into pop psychology, memory researchers have long been suspicious of repression. I lean toward

thinking that the phenomenon is real, for I have seen many clear instances of it in my patients, providing what mathematicians call an "existence proof."

For example, most patients recover from anosognosia after having been in denial for a few days. I had been seeing one such patient who insisted for nine days in a row that his paralyzed arm was "working fine," even with repeated questioning. Then on the tenth day he recovered completely from his denial.

When I questioned him about his condition, he immediately stated, "My left arm is paralyzed."

"How long has it been paralyzed?" I asked, surprised.

He replied, "Why, for the last several days that you have been seeing me."

"What did you tell me when I asked about your arm yesterday?"

"I told you it was paralyzed, of course."

Clearly he was "repressing" his denials!

Anosognosia is a striking illustration of what I have repeatedly stressed in this book—that "belief" is not a single thing. It has many layers that can be peeled away one at a time until the "true" self becomes nothing more than an airy abstraction. As the philosopher Daniel Dennett once said, the self is more akin conceptually to the "center of gravity" of a complicated object, its many vectors intersecting at a single imaginary point.

Thus anosognosia, far from being just another odd syndrome, gives us fresh insights into the human mind. Each time I see a patient with this disorder, I feel like I am looking at human nature through a magnifying glass. I can't help thinking that if Freud had known about anosognosia, he would have taken great delight in studying it. He might ask, for example, what determines which particular defense you use; why use rationalization in some cases and outright denial for others? Does it depend entirely on the particular circumstances or on the patient's personality? Would Charlie always use rationalization and Joe use denial?

Apart from explaining Freudian psychology in evolutionary terms, my model may also be relevant to bipolar disorder (manic-depressive illness). There is an analogy between the coping styles of the left and right hemispheres—manic or delusional for the left, anxious devil's advocate for the right—and the mood swings of bipolar illness. If so, is it possible that such mood swings may actually result from alternation between the hemispheres? As my former teachers Dr. K. C. Nambiar and Jack Pettigrew have shown, even in normal individuals there may be some spontaneous "flipping" between the hemispheres and their corresponding cognitive styles. An extreme exaggeration of this oscillation may be regarded as "dysfunctional" or "bipolar illness" by psychiatrists even though I have known some patients who are willing to tolerate the bouts of depression in order to (for example) continue their brief euphoric communions with God.

OUT OF BODY EXPERIENCE: DOCTOR, I LEFT MY BODY BEHIND

As we saw earlier, one job of the right hemispheres is to take a detached, big-picture view of yourself and your situation. This job also extends to allowing you to "see" yourself from an outsider's point of view. For example, when you are rehearsing a lecture, you may imagine watching yourself from the audience pacing up and down the podium.

This idea can also account for out-of-body experiences. Again, we only need to invoke disruption to the inhibitory circuits that ordinarily keep mirror-neuron activity in check. Damage to the right frontoparietal regions or anesthesia using the drug ketamine (which may influence the same circuits) removes this inhibition. As a result, you start leaving your body, even to the extent of not feeling your own pain; you see your pain "objectively" as if someone else were experiencing it. Sometimes you get the feeling that you have actually left your body and are hovering over it, watching yourself from outside. Note that if these "embodying" circuits are especially vulnerable to lack of oxygen to the brain, this could also explain why such out-of-body sensations are common in near-death experiences.

Odder still than most out-of-body sensations are the symptoms experienced by a patient named Patrick, a software engineer from Utah who had been diagnosed with a malignant brain tumor in his frontoparietal region. The tumor was on the right side of his brain, which was fortunate because he was less worried about it than he would have been had it been on the left. Patrick had been told he had less than two years to live even after the tumor had been removed, but he tended to play it down.

What really intrigued him was much stranger than either he or anyone else could have imagined.

He noticed that he had an invisible but vividly felt "phantom twin" attached to the left side of his body. This was different from the more common sort of out-of-body experience in which a patient feels he is looking down on his own body from above. Patrick's twin mimicked his every action in near-perfect synchrony. Patients like him have been studied extensively by Peter Brugger of the University Hospital Zürich. They remind us that even the congruence between different aspects of your mind such as subjective "ego" and body image can be deranged in brain disease. There must be a specific brain mechanism (or dovetailing suite of mechanisms) that ordinarily preserves such congruence; if there weren't, it could not have been affected selectively in Patrick while leaving other aspects of his mind intact—for indeed, he was emotionally normal, introspective, intelligent, and amiable.[10]

Out of curiosity I irrigated his left ear canal with ice water. This procedure is known to activate the vestibular system and can provide a certain jolt to the body image; it can, for example, fleetingly restore awareness of the paralysis of the body to a patient with anosognosia due to a parietal stroke. When I did this for Patrick, he was astonished to notice the twin shrinking in size, moving, and changing posture. Ah, how little we know about the brain!

Out-of-body experiences are seen often in neurology, but they blend imperceptibly into what we call dissociative states, which are usually seen by psychiatrists. The phrase refers to a condition in which the person mentally detaches herself from whatever is going on in her body during a highly traumatic experience. (Defense lawyers often use the dissociative state diagnosis: that the accused was in a such a state, and that she was watching her body "acting out" the murder without personal involvement.)

The dissociative state involves the deployment of some of the same neural structures already discussed, but in addition two other structures: the hypothalamus and the anterior cingulate.[11] Ordinarily, when confronted with a threat, two outputs flow out from the hypothalamus: a behavioral output, such as running away or fighting; and an emotional output, such as fear or aggression. (We already mentioned the third output: autonomic arousal leading to sweating GSR, blood pressure, and heart-rate elevation.) The anterior cingulate is simultaneously active; it allows you to remain aroused and ever vigilant for new threats and new opportunities for fleeing. But the degree of threat determines the degree to which each of these three subsystems is engaged. When one is confronted with an extreme threat, it is sometimes best to lie still and do nothing at all. This could be regarded as a form of "playing possum," shutting down both the behavioral and emotional output. The possum becomes completely still when a predator is so close that escape is no longer an option, and in fact any attempt would only activate the carnivore's instinct to chase down fleeing prey. Nonetheless, the anterior cingulate remains powerfully engaged the whole time to preserve vigilance, just in case the predator isn't fooled or a quick escape route becomes available.

A vestige of this "possum reflex," or an exaptation of it, may manifest itself as dissociative states in humans in extreme emergencies. You shut down overt behavior as well as emotions and view yourself with objective detachment from your own pain or panic. This sometimes happens in rape, for example, where the woman gets into a paradoxical state: "I was viewing myself being raped as a detached external observer might—feeling the pain but not the agony. And there was no panic." The same thing must have occurred when the explorer David Livingstone was mauled by a lion chewing his arm off; he felt no pain or fear.

The ratio of activation among these circuits and interactions between them can also give rise to less extreme forms of dissociation in which action is not inhibited but emotions are. We have dubbed this the "James Bond reflex": his nerves of steel allow him to remain unperturbed by distracting emotions as he pursues and tackles the villain (or has sex with a woman without paying the "penalty" of love).

Social Embedding

The self defines itself in relation to its social environment. When that environment becomes incomprehensible—for example, when familiar people suddenly seem unfamiliar or vice versa—the self can experience extreme distress or even feel that it is under threat.

THE MISIDENTIFICATION SYNDROMES:
DOCTOR, THAT'S NOT MY MOTHER

A person's brain creates a unified, internally consistent picture of his social world—a stage occupied by different selves like you and me. Seems like a banal statement, but when the self is deranged you begin to realize there are specific brain mechanisms at work to clothe the self with a body and an identity.

In Chapter 2, I offered an explanation for the Capgras syndrome in terms of visual pathways 2 and 3 as they diverge from the fusiform gyrus (Figures 9.1 and 9.2). If pathway 3 (the "so what" stream, which evokes emotions) is compromised while pathway 2 (the "what" stream, which enables identification) remains intact, the patient can recall facts and memories about his nearest and dearest—in a word, he can recognize them—but, jarringly, distressingly, he does not get the warm fuzzy feelings that he "should." The mismatch is either too painful or too bewildering to accept, so he embraces the delusion of an identical imposter. Going further down the path of delusion, he may say things like "my other mother," or even assert that there are several mother-like beings. This is called duplication, or reduplication.

Now think about what happens when the Capgras scenario is reversed: intact pathway 3, compromised pathway 2. The patient loses her ability to recognize faces. She becomes face blind, a condition called prosopagnosia. And yet her pure unconscious discrimination of people's faces continues to be carried out by her intact fusiform gyrus, which can still send signals down her intact "so what" stream (pathway 3) to her amygdala. As a result, she still responds emotionally to familiar faces—she gives a nice big GSR signal when seeing her mother, for example—even though she has no idea who she is looking at. Strangely, her brain—and skin—"knows" something that her mind is unaware of consciously. (This was shown in an elegant series of experiments by Antonio Damasio.) So you can think of the Capgras and prosopagnosia disorders as mirror images of each other, both structurally and in terms of clinical symptoms.[12]

To most of us with our undamaged brains, it seems counterintuitive that identity (facts known about a person) should be segregated from familiarity (emotional reactions to a person). How can you recognize

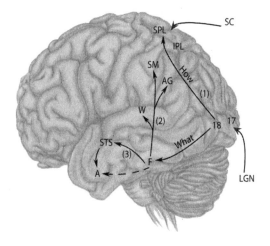

FIGURE 9.1   A highly schematic diagram of the visual pathways and other areas invoked to explain symptoms of mental illness: The superior temporal sulcus (STS) and supramarginal gyrus (SM) are probably rich in mirror neurons. Pathways 1 ("how") and 2 ("what") are identified anatomical pathways. The split of the "what" pathway into two streams—"what" (pathway 2) and "so what" (pathway 3)—is based mainly on functional considerations and neurology. The superior parietal lobule (SPL) is involved in the construction of body image and visual space. The inferior parietal lobule (IPL) is also concerned with body image, but also with pre-hension in monkeys and (probably) apes. The supramarginal gyrus (SM) is unique to humans. During hominin development, it split off from the IPL and became spe-cialized for skilled and semiskilled movements such as tool use. Selection pressure for its split and specialization came from the need to use hands for making tools, wielding weapons, hurling missiles, as well as fine hand and finger manipulation. Another gyrus (AG) is probably unique to us. It split off from IPL and originally subserved cross-modal abstraction capacities, such as tree climbing, and match-ing visual size and orientation with muscle and joint feedback. The AG became exapted for more complex forms of abstraction in humans: reading, writing, lexicon, and arithmetic. Wernicke's area (W) deals with language (semantics). The STS also has connections with the insula (not shown). The amygdaloid complex (A, including the amygdala) deals with emotions. The lateral geniculate nucleus (LGN) of the thalamus relays information from the retina to area 17 (also known as V1, the primary visual cortex). The superior colliculus (SC) receives and pro-cesses signals from the retina that are to be sent via the old pathway to the SPL (after a relay via the pulvinar, not shown). The fusiform gyrus (F) is involved in face and object recognition.

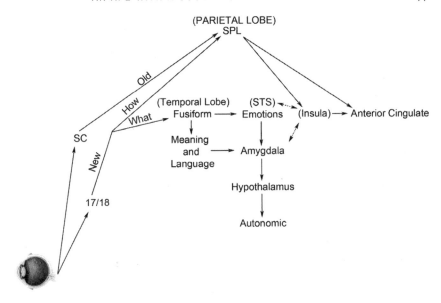

FIGURE 9.2   An abbreviated version of Figure 9.1, showing the distinction between emotions and semantics (meaning).

someone yet not recognize her at the same time? You might get an inkling of what this is like if you think back to an occasion when you ran into an acquaintance somewhere completely out of context, such as an airport in a foreign country, and could not for the life of you remember who he was. You experienced familiarity with lack of identity. The fact that such dissociation can occur at all is proof that separate mechanisms are involved, and in such "airport" moments you experience a miniature, fleeting "syndrome" that is the converse of Capgras. The reason you don't experience this cognitive discrepancy as unpleasant (except briefly as you buy time with small talk while racking your brain) is because such episodes do not last long. If this acquaintance continued to look strange all the time, irrespective of context and no matter how much or how often you spoke with him, he might start looking sinister and you might indeed develop a strong aversion or paranoia.

SELF DUPLICATION: DOCTOR, WHERE IS THE OTHER DAVID?

Astonishingly, we have found that the reduplication seen in Capgras syndrome can even involve the patient's own self. As previously noted, the recursive activity of mirror neurons may result in a representation not only of others' minds but of one's own mind as well.[13] Some mix-up of this mechanism could explain why our patient David pointed to a profile-view photo of himself and said, "That's another David." On other occasions he referred to "the other David" in casual conversation, even asking, poignantly, "Doctor, if the other David comes back, will my real parents disown me?" Of course, we all indulge in role playing from time to time but not to the point where the metaphorical ("I am in two minds," "I'm not the young man that I once was") becomes literal. Again, bear in mind that despite these specific dreamlike misreadings of reality, David was perfectly normal in other respects.

I might add that the Queen of England also refers to herself in the third person, but would hesitate to ascribe this to pathology.

FREGOLI SYNDROME: DOCTOR, EVERYONE LOOKS LIKE AUNT CINDY

In Fregoli syndrome, the patient claims that all people seem to resemble a prototype person he knows. For example, I once met a man who said everyone looked like his aunt Cindy. Perhaps this arises because the emotional pathway 3 (as well as links from pathway 2 to amygdala) has been strengthened by disease. This could happen because of repeated volleys of signals accidentally activating pathway 3, as in epilepsy; it is sometimes called kindling. The outcome is that everyone looks strangely familiar rather than unfamiliar. Why the patient should latch onto a single prototype is unclear, but it may arise from the fact that "diffuse familiarity" makes no sense. By analogy, the diffuse anxiety of the hypochondriac seldom floats free for long, but latches onto a specific organ or disease.

Self-Awareness

Earlier in this chapter I wrote that a self that is not aware of itself is an oxymoron. There are nevertheless certain disorders that can seriously distort one's self-awareness, whether by causing patients to believe that they are dead or by inspiring the delusion that they have become one with God.

COTARD SYNDROME: DOCTOR, I DON'T EXIST

If you do a survey and ask people—whether neuroscientists or Eastern mystics—what the most important puzzling aspect of the self is, the most common answer would be the fact that the self is aware of itself; it can contemplate its own existence and (alas!) its mortality. No nonhuman creature can do this.

I often visit Chennai, India, during the summer to give lectures and see patients at the Institute of Neurology on Mount Road. A colleague of mine, Dr. A. V. Santhanam, often invites me to lecture there and draws my attention to interesting cases. On one particular evening after giving a lecture, I found Dr. Santhanam waiting for me in my office with a patient, a disheveled, unshaven young man of thirty named Yusof Ali. Ali had suffered from epilepsy starting in his late teens. He had periodic bouts of depression, but it was hard to know whether this was related to his seizures or to reading too much Sartre and Heidegger, as many intelligent teenagers do. Ali told me of his deep interest in philosophy.

The fact that Ali was acting strangely was obvious to nearly everyone who knew him long before his epilepsy was diagnosed. His mother had noticed that a couple of times a week there were brief periods when he would become somewhat detached from the world, appear to experience a clouding of consciousness and engage in incessant lip smacking and postural contortions. This clinical history, together with his EEG (electroencephalograph, a record of his brain waves), led us to diagnose Ali's miniseizures as a form of epilepsy called complex partial seizures. Such seizures are different from the dramatic grandmal (whole-body) seizures most people associate with epilepsy; these miniseizures, in contrast, mainly affect the temporal lobes and produce emotional changes. During his long seizure-free intervals Ali was perfectly lucid and intelligent.

"What brings you to our hospital?" I asked.

Ali remained silent, looking a me intently for nearly a minute. He then whispered slowly, "Not much can be done: I am a corpse."

"Ali, where are you?"

"At the Madras Medical College, I think. I used to be a patient at the Kilpauk." (Kilpauk was the only mental hospital in Chennai.)

"Are you saying you are dead?"

"Yes. I don't exist. You could say I am an empty shell. Sometimes I feel like a ghost that exists in an another world."

"Mr. Ali, you are obviously an intelligent man. You are not mentally insane. You have abnormal electrical discharges in certain parts of your brain that can affect the way you think. That's why they moved you here from the mental hospital. There are certain drugs that are very effective for controlling seizures."

"I don't know what you're saying. You know the world is illusory as the Hindus say. Its all *maya* [the Sanskrit word for "illusion"]. And if the world doesn't exist, then in what sense do I exist? We take all that for granted, but it simply isn't true."

"Ali, what are you saying? Are you saying *you* may not exist? How do you explain that you are here talking to me right now?"

Ali appeared confused and a tear started forming in his eye. "Well, I am dead and immortal at the same time."

In Ali's mind—as in the minds of many otherwise "normal" mystics—there is no essential contradiction in his statement. I sometimes wonder whether such patients who have temporal lobe epilepsy have access to another dimension of reality, a wormhole of sorts into a parallel universe. But I usually don't say this to my colleagues, lest they doubt my sanity.

Ali had one of the strangest disorders in neuropsychiatry: Cotard syndrome. It would be all too easy to jump to the conclusion that Ali's delusion was the result of extreme depression. Depression very often accompanies Cotard syndrome. However, depression alone cannot be the cause of it. On the one hand, less extreme forms of depersonalization—in which the patient feels like an "empty shell" but, unlike a Cotard patient, retains insight into his illness—can occur in the complete absence of depression. Conversely, most patients who are severely depressed don't go around claiming they are dead. So something else must be going on in Cotard syndrome.

Dr. Santhanam started Ali on a regimen of the anticonvulsant drug lamotrigine.

"This should help you get better," he said. "We are going to start you on a small dose because in a few rare cases patients develop a very severe allergic skin rash. If you develop such a rash, stop the medicine immediately and come and see us."

Over the next few months Ali's seizures disappeared, and as an added bonus his mood swings diminished and he became less depressed. Yet even three years later he continued to maintain that he was dead.[14]

What would be causing this Kafkaesque disorder? As I noted earlier, pathways 1 (including parts of the inferior parietal lobule) and 3 are both rich in mirror neurons. The former is involved in inferring intentions and the latter, in concert with the insula, is involved in emotional empathy. You have also seen how mirror neurons might not only be involved in modeling other people's behavior—the conventional view—but may also turn "inward" to inspect your own mental states. This could enrich introspection and self-awareness.

The explanation I propose is to think of Cotard syndrome as an extreme and more general form of Capgras syndrome. People with Cotard syndrome often lose interest in viewing art and listening to music, presumably because such stimuli also fail to evoke emotions. This is what we might expect if all or most sensory pathways to the amygdala are totally severed (as opposed to Capgras syndrome, in which just the "face" area in the fusiform gyrus is disconnected from the amygdala). Thus for a Cotard patient, the entire sensory world, not just Mum and Dad, would seem derealized—unreal, as in a dream. If you added to this cocktail a derangement of reciprocal connections between the mirror neurons and the frontal lobe system, you would lose your sense of self as well. Lose yourself and lose the world—that's as close to death in life as you can get. No wonder severe depression frequently, though not always, accompanies Cotard syndrome.

Note that in this framework it is easy to see how a less extreme form of Cotard syndrome could underlie the peculiar states of derealization ("The world looks unreal as in a dream") and depersonalization ("I don't feel real") that are frequently seen in clinical depression. If depressed patients have selective damage to the circuits that mediate empathy and the salience of external objects, but intact circuitry for self-representation, the result could be derealization and a feeling of alienation from the world. Conversely, if self-representation is mainly affected, with normal reactions to the outside world and people, the sense of internal hollowness or emptiness that characterizes depersonalization would be the result. In short, the feeling of unreality is attributed to either oneself or the world depending on differential damage to these closely linked functions.

The extreme sensory-emotional disconnection and diminishment of self I am proposing as an explanation for Cotard syndrome would also explain such patients' curious indifference to pain. They feel pain as a sensation but, like Mikhey (whom we met in Chapter 1), there is no agony. As a desperate attempt to restore the ability to feel something— anything!—such patients may try to inflict pain on themselves in order to feel more "anchored" in their bodies.

It would also explain the paradoxical finding (not proven, but suggestive) that some severely depressed patients commit suicide when first put on antidepressant drugs such as Prozac. It is arguable that in extreme Cotard cases suicide would be redundant, since the self is already "dead"; there is no one there who can or should be put out of her suffering. On the other hand, an antidepressant drug may restore just enough self-awareness for the patient to recognize that her life and world are meaningless; now that it matters that the world is meaningless, suicide may seem the only escape. In this scheme, Cotard syndrome is apotemnophilia for one's entire self, rather than just one arm or leg, and suicide is its successful amputation.[15]

DOCTOR, I AM ONE WITH GOD

Now consider what would happen if the extreme opposite were to occur—if there were a tremendously overactivation of pathway 3 caused by the kind of kindling one sees in temporal lobe epilepsy (TLE). The result would be an extreme heightening of empathy for others, for the self, and even for the inanimate world. The universe and everything in it become deeply significant. It would feel like union with God. This, too, is frequently reported in TLE.

Now, as in Cotard syndrome, imagine adding into this cocktail some damage to the system in the frontal lobes that inhibits mirror-neuron activity. Ordinarily this system preserves empathy while preventing "overempathy," thus preserving your sense of identity. The result of damaging this system would be a second, even deeper sense of merging with everything.

This sense of transcending your body and achieving union with some immortal, timeless essence is also unique to humans. To their credit, apes are not preoccupied with theology and religion.

DOCTOR, I'M ABOUT TO DIE

Incorrect "attribution" of our internal mental states to the wrong trigger in the external world is very much a part of the complex web of interactions that lead to mental illness in general. Cotard syndrome and "merging with God" are extreme forms of this.[16] A far more common form is the syndrome of panic attacks.

A certain proportion of otherwise normal people are seized for forty to sixty seconds by a sudden feeling of impending doom—a sort of transient Cotard syndrome (combined with a strong emotional component). The heart starts beating faster (felt as palpitations, an intensification of heartbeats), palms sweat, and there is an extreme sense of helplessness. Such attacks can occur several times a week.

One possible source of panic attacks might be brief miniseizures affecting pathway 3, especially the amygdala and its emotional and autonomic arousal outflow through the hypothalamus. In such a case, a powerful fight-or-flight reaction would be triggered, but since there is nothing external you can ascribe the changes to, you internalize it and start to feel as if you're dying. It's the brain's aversion to discrepancy again—this time between the neutral external input and the far-from-neutral internal physiological feelings. The only way your brain can account for this combination is to ascribe the changes to some indecipherable and terrifying internal source. The brain finds free-floating (inexplicable) anxiety less tolerable than anxiety which can be clearly attributed to a source.

If this is correct, one wonders if it might be possible to "cure" panic attacks by taking advantage of the fact that the patient often knows a few seconds ahead of time that an attack is about to occur. If you are the patient, then as soon as you sense the attack coming on, you could quickly start watching a horror movie on your iPhone, for example. This might abort the attack by allowing your brain to ascribe the physiological arousal to the external horror, rather than to some terrifying but intangible inner cause. The fact that you "know" that it's only a movie at some higher intellectual level doesn't necessarily rule out this treatment; after all, you do feel fear when watching a horror movie even while recognizing that it's "only a movie." Belief is not monolithic; it exists in many layers whose interactions one can manipulate clinically using the right trick.

Continuity

Implicit in the idea of the self is the notion of sequentially organized memories accumulated over a lifetime. There are syndromes that can profoundly affect different aspects of memory formation and retrieval. Psychologists classify memory (the word is used loosely synonymous with learning) into three distinct types that might have separate neural substrates. The first of these, called procedural memory, allows you to acquire new skills, such as riding a bicycle or brushing your teeth. Such memories are summoned up instantly when the occasion demands; no conscious recollection is involved. This type of memory is universal to all vertebrates and some invertebrates; it certainly isn't unique to humans. Second, there are memories that comprise your semantic memory, your factual knowledge of objects and events in the world. For example, you know that winter is cold and bananas are yellow. This form of memory, too, is not unique to humans. The third category, first recognized by Endel Tulving, is called episodic memory, memories for specific events, such as your prom night, or the day you broke your ankle playing basketball, or as the psycholinguist Steve Pinker puts it, "When and where who did what to whom." Semantic memories are like a dictionary whereas episodic ones are like a diary. Psychologists also refer to them as "knowing" versus "remembering"; only humans are capable of the latter.

Harvard psychologist Dan Schacter has made the ingenious suggestion that episodic memories may be intimately linked to your sense of self: you need a self to which you attach the memories, and the memories in turn enrich your self. In addition to this we tend to organize episodic memories in approximately the correct sequence and can engage in a sort of mental time travel, conjuring them up in order to "visit" or "relive" episodes in our lives in vivid nostalgic detail. These abilities are almost certainly unique to humans. More paradoxical is our ability to engage in more open-ended forward time travel to anticipate and plan the future. This ability is probably also unique to us (and may require well-developed frontal lobes). Without such planning, our ancestors couldn't have made stone tools in advance of a hunt or sown seeds for the next harvest. Chimpanzees and orangutans engage in opportunistic tool making and tool use (stripping leaves from twigs in order to fish termites from their mounds) but they cannot make tools with the intent to store them for future use.

DOCTOR, WHEN AND WHERE DID MY MOTHER DIE?

All of this makes intuitive sense but there is also evidence from brain disorders—some common, others rare—in which the different components of memory are selectively compromised. These syndromes vividly illustrate the different subsystems of memory, including ones that have evolved only in humans. Almost everyone has heard of amnesia following head trauma: The patient has difficulty recollecting specific incidents that took place during the weeks or months preceding the injury, even though he is smart, recognizes people and is able to acquire new episodic memories. This syndrome—retrograde amnesia—is quite common, seen as often in real life as in Hollywood.

Far rarer is a syndrome described by Endel Tulving, whose patient Jake had damage to parts of both his frontal and temporal lobes. As a result Jake had no episodic memories of any kind, whether from childhood or from the recent past. Nor could he form new episodic memories. However, his semantic memories about the world remained intact; he knew about cabbages, kings, love, hate, and infinity. It is very hard for us to imagine Jake's inner mental world. Yet despite what you would expect from Schacter's theory, there was no denying that he had a sense of self. The various attributes of self, it would seem, are like arrows pointing toward an imaginary point: the mental "center of gravity" of the self that I mentioned earlier. Losing any one arrow might impoverish the self but does not destroy it; the self valiantly defies the slings and arrows of outrageous fortune. Even so, I would agree with Schacter that the autobiography we each carry around in our minds based on a lifetime of episodic memories is intimately linked to our sense of self.

Tucked away in the lower, inner portion of the temporal lobes is the hippocampus, a structure required for the acquisition of new episodes. When it is damaged on both sides of the brain, the result is a striking memory disorder called anterograde amnesia. Such patients are mentally alert, talkative, and intelligent but cannot acquire any new episodic memories. If you were introduced to such a patient for the first time, walked out, and returned after five minutes, there would be no glimmer of recognition on her part; it's as if she had never seen you before. She could read the same detective novel again and again and never get bored. Yet, unlike Tulving's patient, her old memories, acquired prior to the damage, are for the most part intact: she remembers the boy she was dating in the year of her accident, her fortieth birthday party, and so on. So you need your hippocampus to create new memories, but not to retrieve old memories. This suggests that memories are not actually stored in the hippocampus. Furthermore, the patient's semantic memories are unaffected. She still knows facts about people, history, word meanings and so forth. A great deal of pioneering work has been done on these disorders by my colleagues Larry Squire and John Wixted at UC San Diego and by Brenda Milner at McGill University, Montreal.

What would happen if someone were to lose both his semantic and episodic memories, so that he had neither factual knowledge of the world nor episodic memories of a lifetime? No such patient exists, and even if you were to stumble on one who had the right combination of brain lesions, what would you expect him to say about his sense of self? In fact, if he really had neither factual nor episodic memories, it is unlikely that he could even talk to you or understand your question, let alone understand the meaning of "I." However, his motor skills would be unaffected; he might surprise you by cycling home.

Free Will

One attribute of the self is your sense of "being in charge" of your actions and, as a corollary, of your belief that you could have acted otherwise if you had chosen to. This may seem like an abstract philosophical issue but it plays an important role in the criminal justice system. You can deem someone guilty only if he (1) could fully envisage alternate courses of action available to him; (2) he was fully aware of the potential consequences of his actions, both short- and long-term; (3) he could have chosen to withhold the action; and (4) he wanted the result that ensued.

The upper gyrus branching from the left inferior parietal lobule, which I earlier referred to as the supramarginal gyrus, is very much involved in this ability to create a dynamic internal image of anticipated actions. This structure is highly evolved in humans; damage to it results in a curious disorder called apraxia, defined as an inability to carry out skilled actions. For example, if you ask an apraxic patient to wave good-bye, she will simply stare at her hand and start wiggling her fingers. But if you ask her, "What does 'goodbye' mean?" she will reply, "Well, you wave your hand when parting company." Furthermore, her hand and arm muscles are fine; she can untie a knot. Her thinking and language are unaffected and so is her motor coordination, but she cannot translate thought into action. I have often wondered whether this gyrus, which exists only in humans, evolved initially for the manufacture and deployment of multicomponent tools, such as hafting an axe head on a suitably carved handle.

All of this is only part of the story. We usually think of free will as the drive to perform that is linked to your sense of being a purposeful agent with multiple choice options. We have only a few clues as to where this sense of agency—your desire to act, and belief in your ability—emerges from. Strong hints come from studying patients with damage to the anterior cingulate in the frontal lobes, which in turn gets a major input from the parietal lobes, including supramarginal gyrus. Damage here can result in the akinetic mutism, or vigilant coma, we saw in Jason at the beginning of this chapter. A few patients recover after some weeks and say things like, "I was fully conscious and aware of what was going on, Doctor. I understood all your questions but I simply didn't want to reply or do anything." Wanting, it turns out, is crucially dependent on the anterior cingulate.

Another consequence of damage to the anterior cingulate is the alien-hand syndrome, in which the person's hand does something he doesn't "will" it to do. I saw a woman with this disorder in Oxford (together with Peter Halligan). The patient's left hand would reach out and grab objects without her intending to, and she had to use her right hand to pry loose her fingers to let go of the object. (Some of the male graduate students in my lab have dubbed this the "third-date syndrome.") Alien-hand syndrome underscores the important role of the anterior cingulate in free will, transforming a philosophical problem into a neurological one.

Philosophy has set up a way of looking at the consciousness problem by considering abstract questions such as qualia and their relationship to the self. Psychoanalysis, while able to frame the problem in terms of conscious and unconscious brain processes, hasn't formulated clearly testable theories nor do they have the tools to test them. My goal in this chapter has been to demonstrate that neuroscience and neurology provide us with a new and unique opportunity to understand the structure and function of the self, not only from the outside by observing behavior, but also from studying the inner workings of the brain.[17] By studying patients such as those in this chapter, who have deficits and disturbances in the unity of self, we can gain deeper insight into what it means to be human.[18]

If we succeed in this, it will be the first time in evolution that a species has looked back on itself and not only understood its own origins but also figured out what or who is the conscious agent doing the understanding. We don't know what the ultimate outcome of such a journey will be, but surely it is the greatest adventure humankind has ever embarked on.

EPILOGUE

———

. . . gives to airy nothing a local habitation and a name . . .

—WILLIAM SHAKESPEARE

ONE OF THE MAJOR THEMES IN THE BOOK—WHETHER TALKING about body image, mirror neurons, language evolution, or autism—has been the question of how your inner self interacts with the world (including the social world) while at the same time maintaining its privacy. The curious reciprocity between self and others is especially well developed in humans and probably exists only in rudimentary form in the great apes. I have suggested that many types of mental illness may result from derangements in this equilibrium. Understanding such disorders may pave the way not only for solving the abstract (or should I say philosophical) problem of the self at a theoretical level, but also for treating mental illness.

My goal has been to come up with a new framework to explain the self and its maladies. The ideas and observations I have presented will hopefully inspire new experiments and set the stage for a more coherent theory in the future. Like it or not, this is the way science often works in its early stage: Discover the lay of the land first before attempting all-encompassing theories. Ironically it's also the stage when science is most fun; every little experiment you do, you feel like Darwin unearthing a new fossil or Richard Burton turning another bend of the Nile to discover its source. You may not share their lofty stature, but in trying to emulate their style you feel their presence as guardian angels.

To use an analogy from another discipline, we are now at the same stage that chemistry was in the nineteenth century: discovering the basic elements, grouping them into categories, and studying their interactions. We are still grouping our way toward the equivalent of the periodic table but are not anywhere near atomic theory. Chemistry had many false leads—such as the postulation of a mysterious substance, phlogiston, which seemed to explain some chemical interactions until it was discovered that to do so phlogiston had to have a negative weight! Chemists also came up with spurious correlations. For example, John Newlands's law of octaves, which claimed that elements came in clusters of eight like the eight notes in one octave of the familiar *do-re-mi-fa-so-la-ti-do* scale of Western music. (Though wrong, this idea paved the way for the periodic table.) One hopes the self isn't like phlogiston!

I started by outlining an evolutionary and anatomical framework for understanding many strange neuropsychiatric syndromes. I suggested that these disorders could be regarded as disturbances of consciousness and self-awareness, which are quintessentially human attributes. (It's hard to imagine an ape suffering from Cotard syndrome or God delusions.) Some of the disorders arise from the brain's attempts to deal with intolerable discrepancies among the outputs of different brain modules (as in Capgras syndrome and apotemnophilia) or inconsistencies between internal emotional states and a cognitive appraisal of the external circumstances (as in panic attacks). Other disorders arise from derangement of the normally harmonious interplay of self-awareness and other-awareness that partly involves mirror neurons and their regulation by the frontal lobes.

I began this book with Disraeli's rhetorical question, "Is man an ape or angel?" I discussed the clash between two Victorian scientists, Huxley and Owen, who argued over this issue for three decades. The former emphasized continuity between the brains of apes and humans, and the latter emphasized human uniqueness. With our increasing knowledge of the brain, we need not take sides on this issue anymore. In a sense they were both right, depending on how you ask the question. Aesthetics exists in birds, bees, and butterflies, but the word "art" (with all its cultural connotations) is best applied to humans—even though, as we have seen, art taps into much of the same circuitry in us as in other animals. Humor is exclusively human but laughter isn't. No one would ascribe humor to a hyena or even to an ape that "laughs" when tickled. Rudimentary imitation (such as opening a lock) can be also accomplished by orangutans, but imitation of more demanding skills such as spearing an antelope or hafting a hand axe—and in the wake of such imitation the rapid assimilation and spread of sophisticated culture—is seen only in humans. The kind of imitation humans do may have required, among other things, a more complexly evolved mirror-neuron system than what exists in lower primates. A monkey can learn new things, of course, and retain memory. But a monkey cannot engage in conscious recollection of specific events from its past in order to construct an autobiography, imparting a sense of narrative and meaning to its life.

Morality—and its necessary antecedent "free will," in the sense of envisioning consequences and choosing among them—requires frontal lobe structures that embody values on the basis of which choices are made via the anterior cingulate. This trait is seen only in humans, although simpler forms of empathy are surely present in the great apes.

Complex language, symbol juggling, abstract thought, metaphor, and self-awareness are all almost certainly unique to humans. I have offered some speculation on their evolutionary origins, and suggested also that these functions are mediated partly by specialized structures, such as the angular gyrus and Wernicke's area. The manufacture and deployment of multicomponent tools intended for future use probably requires yet another uniquely human brain structure, the supramarginal gyrus, which branched off from its ancestor (the inferior parietal lobule) in apes. Self-awareness (and the interchangeably used word "consciousnesses") has proved to be an especially elusive quarry, but we have seen how it can be approached through studying the inner mental life of neurological and psychiatric patients. Self-awareness is a trait that not only makes us human but also paradoxically makes us want to be more than merely human. As I said in my BBC Reith Lectures, "Science tells us we are merely beasts, but we don't feel like that. We feel like angels trapped inside the bodies of beasts, forever craving transcendence." That's the essential human predicament in a nutshell.

We have seen that the self consists of many strands, each of which can be unraveled and studied by doing experiments. The stage is now set for understanding how these strands harmonize in our normal day-to-day consciousness. Moreover, treating at least some forms of mental illness as disorders of self might enrich our understanding of them and help us devise new therapies to complement traditional ones.

The real drive to understand the self, though, comes not from the need to develop treatments, but from a more deep-seated urge that we all share: the desire to understand ourselves. Once self-awareness emerged through evolution, it was inevitable that an organism would ask, "Who am I?" Across vast stretches of inhospitable space and immeasurable time, there suddenly emerged a person called Me or I. Where does this person come from? Why here? Why now? You, who are made of stardust, are now standing on a cliff, gazing at the starlit sky pondering your own origins and your place in the cosmos. Perhaps another human stood in that very same spot fifty thousand years ago, asking the very same question. As the mystically inclined, Nobel Prize–winning physicist Erwin Schrödinger once asked, Was he really another person? We wander—to our peril—into metaphysics, but as human beings we cannot avoid doing so.

When informed that their conscious self emerges "simply" from the mindless agitations of atoms and molecules in their brains, people often feel let down, but they shouldn't. Many of the greatest physicists of this century—Werner Heisenberg, Erwin Schrödinger, Wolfgang Pauli, Arthur Eddington, and James Jeans—have pointed out that the basic constituents of matter, such as quanta, are themselves deeply mysterious if not downright spooky, with properties bordering on the metaphysical. So we need not fear that the self might be any less wonderful or awe inspiring for being made of atoms. You can call this sense of awe and perpetual astonishment God, if you like.

Charles Darwin himself was at times ambivalent about these issues:

> I feel most deeply that this whole question of Creation is too profound for human intellect. A dog might as well speculate on the mind of Newton! Let each man hope and believe what he can.

And elsewhere:

> I own that I cannot see as plainly as others do, and as I should wish to do, evidence of design and beneficence on all sides of us. There seems to me too much misery in the world. I cannot persuade myself that a beneficent and omnipotent God would have designedly created the Ichneumonidae [a family of parasitic wasps] with the express intention of their feeding within the living bodies of caterpillars or that a cat should play with mice . . .

On the other hand, I cannot anyhow be contented to view this wonderful universe, and especially the nature of man, and to conclude that everything is the result of brute force.

These statements[1] are pointedly directed against creationists, but Darwin's qualifying remarks are hardly the kind you would expect from the hard-core atheist he is often portrayed to be.

As a scientist, I am one with Darwin, Gould, Pinker, and Dawkins. I have no patience with those who champion intelligent design, at least not in the sense that most people would use that phrase. No one who has watched a woman in labor or a dying child in a leukemia ward could possibly believe that the world was custom crafted for our benefit. Yet as human beings we have to accept—with humility—that the question of ultimate origins will always remain with us, no matter how deeply we understand the brain and the cosmos that it creates.

GLOSSARY

———

Words and terms in italics have their own entries.

AGNOSIA   A rare disorder characterized by an inability to recognize and identify objects and people even though the specific sensory modality (such as vision or hearing) is not defective nor is there any significant loss of memory or intellect.

ALIEN-HAND SYNDROME   The feeling that one's hand is possessed by an uncontrollable outside force resulting in its actual movement. The syndrome usually stems from an injury to the corpus collosum or *anterior cingulate*.

AMES ROOM ILLUSION   A distorted room used to create the optical illusion that a person standing in one corner appears to be a giant while a person standing in another corner appears to be a dwarf.

AMNESIA   A condition in which memory is impaired or lost. Two of the most common forms are anterograde amnesia (the inability to acquire new memories) and retrograde amnesia (the loss of preexisting memories).

AMYGDALA   A structure in the front end of the *temporal lobes* that is an important component of the limbic system. It receives several parallel inputs including two projections arriving from the *fusiform gyrus*. The amygdala helps activate the *sympathetic nervous system* (fight-or-flight responses). The amygdala sends outputs via the *hypothalamus* to trigger appropriate reactions to objects—namely, feeding, fleeing, fighting, and sex. Its affective component (the subjective emotions) partly involves connections with the *frontal lobes*.

ANGULAR GYRUS   A brain area situated in the lower part of the *parietal lobe* near its junction with the *occipital* and *temporal lobes*. It is involved in high-level abstraction and abilities such as reading, writing, arithmetic, left-right discrimination, word representation, the representation of fingers, and possibly also comprehension of metaphor and proverbs. The angular gyrus is possibly unique to humans. It is also probably rich in *mirror neurons* that allow you to see the world from another's point of view spatially and (perhaps) metaphorically—a key ingredient in morality.

ANOSOGNOSIA   A syndrome in which a person who suffers a disability seems unaware of, or denies the existence of, the disability. (*Anosognosia* is Greek for "denial of illness.")

ANTERIOR CINGULATE   A C-shaped ring of cortical tissue abutting and partially encircling the front part of the large bundle of nerve fibers, called the corpus callosum, that link the left and right hemispheres of the brain. The anterior cingulate "lights up" in many—almost too many—brain-imaging studies. This structure is thought to be involved in free will, vigilance, and attention.

APHASIA   A disturbance in language comprehension or production, often as a result of a stroke. There are three main kinds of aphasia: anomia (difficulty finding words), Broca's aphasia (difficulty with grammar, more specifically the deep structure of language), and Wernicke's aphasia (difficulty with comprehension and expression of meaning).

APOTEMNOPHILIA   A neurological disorder in which an otherwise mentally competent person desires to have a healthy limb amputated in order to "feel whole." The old Freudian explanation was that the patient wants a large amputation stump resembling a penis. Also called body integrity identity disorder.

APRAXIA   A neurological condition characterized by an inability to carry out learned purposeful movements despite knowing what is expected and having the physical ability and desire to do so.

ASPERGER SYNDROME   A type of *autism* in which people have normal language skills and cognitive development but have significant problems with social interaction.

ASSOCIATIVE LEARNING   A form of learning in which the mere exposure to two phenomena that always occur together (such as Cinderella and her carriage) leads subsequently to one of the two things spontaneously evoking the memory of the other. Often invoked, incorrectly, as an explanation of *synesthesia*.

AUTISM   One of a group of serious developmental problems called autism spectrum disorders that appear early in life, usually before age three. While symptoms and severity vary, autistic children have problems communicating and interacting with others. The disorder may be related to defects in the mirror-neuron system or the circuits it projects to, although this has yet to be clearly established.

AUTONOMIC NERVOUS SYSTEM   A part of the peripheral nervous system responsible for regulating the activity of internal organs. It includes the *sympathetic* and *parasympathetic nervous systems*. These originate in the *hypothalamus*; the sympathetic component also involves the *insula*.

AXON   The fiber-like extension of a neuron by which the cell sends information to target cells.

BASAL GANGLIA   Clusters of neurons that include the caudate nucleus, the putamen, the globus pallidus, and the substantia nigra. Located deep in the brain, the basal ganglia play an important role in movement, especially control of posture and equilibrium and unconscious adjustments of certain muscles for execution of more voluntary movements regulated by the motor cortex (see *frontal lobe*). The finger and wrist movements for screwing a bolt are mediated by the motor cortex, but adjusting the elbow and shoulder to carry this out requires the basal ganglia. Cell death in the substantia nigra contributes to signs of Parkinson's disease, including a stiff gait and the absence of postural adjustments.

BIPOLAR DISORDER   A psychiatric disorder characterized by wild mood swings. Individuals experience manic periods of high energy and creativity and depressed periods of low energy and sadness. Also called manic depressive disorder.

BLACK BOX   Before the advent of modern imaging technologies in the 1980s and 1990s, there was no way to peer inside the brain, hence it was likened to a black box. (The phrase is borrowed from electrical engineering.) The black-box approach is also one favored by cognitive psychologists and perceptual psychologists, who draw flow diagrams, or charts that indicate purported stages of information processing in the brain without being burdened by knowledge of brain anatomy.

BLINDSIGHT   A condition in some patients who are effectively blind because of damage to the visual cortex but can carry out tasks which would ordinarily appear to be impossible unless they can see the objects. For instance they can point out an object and accurately describe whether a stick is vertical or horizontal, even though they can't consciously perceive the object. The explanation appears to be that visual information travels along two pathways in the brain: the *old pathway* and the *new pathway*. If only the new pathway is damaged, a patient may lose the ability to see an object but still be aware of its location and orientation.

BRAINSTEM   The major route by which the *cerebral hemispheres* send information to and receive information from the spinal cord and peripheral nerves. It also gives rise directly to cranial nerves that go out to muscles of facial expression (frowning, winking, smiling, biting, kissing, pouting, and so forth) and facilitates swallowing and shouting. The brainstem also controls, among other things, respiration and the regulation of heart rhythms.

BROCA'S AREA   The region that is located in the left *frontal lobe* and is responsible for the production of speech that has syntactic structure.

CAPGRAS SYNDROME   A rare syndrome in which the person is convinced that close relatives—usually parents, spouse, children or siblings—are imposters. It may be caused by damage to connections between areas of the brain dealing with face recognition and those handling emotional responses. Someone with Capgras syndrome might recognize the faces of loved ones but not feel the emotional reaction normally associated with that person. Also called Capgras delusion.

CEREBELLUM   An ancient region of the brain that plays an important role in motor control and in some aspects of cognitive functioning. The cerebellum (Latin for "little brain") contributes to the coordination, precision, and accurate timing of movements.

CEREBRAL CORTEX   The outermost layer of the *cerebral hemispheres* of the brain. It is responsible for all forms of high(er)-level functions, including perception, nuanced emotions, abstract thinking, and planning. It is especially well developed in humans and to a lesser extent in dolphins and elephants.

CEREBRAL HEMISPHERES   The two halves of the brain partially specialized for different things—the left hemisphere for speech, writing, language, and calculation; the right hemisphere for spatial abilities, face recognition in vision, and some aspects of music perception (scales rather than rhythm or beat). A speculative conjecture holds that the left hemisphere is the "conformist," trying to make everything fit in order to forge ahead, whereas the right hemisphere is your devil's advocate, or reality check. Freudian *defense mechanisms* probably evolved in the left hemisphere to confer coherence and stability on behavior.

CLASSICAL CONDITIONING   Learning in which a stimulus that naturally produces a specific response (an unconditioned stimulus) is repeatedly paired with a neutral stimulus (a conditioned stimulus). As a result, the conditioned stimulus starts evoking a response similar to that of the unconditioned stimulus. Related to *associative learning*.

COGNITION   The process or processes by which an organism gains knowledge of, or becomes aware of, events or objects in its environment and uses that knowledge for comprehension and problem solving.

COGNITIVE PSYCHOLOGY   The scientific study of information processing in the brain. Cognitive psychologists often do experiments to isolate the stages of information processing. Each stage can be described as a *black box* within which certain specialized computations are performed before the output goes to the next box, so the researcher can construct a flow diagram. The British psychologist Stuart Sutherland defined cognitive psychology as the "ostentatious display of flow diagrams as a substitute for thought."

COGNITIVE NEUROSCIENCE   The discipline that attempts to provide neurological explanations of cognition and perception. The emphasis is on basic science, although there may be clinical spin-offs.

CONE   A primary receptor cell for vision located in the retina. Cones are sensitive to color and used primarily for daytime vision.

COTARD SYNDROME   A disorder in which a patient asserts that he or she is dead, even claiming to smell rotting flesh or worms crawling over the skin (or some other equally absurd delusion). It may be an exaggerated form of the *Capgras syndrome*, in which not just one sensory area (such as face recognition) but all sensory areas are cut off from the limbic system, leading to a complete lack of emotional contact with the world and with oneself.

CROSS-MODAL   Describes interactions across different sensory systems, such as touch, hearing, and vision. If I showed you an unnameable, irregularly shaped object, then blindfolded you and asked you to pick out the object with your hands from a collection of similar objects, you would use cross-modal interactions to do so. These interactions occur especially in the *inferior parietal lobule* (especially the *angular gyrus*) and in certain other structures such as the claustrum (a sheet of cells buried in the sides of the brain that receives inputs from many brain regions) and the *insula*.

DEFENSE MECHANISMS   Term coined by Sigmund and Anna Freud. Information that is potentially threatening to the integrity of one's "ego" is deflected unconsciously by various psychological mechanisms. Examples include repression of unpleasant memories, denial, rationalization, projection, and reaction formation.

DENDRITE   A treelike extension of the neuron cell body. Along with the cell body, it receives information from other neurons.

ELECTROENCEPHALOGRAPHY (EEG)   A measure of the brain's electrical activity in response to sensory stimuli. This is obtained by placing electrodes on the surface of the scalp (or, more rarely, inside the head), repeatedly administering a stimulus, and then using a computer to average the results. The result is an electroencephalogram (also abbreviated EEG).

EPISODIC MEMORY   Memory for specific events from your personal experience.

EXAPTATION   A structure evolved through natural selection for a particular function that becomes subsequently used—and refined through further natural selection—for a completely novel unrelated function. For example, bones of the ear that evolved for amplifying sound were exapted from reptilian jaw bones used for chewing. Computer scientists and evolutionary psychologists find the idea irritating.

EXCITATION   A change in the electrical state of a neuron that is associated with an enhanced probability of action potentials (a train of electrical spikes that occurs when a neuron sends information down an *axon*).

FRONTAL LOBE   One of the four divisions of each *cerebral hemisphere*. (The other three divisions are the *parietal, temporal,* and *occipital lobes*.) The frontal lobes include the motor cortex, which sends commands to muscles on the opposite side of the body; the premotor cortex, which orchestrates these commands; and the prefrontal cortex, which is the seat of morality, judgment, ethics, ambition, personality, character, and other uniquely human attributes.

FUNCTIONAL MAGNETIC RESONANCE IMAGINING (FMRI)   A technique—in which the baseline activity of the brain (with the person doing nothing) is subtracted from the activity during task performance—that determines which anatomical regions of the brain are active when a person engages in a specific motor, perceptual, or cognitive task. For example, subtracting a German brain's activity from that of an Englishman might reveal the "humor center" of the brain.

FUSIFORM GYRUS A gyrus near the bottom inner part of the *temporal lobe* that has subdivisions specialized for recognizing color, faces, and other objects.

GALVANIC SKIN RESPONSE (GSR) When you see or hear something exciting or significant (such as a snake, a mate, prey, or a burglar), your *hypothalamus* is activated; this causes you to sweat, which changes your skin's electrical resistance. Measuring this resistance provides an objective measure of emotional arousal. Also called skin conductance response (SCR).

HEMISPHERES See *Cerebral hemispheres*.

HIPPOCAMPUS A seahorse-shaped structure located within the *temporal lobes*. It functions in memory, especially the acquisition of new memories.

HOMININS Members of the Hominini tribe, a taxonomic group recently reclassified to include chimpanzees (*Pan*), human and extinct protohuman species (*Homo*), and some ancestral species with a mix of human and apelike features (such as *Australopithecus*). The hominins are thought to have diverged from the gorillas (Gorillini tribe).

HORMONES Chemical messengers secreted by endocrine glands to regulate the activity of target cells. They play a role in sexual development, calcium and bone metabolism, growth, and many other activities.

"HOW" STREAM The pathway from the visual cortex to the *parietal lobe* that guides muscle twitch sequences that determine how you move your arm or leg in relation to your body and environment. You need this pathway to accurately reach for an object, and for grasping, pulling, pushing, and other types of object manipulation. To be distinguished from the *"what" stream* in the *temporal lobes*. Both "what" and "how" streams diverge from the *new pathway*, whereas the *old pathway* starts from the superior colliculus and projects onto the *parietal lobe*, converging on it with the "how" stream. Also called pathway 1.

HYPOTHALAMUS A complex brain structure composed of many cell clusters with various functions. These include emotions, regulating the activities of internal organs, monitoring information from the *autonomic nervous system*, and controlling the pituitary gland.

INFERIOR PARIETAL LOBULE (IPL) A cortical region in the middle part of the *parietal lobe*, just below the *superior parietal lobule*. It became several times bigger in humans compared with apes, especially on the left. In humans the IPL split into two entirely new structures: the *supramarginal gyrus* (on top), which is involved in skilled actions such as tool use; and the *angular gyrus*, involved in arithmetic, reading, naming, writing, and possibly also in metaphorical thinking.

INHIBITION   In reference to neurons, a synaptic message that prevents the recipient cell from firing.

INSULA   An island of cortex buried in the folds on the side of the brain, divided into anterior, middle, and posterior sections, each of which has many subdivisions. The insula receives sensory input from the viscera (internal organs) as well as taste, smell, and pain inputs. It also gets inputs from the somatosensory cortex (touch, muscle and joint, and position sense) and the vestibular system (organs of balance in the ear). Through these interactions, the insula helps construct a person's "gut level," but not fully articulated, sense of a rudimentary "body image." In addition, the insula has *mirror neurons* that both detect disgusting facial expressions and express disgust toward unpleasant food and smells. The insula is connected via the parabrachial nucleus to the *amygdala* and the *anterior cingulate*.

KORO   A disorder that purportedly afflicts young Asian men who develop the delusion that their penises are shrinking and may eventually drop off. The converse of this syndrome—aging Caucasian men who develop the delusion that their penises are expanding—is much more common (as noted by our colleague Stuart Anstis). But it has not been officially given a name.

LIMBIC SYSTEM   A group of brain structures—including the *amygdala*, *anterior cingulate*, fornix, *hypothalamus*, hippocampus, and septum—that work to help regulate emotion.

MIRROR NEURONS   Neurons that were originally identified in the *frontal lobes* of monkeys (in a region homologous to the Broca's language area in humans). The neurons fire when the monkey reaches for an object or merely watches another monkey start to do the same thing, thereby simulating the other monkey's intentions, or reading its mind. Mirror neurons have also been found for touch; that is, sensory touch mirror neurons fire in a person when she is touched and also when she watches another person being stroked. Mirror neurons also exist for making and recognizing facial expressions (in the *insula*) and for pain "empathy" (in the *anterior cingulate*).

MOTOR NEURON   A neuron that carries information from the central nervous system to a muscle. Also loosely used to include motor-command neurons, which program a sequence of muscle contractions for actions.

MU WAVES   Some specific brain waves that are affected in *autism*. Mu waves may or may not be an index of mirror-neuron function, but they get suppressed both during action performance and action observation, suggesting a close link with the mirror-neuron system.

NATURAL SELECTION   Sexual reproduction results in shuffling genes into novel combinations. Nonlethal mutations arise spontaneously. Those mutations or gene combinations that make some species better adapted to their current environment are the ones that survive more often because the parents survive and reproduce more often. The term is used in opposition to creationism (which holds that all species were created at once) and in contrast to artificial selection by humans to improve livestock and plants. Natural selection is not synonymous with evolution; it is a mechanism that drives evolutionary change.

NEURON   Nerve cell. It is specialized for the reception and transmission of information, and is characterized by long fibrous projections called *axons* and shorter, branchlike projections called *dendrites*.

NEUROTRANSMITTER   A chemical released by *neurons* at a *synapse* for the purpose of relaying information via receptors.

NEW PATHWAY   Passes information from visual areas to the *temporal lobes*, via the *fusiform gyrus*, to help with the recognition of objects as well as with their meaning and emotional significance. The new pathway diverges into the *"what" stream* and the *"how" stream*.

OCCIPITAL LOBE   One of the four subdivisions (the others being *frontal*, *temporal*, and *parietal lobes*) of each *cerebral hemisphere*. The occipital lobes play a role in vision.

OLD PATHWAY   The older of two main pathways in the brain for visual processing. This pathway goes from the superior colliculus (a primitive brain structure in the brain stem) via the *thalamus* to the *parietal lobes*. The old pathway converges on the *"how" stream* to help move eyes and hands toward objects even when the person does not consciously recognize them. The old pathway is involved in mediating *blindsight*, when the *new pathway* alone is damaged.

PARASYMPATHETIC NERVOUS SYSTEM   A branch of the *autonomic nervous system* concerned with the conservation of the body's energy and resources during relaxed states. This system causes pupils to constrict, blood to be diverted to the gut for leisurely digestion, and heart rate and blood pressure to fall in order to diminish the load on the heart.

PARIETAL LOBE   One of the four subdivisions (the others being *frontal*, *temporal*, and *occipital lobes*) of each *cerebral hemisphere*. A portion of the parietal lobe in the right hemisphere plays a role in sensory attention and body image, while the left parietal is involved in skilled movements and in aspects of language (object naming, reading, and writing). Ordinarily the parietal lobes have no role in the comprehension of language, which happens in the *temporal lobes*.

PERIPHERAL NERVOUS SYSTEM   A division of the nervous system consisting of all nerves not part of the central nervous system (in other words, not part of the brain or spinal cord).

PHANTOM LIMB   The perceived existence of a limb lost through accident or amputation.

PONS   A part of the stalk on which the brain sits. Together with other brain structures, it controls respiration and regulates heart rhythms. The pons is a major route by which the *cerebral hemispheres* send information to and receive information from the spinal cord and the *peripheral nervous system*.

POPOUT TEST   A test visual psychologists use to determine whether or not a particular visual feature is extracted early in visual processing. For example, a single vertical line will "pop out" in a matrix of horizontal lines. A single blue dot will "pop out" against a collection of green dots. There are cells tuned to orientation and color in low-level (early) visual processing. On the other hand, a female face will not pop out from a matrix of male faces, because cells responding to the sex of a face occur at a much higher level (later) in visual processing.

PREFRONTAL CORTEX   See *Frontal lobe*.

PROCEDURAL MEMORY   Memory for skills (such as learning to ride a bicycle), as opposed to declarative memory, which is storage of specific information that can be consciously retrieved (such as Paris being the capital of France).

PROTOLANGUAGE   Presumed early stages of language evolution that may have been present in our ancestors. It can convey meaning by stringing together words in the right order (for example, "Tarzan kill ape") but has no *syntax*. The word was introduced by Derek Bickerton of the University of Hawaii.

QUALIA   Subjective sensations. (Singular: quale.)

RECEPTOR CELL   Specialized sensory cells designed to pick up and transmit sensory information.

RECEPTOR MOLECULE   A specific molecule on the surface or inside of a cell with a characteristic chemical and physical structure. Many *neurotransmitters* and *hormones* exert their effects by binding to receptors on cells. For example, insulin released by islet cells in the pancreas acts on receptors on target cells to facilitate glucose intake by the cells.

REDUCTIONISM   One of the most successful methods used by scientists to understand the world. It only makes the innocuous claim that the whole can be explained in terms of *lawful interactions between* (not simply the sum of) the component parts. For example, heredity was "reduced" to the genetic code and complementarity of DNA strands. Reducing a complex phenomenon to its component parts does not negate the existence of the complex phenomenon. For ease of human comprehension, complex phenomena can also be described in terms of lawful interactions between causes and effects that are at the "same level" of description as the phenomenon (such as when your doctor tells you, "Your illness is caused by a reduction in vitality"), but this rarely gets us very far. Many psychologists and even some biologists resent reductionism, claiming, for example, that you cannot explain sperm if you know only its molecular constituents but not about sex. Conversely, many neuroscientists are mesmerized by reductionism for its own sake, quite independent of whether it helps explain higher-level phenomena.

REUPTAKE   A process by which released *neurotransmitters* are absorbed at the *synapse* for subsequent reuse.

SEIZURES   A brief paroxysmal discharge of a small group of hyperexcitable brain cells that results in a loss of consciousness (grand mal seizure) or disturbances in consciousness, emotions, and behavior without loss of consciousness (*temporal lobe epilepsy*). Petit mal seizures are seen in children as a brief "absence." Such seizures are completely benign and the child almost always outgrows them. Grand mal is often familial and begins in the late teens.

SELF-OTHER DISTINCTION   The ability to experience yourself as a self-conscious being whose inner world is separate from the inner worlds of others. Such separateness does not imply selfishness or lack of empathy for others, although it may confer a propensity in that direction. Disturbances of self-other distinctions, as we have argued in Chapter 9, may underlie many strange types of neuropsychiatric illness.

SEMANTIC MEMORY   Memory for the meaning of an object, event, or concept. Semantic memory for a pig's appearance would include a cluster of associations: ham, bacon, *oink oink*, mud, obesity, Porky the Pig cartoons, and so on. The cluster is bound together by the name "pig." But our research on patients with anomia and Wernicke's *aphasia* suggests that the name is not merely another association; it is a key that opens a treasury of meanings and a handle that can be used for juggling the object or concept around in accordance with certain rules, such as those required for thinking. I have noticed that if an intelligent person with anomia or Wernicke's *aphasia*, who can recognize objects but names them incorrectly, initially misnames an object (such as calling a paintbrush a comb), she often proceeds to use it as a comb. She is forced to head up the wrong semantic path by the mere act of mislabeling the object. Language, visual recognition, and thought are more closely interlinked than we realize.

SEROTONIN   A monoamine *neurotransmitter* believed to play many roles including, but not limited to, temperature regulation, sensory perception, and inducing the onset of sleep. Neurons using serotonin as a transmitter are found in the brain and in the gut. A number of antidepressant drugs are used to target serotonin systems in the brain.

"SO WHAT" STREAM   Not well defined or anatomically delineated, this pathway involves parts of the *temporal lobes* concerned with the biological significance of what you are looking at. Includes connections with the *superior temporal sulcus*, the *amygdala*, and the *insula*. Also called pathway 3.

STIMULUS   A highly specific environmental event capable of being detected by sensory receptors.

STROKE   An impeded blood supply to the brain, caused by a blood clot forming in a blood vessel, the rupture of a blood vessel wall, or an obstruction of flow caused by a clot or fat globule released from injury elsewhere. Deprived of oxygen (which is carried by the blood), nerve cells in the affected area cannot function and thus die, leaving the part of the body controlled by these cells also unable to function. A major cause of death in the West, stroke can result in loss of consciousness and brain function, and in death. During the last decade, studies have shown that feedback from a mirror can accelerate recovery of sensory and motor function in the arm in some stroke patients.

SUPERIOR PARIETAL LOBULE (SPL)   A brain region that lies near the top of the *parietal lobe*. The right SPL is partially concerned with creating one's body image using inputs from vision and area S2 (joint and muscle sense). The *inferior parietal lobule* is also involved in this function.

SUPERIOR TEMPORAL SULCUS (STS)   The topmost of two horizontal furrows, or sulci, in the *temporal lobes*. The STS has cells that respond to changing facial expressions, biological movements such as gait, and other biologically salient inputs. The STS sends its output to the *amygdala*.

SUPRAMARGINAL GYRUS   An evolutionarily recent gyrus that split off from the *inferior parietal lobule*. The supramarginal gyrus is involved in the contemplation and execution of skilled or semiskilled movements. It is unique to humans, and damage to it leads to *apraxia*.

SYMPATHETIC NERVOUS SYSTEM   A branch of the *autonomic nervous system*, responsible for mobilizing the body's energy and resources during times of stress and arousal. It does this by regulating temperature as well as increasing blood pressure, heart rate, and sweating in anticipation of exertion.

SYNAPSE   A gap between two neurons that functions as the site of information transfer from one neuron to another.

SYNESTHESIA   A condition in which a person literally perceives something in a sense besides the sense being stimulated, such as tasting shapes or seeing colors in sounds or numbers. Synesthesia is not just a way of describing experiences as a writer might use metaphors; some synesthetes actually experience the sensations.

SYNTAX   Word order that enables compact representation of complex meaning for communicative intent; loosely synonymous with grammar. In the sentence "The man who hit John went to the car," we recognize instantly that "the man" went to the car, not John. Without syntax we could not arrive at this conclusion.

TEMPORAL LOBE   One of the four major subdivisions (the others being *frontal, parietal*, and *occipital lobes*) of each *cerebral hemisphere*. The temporal lobe functions in perception of sounds, comprehension of language, visual perception of faces and objects, acquisition of new memories, and emotional feelings and behavior.

TEMPORAL LOBE EPILEPSY (TLE)   *Seizures* confined mainly to the *temporal lobes* and sometimes the *anterior cingulate*. TLE may produce a heightened sense of self and has been linked to religious or spiritual experiences. The person may undergo striking personality changes and/or become obsessed with abstract thoughts. People with TLE have a tendency to ascribe deep significance to everything around them, including themselves. One explanation is that repeated seizures may strengthen the connections between two areas of the brain: the temporal cortex and the *amygdala*. Interestingly, people with TLE tend to be humorless, a characteristic also seen in seizure-free religious people.

THALAMUS   A structure consisting of two egg-shaped masses of nerve tissue, each about the size of a walnut, deep within the brain. The thalamus is the key "relay station" for sensory information, transmitting and amplifying only information of particular importance from the mass of signals entering the brain.

THEORY OF MIND   The idea that humans and some higher primates can construct a model in their brains of the thoughts and intentions of other people. The more accurate the model, the more accurately and rapidly the person can predict the other person's thoughts, beliefs, and actions. The idea is that there are specialized brain circuits in human (and some apes') brains that allow for theory of mind. Uta Frith and Simon Baron-Cohen have suggested that autistic children may have a deficient theory of mind, which complements our view that a dysfunction of *mirror neurons* or their targets may underlie *autism*.

WERNICKE'S AREA   A brain region responsible for the comprehension of language and the production of meaningful speech and writing.

"WHAT" STREAM   The *temporal lobe* pathway concerned with recognizing objects and their meaning and significance. Also called pathway 2. See also *new pathway* and *"how" stream*.

NOTES

PREFACE

1. I have since learned that this observation has resurfaced from time to time, but for obscure reasons isn't part of mainstream oncology research. See, for example, Havas (1990), Kolmel et al. (1991), or Tang et al. (1991).

INTRODUCTION: NO MERE APE

1. This basic method for studying the brain is how the whole field of behavioral neurology got started back in the nineteenth century. The major difference between then and now is that in those days there was no brain imaging. The doctor had to wait around for a decade or three for the patient to die, then dissect his brain.

2. In contrast to the hobbits, African pigmies, who are also extraordinarily short, are modern humans in every way, from their DNA right on up through their brains, which are the same size as those of all other human groups.

CHAPTER 2   SEEING AND KNOWING

1. Strictly speaking, the fact that octopuses and humans both have complex eyes is probably not an example of true convergent evolution (unlike the wings of birds, bats, and pterosaurs). The same master control genes are at work in "primitive" eyes as in our own. Evolution sometimes reuses genes that have been stored away in the attic.

2. John was originally studied by Glyn Humphreys and Jane Riddoch, who wrote a beautiful monograph about him: *To See but Not to See: A Case Study of Visual Agnosia* (Humphreys & Riddoch, 1998). What follows is not a literal transcript but for the most part preserves the patient's original comments. John suffered from an embolus following appendectomy as indicated, but the circumstances leading up to the appendectomy are a reenactment of the way things might have occurred during a routine diagnosis of appendicitis. (As mentioned in the Preface, to preserve patient confidentiality, throughout the book I often use fictitious names for patients and alter circumstances of hospital admission that are not relevant to the neurological symptoms.)

3. Can you see the Dalmatian dog in Figure 2.7?

4. The distinction between the "how" and "what" pathways is based on the pioneering work of Leslie Ungerleider and Mortimer Mishkin working at the National Institutes of Health. Pathways 1 and 2 ("how" and "what") are clearly defined anatomically. Pathway 3 (dubbed "so what," or the emotional pathway) is currently considered a functional pathway, as inferred from physiological and brain lesion studies (such as studies on the double dissociation between the Capgras delusion and prosopagnosia; see Chapter 9).

5. Joe LeDoux has discovered there is also a small, ultra-shortcut pathway from the thalamus (and possibly the fusiform gyrus) directly to the amygdala in rats, and quite possibly in primates. But we won't concern ourselves with that here. The details of neuroanatomy are unfortunately far messier than we would like, but that shouldn't stop us from looking for overall patterns of functional connectedness, as we've been doing.

6. This idea about the Capgras syndrome was proposed independently of us by Hadyn Ellis and Andrew Young. However, they postulate a preserved "how" stream (pathway 1) and combined damage to the two components of the "what" stream (pathways 2 plus 3), whereas we postulate a selective damage to the emotional stream (pathway 3) alone with sparing of pathway 2.

CHAPTER 3   LOUD COLORS AND HOT BABES: SYNESTHESIA

1. Several experiments point to the same conclusion. In our very first paper on synesthesia, published in 2001 in the *Proceedings of the Royal Society of London*, Ed Hubbard and I noted that in some synesthetes the strength of color induced seemed to depend not just on the number but on where in the visual field it was presented (Ramachandran & Hubbard, 2001a). When the subject looked straight, then numbers or letters presented off to one side (but made larger to be equally visible) seemed less vividly colored than ones presented in central vision. This, in spite of the fact that they were equally identifiable as particular numbers and in spite of the fact that real colors are just as vividly visible in off-axis (peripheral) vision. Again, these results exclude high-level memory associations as the source of synesthesia. Visual memories are spatially invariant. By that I mean that when you learn something in one region of your visual field—recognizing a particular face, for instance—you can recognize the face presented in a completely new visual location. The fact that the evoked colors are *different* in different regions argues strongly against memory associations. (I should add that even for the same eccentricity the color is sometimes different for left and right halves of the visual field; possibly because the cross-activation is more pronounced in one hemisphere than the other.)

2. This basic result—that the 2s are more quickly segregated from the 5s in synesthetes than in nonsynesthetes—has been confirmed by other scientists, especially Randolph Blake and Jamie Ward. In a meticulously controlled experiment, Ward and his colleagues found that synesthetes as a group are significantly better than control subjects at seeing the embedded shape made of 2s. Intriguingly, some of them perceived the shape even before any color was evoked! This lends credibility to our early cross-activation model; it's possible that during brief presentations the colors are evoked sufficiently strongly to permit segregation to occur but not strongly enough to evoke consciously perceived colors.

3. In lower, "projection," synesthetes there are several lines of evidence (in addition to segregation) supporting the low-level perceptual cross-activation model as opposed to the notion that synesthesia is based entirely on high-level associative learning and memories:

    (a) In some synesthetes, different parts of a single number or letter are seen as colored differently. (For example, the V part of an M might be colored red, whereas the vertical lines might be green.)

    Soon after the popout/segregation experiment had been done, I noticed something strange in one of the many synesthetes we had been recruiting. He saw numbers as being colored—nothing unusual so far—but what surprised me was his claim that some of the numbers (for example, 8) had different portions colored differently. To make sure he wasn't making this up, we showed him the same numbers a few months later—without letting him know ahead of time that he would be retested. The new drawing he produced was virtually identical to the first, making it unlikely that he was fibbing.

    This observation provides further evidence that, at least in some synesthetes, the colors should be seen as emerging from (to use a computer metaphor) a glitch in neural hardware rather than from an exaggeration of memories or metaphors (a software glitch) Associative learning cannot explain this observation; for example, we don't play with multicolored magnets. On the other hand, there may be "form primitives" such as line orientation, angles, and curves that get linked to color neurons that execute an earlier stage of form processing within the fusiform than the one at which full-fledged graphemes are assembled.

    (b) As previously noted, in some synesthetes the evoked color becomes less vivid when the number is viewed off-axis (in peripheral vision). This probably reflects the greater emphasis on color in central vision (Ramachandran & Hubbard, 2001a; Brang & Ramachandran, 2010). In some of these synesthetes the color is also more saturated in one visual field (left or right) relative to the other. Neither of these observations supports the high-level associative learning model for synesthesia.

(c) An actual increase in anatomical connectivity within the fusiform area of lower synesthetes has been observed by Rouw and Scholte (2007) using diffusion tensor imaging.

(d) The synesthetically evoked color can provide an input to apparent motion perception (Ramachandran & Hubbard, 2002; Kim, Blake, Palmeri, 2006; Ramachandran & Azoulai, 2006).

(e) If you have one type of synesthesia, then you are more likely to have a second unrelated one as well. This supports my "increased cross activation model" of synesthesia; with the mutated gene being more prominently expressed in certain brain regions (in addition to making some synesthetes more creative).

(f) The existence of color-blind (strictly speaking, color anomalous) synesthetes who can see colors in numbers that they can't see in the real world. The subject couldn't have learned such associations.

(g) Ed Hubbard and I showed in 2004 that letters that are similar in shape (e.g., curvy rather than angular) tend to evoke similar colors in "lower" synesthetes. This shows that certain figural primitives that define the letters cross-activate colors even before they are fully processed. We suggested that the technique might be used to map an abstract color-space in a systematic manner onto form-space. More recently David Brang and I confirmed this using brain imaging (MEG or magnetoencephalography) in collaboration with Ming Xiong Huang, Roland Lee, and Tao Song.

Taken collectively these observations strongly support the sensory cross-activation model. This is not to deny that learned associations and high-level rules of cross-domain mapping are not also involved (see Notes 8 and 9 for this chapter). Indeed, synesthesia may help us discover such rules.

4. The model of cross-activation—either through disinhibition (a loss or lessening of inhibition) of back projections, or through sprouting—can also explain many forms of "acquired" synesthesia that we have discovered. One blind patient with retinitis pigmentosa whom we studied (Armel and Ramachandran, 1999) vividly experienced visual phosphenes (including visual graphemes) when his fingers were touched with a pencil or when he was reading Braille. (We ruled out confabulation by measuring thresholds and demonstrating their stability across several weeks; there is no way he could have memorized the thresholds.) A second blind patient, whom I tested with my student Shai Azoulai, could quite literally see his hand when he waved it in front of his eyes, even in complete darkness. We suggest that this is caused either by hyperactive back projections or by disinhibition caused by visual loss, so that the moving hand is not merely felt but is also seen. Cells with multimodal receptive fields in the parietal lobes may also be involved in mediating this phenomenon (Ramachandran and Azoulai, 2004).

5. Although synesthesia often involves adjacent brain areas (an example is grapheme-color synesthesia in the fusiform), it doesn't have to. Even far-flung brain regions, after all, may have preexisting connections that could be amplified (through disinhibition, say). Statistically speaking, however, adjacent brain areas tend to be more "cross-wired" to begin with, so synesthesia is likely to involve those more often.

6. The link between synesthesia and metaphor has already been alluded to. The nature of the link remains elusive given that synesthesia involves arbitrarily connecting two unrelated things (such as color and number), whereas in metaphor there is a nonarbitrary conceptual connection between two things (for example, Juliet and the sun).

    One potential solution to this problem emerged from a conversation I had with the eminent polymath Jaron Lanier: We realized that any given word has only a *finite* set of strong, first-order associations (sun = warm, nurturing, radiant, bright) surrounded by a penumbra of weaker, second-order associations (sun = yellow, flowers, beach) and third- and fourth-order associations that fade way like an echo. It is the overlapping region between two halos of associations that forms the basis of metaphor. (In our example of Juliet and the sun, this overlap derives from observations that both are radiant, warm, and nurturing). Such overlap in halos of associations exists in all of us, but the overlaps are larger and stronger in synesthetes because their the cross-activation gene produces larger penumbras of associations.

    In this formulation, synesthesia is not synonymous with metaphor, but the gene that produces synesthesia confers a propensity toward metaphor. A side effect of this may be that associations that are only vaguely felt in all of us (for example, masculine or feminine letters, or good and bad shapes produced by subliminal associations) become more explicitly manifest in synesthetes, a prediction that can be tested experimentally. For instance, most people consider certain female names (Julie, Cindy, Vanessa, Jennifer, Felicia, and so on) to be "sexier" than others (such as Martha and Ingrid). Even though we may not be consciously aware of it, this may be because saying the former involves pouting and other tongue and lip movements with unconscious sexual overtones. The same argument would explain why the French language is often thought of as being more sexy than German. (Compare *Busten-halten* with *brassière*.) It might be interesting to see if these spontaneously emerging tendencies and classifications are more pronounced in synesthetes.

    Finally, my student David Brang and I showed that completely new associations between arbitrary new shapes and colors are also learned more readily by synesthetes.

    Taken collectively, these results show that the different forms of synesthesia span the whole spectrum from sensation to cognition, and indeed this is precisely why synesthesia is so interesting to study.

Another familiar yet intriguing kind of visual metaphor, where meaning resonates with form, is the use (in advertising, for example) of type that mirrors the meaning of the word; for example, using tilted letters to print "tilt," and wiggly lines to print "fear," "cold," or "shiver." This form of metaphor hasn't yet been studied experimentally.

7. Effects similar to this were originally studied by Heinz Werner, although he didn't put it in the broader context of language evolution.

8. We have observed that chains of associations, which would normally evoke only memories in normal individuals, would sometimes seem to evoke qualia-laden sense impressions in some higher synesthetes. So the merely metaphorical can become quite literal. For example, *R* is red and red is hot so *R* is hot, and so forth. One wonders whether the hyperconnectivity (either the sprouting or disinhibition) has affected back projections between different areas in the neural hierarchy in these subjects. This would also explain an observation David Brang and I made—that eidetic imagery (photographic memory) is more common in synesthetes. (Back projections are thought to be involved in visual imagery.)

9. The introspections of some higher synesthetes are truly bewildering in their complexity; as they go completely "open loop." Here is a quotation from one of them: "Most men are shades of blue. Women are more colorful. Because people and names both have color associations, the two don't necessarily match." Such remarks imply that any simple phrenological model of synesthesia is bound to be incomplete, although it is not a bad place to start.

In doing science one is often forced to choose between providing precise answers to boring (or trivial) questions such as, How many cones are there in the human eye? or vague answers to big questions such as, What is consciousness? or, What is a metaphor? Fortunately, every now and then we get a precise answer to a big question and hit the jackpot (like DNA being the answer to the riddle of heredity). So far, synesthesia seems to lie halfway between those two extremes.

10. For up-to-date information, see the entry "Synesthesia," by David Brang and me, at *Scholarpedia* (www.scholarpedia.org/article/Synesthesia). *Scholarpedia* is an open-access online encyclopedia written and peer-reviewed by scholars from around the world.

CHAPTER 4   THE NEURONS THAT SHAPED CIVILIZATION

1. A young orangutan in the London zoo once watched Darwin play a harmonica, grabbed it from him, and started to mime him; Darwin had

already been thinking of the imitative capacities of apes in the nineteenth century.

2. Since their original discovery, the concept of mirror neurons has been confirmed repeatedly in experiments and has had tremendous heuristic value in our understanding the interface between structure and function in the brain. But it has also been challenged on various grounds. I will list the objections and reply to each.

    (a) *"Mirroritis": There is a great deal of media hype surrounding the mirror-neuron system (MNS), with anything and everything being attributed to them.* This is true, but the existence of hype doesn't by itself negate the value of a discovery.

    (b) *The evidence for their existence in humans is unconvincing.* This criticism seems odd to me given that we are closely related to monkeys; the default assumption should be that human mirror neurons *do* exist. Furthermore, Marco Iacoboni has shown their presence by directly recording from nerve cells in human patients (Iacoboni & Dapretto, 2006).

    (c) *If such a system exists, why isn't there a neurological syndrome in which damage to a small region leads to difficulty in BOTH performing and miming skilled or semiskilled actions (such as combing your hair or hammering a nail) AND recognizing the same action performed by someone else?* Answer: Such a syndrome does exist, although most psychologists are unaware of it. It is called ideational apraxia and it's seen after damage to the left supramarginal gyrus. Mirror neurons have been shown to exist in this region.

    (d) *The antireductionist stance: "Mirror neurons" is just a sexy phrase synonymous with what psychologists have long called "theory of mind." There's nothing new about them.* This argument confounds metaphor with mechanism: It's like saying that, since we know what the phrase "passage of time" means, there is no need to understand how clocks work. Or that, since we already knew Mendel's laws of heredity during the first half of the twentieth century, understanding DNA structure and function would have been superfluous. Analogously, the idea of mirror neurons doesn't negate the concept of theory of mind. On the contrary, the two concepts complement each other and allow us to home in on the underlying neural circuitry.

    This power of having a mechanism to work with can be illustrated with many examples; here are three: In the 1960s, John Pettigrew, Peter Bishop, Colin Blakemore, Horace Barlow, David Hubel, and Torsten Wiesel discovered disparity-detecting neurons in the visual cortex; this finding alone provides an explanation for stereoscopic vision. Second, the discovery that the hippocampus is involved in memory allowed Eric Kandel to discover long-term potentiation (LTP), one of the key mechanisms of memory storage. And finally, one could argue that more was learned about memory in five years of research by Brenda Milner on the single patient "HM," who had hippocampal damage, than in the previous hundred years of purely psychological approaches to memory. The falsely constructed antithesis between reductionist and holistic views of brain function is detrimental to science, something I discuss at length in Note 16 of Chapter 9.

(e) *The MNS is not a dedicated set of hardwired neural circuits; it may be constructed through associative learning. For instance, every time you move your hand, there is activation of motor-command neurons, with simultaneous activation of visual neurons by the appearance of the moving hand. By Hebb's rule, such repeated coactivations will eventually result in the visual appearance itself triggering these motor neurons, so that they become mirror neurons.*

I have two response to this criticism: First, *even if* the MNS is set up partially through learning, that wouldn't diminish its importance. The question of how the system works is logically orthogonal to how it is set up (as already mentioned under point d above). Second, if this criticism were true, why wouldn't *all* the motor-command neurons become mirror neurons through associative learning? Why only 20 percent? One way to settle this would be to see if there are touch mirror neurons for the back of your head that you have never seen. Since you don't often touch the back of your head or see the back of it being touched, you aren't likely to construct an internal mental model of the back of your head in order to deduce that it's being touched. So you should have far fewer mirror neurons, if any, on this part of your body.

3.   The basic idea of the coevolution between genes and culture isn't new. Yet my claim that a sophisticated mirror-neuron system—conferring an ability to imitate complex actions—was a turning point in the emergence of civilization might be construed as an overstatement. So let's see how the events may have played out.

Assume that a large population of early hominins (such as *Homo erectus* or early *H. sapiens*) had some degree of genetic variation in innate creative talent. If one rare individual through his or her special intellectual gifts had invented something useful, then without the concomitant emergence of sophisticated imitative ability among peers (which requires adopting the other's point of view and "reading" that person's intentions), the invention would have died with the inventor. But as soon as the ability to imitate emerged, such one-of-a-kind innovations (including "accidental" ones) would have spread rapidly through the population, both horizontally through kin and vertically through offspring. Then, if any new "innovative ability" mutation later appeared in another individual, she could instantly capitalize on the preexisting inventions in novel ways, leading to the selection and stabilization of the "innovatability" gene. The process would have spread exponentially, setting up an avalanche of innovations that transforms evolutionary change from Darwinian to Lamarckian, culminating in modern civilized humans. Thus the great leap forward was indeed propelled by genetically selected circuits, but ironically the circuits were specialized for learnability—that is, for liberating us from genes! Indeed, cultural diversity is so vast in modern humans that there is probably a greater difference in mental quality and behavior between a university professor and (say) a Texan cowboy (or president) than between the latter and early *H. sapiens*. Not only is the human brain phylogenetically unique as a whole, but the "brain" of each different culture is unique (through "nurture")—much more so than in any other animal.

CHAPTER 5   WHERE IS STEVEN? THE RIDDLE OF AUTISM

1. Another way of testing the mirror-neuron hypothesis would be to see if autistic children do not show unconscious subvocalization when listening to others talking. (Laura Case and I are testing this.)

2. Many studies have confirmed my original observation (made with Lindsay Oberman, Eric Altschuler, and Jaime Pineda) of a dysfunctional mirror-neuron system (MNS) in autism (which we accomplished by using mu-wave suppression and fMRI). There is an fMRI study, however, claiming that in one specific brain region (the ventral premotor area, or Broca's area), autistic children have normal mirror-neuron-like activity. Even if we accept this observation at face value (despite the inherent limitations of fMRI), my theoretical reasons for postulating such a dysfunction will still stand. More important, such observations highlight the fact that the MNS is composed of many far-flung subsystems in the brain that are interconnected for a common function: action and observation. (As an analogy, consider the lymphatic system of the body, which is distributed throughout the body but is functionally a distinct system.)

    It is also possible that this part of the MNS itself is normal but its projections or recipient zones in the brain are abnormal. The net result would be the same kind of dysfunction that I originally suggested. In another analogy, consider the fact that diabetes is fundamentally a disturbance of carbohydrate metabolism; no one disputes that. While it is sometimes caused by damage to the pancreatic islet cells, causing a reduction of insulin and an elevation of blood glucose, it can also be caused by a reduction of insulin receptors on cell surfaces throughout the body. This would produce the same syndrome as diabetes *without* damage to the islets (for islets in the pancreas, think "mirror neurons in the brain's premotor area called F5"), but the logic of the original argument is unaffected.

    Having said all this, let me emphasize that the evidence for MNS dysfunction in autism is, at this point, compelling but not conclusive.

3. The treatments I have proposed for autism in this chapter were inspired in part by the mirror-neuron hypothesis. But their plausibility does not in itself depend on the hypothesis; they would be interesting to try anyway.

4. To further test the mirror-neuron hypothesis of autism, it would be interesting to monitor the activity of the mylohyoid muscle and vocal cords to determine whether autistic children do not show unconscious subvocalization when listening to others talking (unlike normal children, who do). This might provide an early diagnostic tool.

CHAPTER 6   THE POWER OF BABBLE: THE EVOLUTION OF LANGUAGE

1. This approach was pioneered by Brent Berlin. For cross-cultural studies similar to Berlin's, see Nuckolls (1999).
2. The gestural theory of language origins is also supported by several other ingenious arguments. See Corballis (2009).
3. Even though Wernicke's area was discovered more than a century ago, we know very little about how it works. One of our main questions in this chapter has been, What aspects of thought require Wernicke's language area? In collaboration with Laura Case, Shai Azoulai, and Elizabeth Seckel, I examined two patients (LC and KC) on whom I did several experiments (in addition to the ones described in the chapter); here is a brief description of these and other casual observations that are revealing:

    (a) LC was shown two boxes: one with a cookie, one without. A student volunteer entered the room and looked at each box expectantly, hoping to open the one with the cookie. I had previously winked to the patient, gesturing him to "lie." Without hesitation LC pointed out the empty box to the student. (KC responded to this situation the same way.) This experiment shows you don't need language for a theory-of-mind task.

    (b) KC had a sense of humor, laughing at nonverbal Gary Larson cartoons and playing a practical joke on me.

    (c) Both KC and LC could play a reasonable game of chess and tic-tac-toe, implying that they have at least a tacit knowledge of if-then conditionals.

    (d) Both could understand visual analogy (for example, airplane is to bird as submarine is to fish) when probed nonverbally using pictorial multiple choice.

    (e) Both could be trained to use symbols designating the abstract idea "similar but not identical" (wolf and dog, for example).

    (f) Both were blissfully unaware of their profound language problem, even though they were producing gibberish. When I spoke to them in Tamil (a south Indian language), one of them said, "Spanish," while the other nodded as if in understanding and replied in gibberish. When we played a DVD recording of LC's own utterances back to him, LC nodded and said, "It's okay."

    (g) LC had profound dyscalculia (for example, reporting 14 minus 5 as 3). Yet he could do nonverbal subtraction. We showed him two opaque cups A and B, and dropped three cookies in A and four in B while he watched. When we removed two cookies from B (as he watched), LC subsequently went straight for A. (KC was not tested.)

(h) LC had a profound inability to understand even simple gestures such as "okay," "hitchhike," or "salute," Nor could he comprehend iconic signs like the restroom sign. He couldn't match a dollar with four quarters. And preliminary tests showed he was poor at transitivity.

A paradox arises: Given that LC was okay at learning paired associations (for example, *pig* = *nagi*) after extensive training, why can't he relearn his own language? Perhaps the very attempt to engage his preexisting language introduces a software "bug" that forces the malfunctioning language system to go on autopilot. If so, then teaching the patient a completely new language may, paradoxically, be easier than retraining the patient to the original.

Could he learn pidgin, which requires only that words be strung together in the right order (given that his *concept* formation is unimpaired)? And if he could be taught something as complex as "similar but not same," why can't he be taught to attach arbitrary Sassurian *symbols* (that is, words) to other concepts such as "big," small," "on," "if," "and," and "give"? Would this not enable him to understand a new language (such as French or American Sign Language), which would allow him to at least converse with French people or signers? Or if the problem is in linking heard sounds with objects and ideas, why not use a language based on visual tokens (as was done with Kanzi, the bonobo)?

The oddest aspects of Wernicke's aphasia are the patients' complete lack of insight into their own profound inability to comprehend or produce language, whether written or spoken, and their total lack of any frustration. We once gave LC a book to read and walked out of the room. Even though he couldn't understand a single word, he kept scanning the print and turning the pages for fifteen minutes. He even bookmarked some pages! (He was unaware of the fact that the video camera filming him had been left on during our absence.)

CHAPTER 7 BEAUTY AND THE BRAIN: THE EMERGENCE OF AESTHETICS

1. One has to be careful to not overdo this type of reductionist thinking about art and the brain. I recently heard an evolutionary psychologist give a lecture about why we like kinetic art, which includes pieces like Calder mobiles made up of moving cutout shapes dangling from the ceiling. With a perfectly straight face he proclaimed that we like such art because an area in our brain called the MT (middle temporal) area possesses cells that

are specialized for detecting the direction of motion. This claim is nonsense. Kinetic art obviously excites such cells, but so would a snowstorm. So would a copy of the *Mona Lisa* set spinning on a peg. Neural circuitry for motion detection is certainly necessary for kinetic art but it's not sufficient: It doesn't explain the appeal of kinetic art by any stretch of logic. This chap's explanation is like saying that the existence of face-sensitive cells in the fusiform gyrus of your brain explains why you like Rembrandt. Surely to explain Rembrandt you need to show how he enhanced his images and why such embellishments elicit responses from the neural circuits in your brain more powerfully than a realistic photograph does. Until you do that, you have explained nothing.

2. Note that peak shift should also be applicable in animation. For example, you can create a striking perceptual illusion by mounting tiny LEDs (light-emitting diodes) on a person's joints and having her walk around in a dark room. You might expect to see just a bunch of LEDs moving around randomly, but instead you get a vivid sense of seeing a whole person walking, even though all her other features—face, skin, hair, outline, and so forth—are invisible. If she stops moving, you suddenly cease to see the person. This implies that the information about her body is conveyed entirely by the motion trajectories of the light spots. It's as though your visual areas are exquisitely sensitive to the parameters that distinguish this type of biological motion from random motion. It's even possible to tell if the person is a man or woman by looking at the gait, and a couple dancing provides an especially amusing display.

    Can we exploit our laws to heighten this effect? Two psychologists, Bennett Bertenthal of Indiana University and James Cutting of Cornell University, mathematically analyzed the constraints underlying biological motion (which depend on permissible joint motions) and wrote a computer program that incorporates the constraints. The program generates a perfectly convincing display of a walking person. While these images are well known, their aesthetic appeal has rarely been commented on. In theory it should be possible to amplify the constraints so that the program could produce an especially elegant feminine gait caused by a large pelvis, swaying hips and high heels as well as an especially masculine gait caused by erect posture, stiff stride, and tight buttocks. You'd create a peak shift with a computer program.

    We know the superior temporal sulcus (STS) has dedicated circuitry for extracting biological motion, so a computer manipulation of human gait might hyperactivate those circuits by exploiting two aesthetic laws in parallel: isolation (isolating the biological motion cues from other static cues) and peak shift (amplifying the biological characteristics of the motion). The result might end up being an evocative work of kinetic art that surpasses any Calder mobile. I predict that STS cells for biological motion could react even more strongly to "peak-shifted" point-light walkers.

CHAPTER 8 THE ARTFUL BRAIN: UNIVERSAL LAWS

1. Indeed, peekaboo in children may be enjoyable for precisely the same reason. In early primate evolution while still primarily inhabiting the treetops, most juveniles often became temporarily occluded completely by foliage. Evolution saw fit to make peekaboo visually reinforcing for offspring and mother, as they periodically glimpsed each other, thereby ensuring that the child was kept safe and within a reasonable distance. Additionally, the smile and laugh of parent and offspring would have mutually reinforced each other. One wonders whether apes enjoy peekaboo.

    The laughter seen after peekaboo is also explained by my ideas on humor (see Chapter 1), that it results from; a buildup of expectation followed by a surprising deflation. Peakaboo could be regarded as a cognitive tickle.

2. See also Note 6 of Chapter 3, where the effect of altering type to match the meaning of the words was discussed—there from the standpoint of synesthesia rather than humor and aesthetics.

3. To these nine laws of aesthetics we may add a tenth law that overarches the others. Let's call it "resonance" because it involves the clever use of multiple laws enhancing each other in a single image. For example, in many Indian sculptures, a sexy nymph is portrayed languorously standing beneath the arched branch of a tree which has ripe fruits dangling from it. There are the peak shifts in posture and form (for example, large breasts) that make her exquisitely feminine and voluptuous. Additionally, the fruits are a *visual* echo of her breasts, but they also conceptually symbolize the fecundity and fertility of nature just as the nymph's breasts do; so the perceptual and conceptual elements resonate. The sculptor will also often add baroque ornate jewelry on her otherwise naked torso to enhance, by *contrast*, the smoothness and suppleness of her youthful estrogen-charged skin. (I mean contrast of texture rather than of luminance here.) A more familiar example would be a Monet in which peekaboo, peak shift, and isolation are all combined in a single painting.

CHAPTER 9 AN APE WITH A SOUL: HOW INTROSPECTION EVOLVED

1. Two questions may legitimately be raised about metarepresentations. First, isn't this just a matter of degree? Perhaps a dog has a metarepresentation of sorts that's richer than what a rat has but not quite as rich as a human's (the

"When to you start calling a man bald" issue). This question was raised and answered in the Introduction, where we noted that nonlinearities are common in nature—especially in evolution. A fortuitous coemergence of attributes can produce a relatively sudden, qualitative jump, resulting in a novel ability. A metarepresentation doesn't merely imply richer associations; it also requires the ability to intentionally summon up these associations, attend to them at will, and manipulate them mentally. These abilities require frontal lobe structures, including the anterior cingulate, to direct attention to different aspects of the internal image (although concepts such as "attention" and "internal image" conceal vast depths of ignorance). An idea similar to this was originally proposed by Marvin Minsky.

Second, doesn't postulating a metarepresentation make us fall into the homunculus trap? (See Chapter 2, where the homunculus fallacy was discussed.) Doesn't it imply a little man in the brain watching the metarepresentation and creating a meta-metarepresentation in *his* brain? The answer is no. A metarepresentation is not a picture-like replica of sensory representation; it results from further processing of early sensory representations and packaging them into more manageable chunks for linking to language and symbol juggling.

The telephone syndrome, which Jason had, has been studied by Axel Klee and Orrin Devinsky.

2. I recall a lecture given at the Salk Institute by Francis Crick, who with James Watson codiscovered the structure of DNA and deciphered the genetic code, thereby unraveling the physical basis of life. Crick's lecture was on consciousness, but before he could begin, a philosopher in the audience (from Oxford, I believe) raised his hand and protested, "But Professor Crick, you say you are going to talk about the neural mechanisms of consciousness, but you haven't even bothered to define the word properly." Crick's response: "My dear chap, there was never a time in the history of biology when a group of us sat around the table saying let's define life first. We just went out there and found out what it was—a double helix. We leave matters of semantic distinctions and definitions to you philosophers."

3. Almost everyone knows of Freud as the father of psychoanalysis, but few realize that he began his career as a neurologist. Even as a student he published a paper on the nervous system of a primitive fishlike creature called a lamprey, convinced that the surest way to understand the mind was to approach it through neuroanatomy. But he soon became bored with lampreys and began to feel that his attempts to bridge neurology and psychiatry were premature. So he switched to "pure" psychology, inventing all the ideas we now associate with his name: id, ego, superego, Oedipus complex, penis envy, *thanatos*, and the like.

In 1896 he became disillusioned once again and wrote his now famous

"Manifesto for a Scientific Psychology" urging a neuroscientific approach to the human mind. Unfortunately he was way ahead of his time.

4. Although we intuitively understand what Freud meant, one could argue that the phrase "unconscious self" is an oxymoron since self-awareness (as we shall see) is one of the defining characteristics of the self. Perhaps the phrase "unconscious mind(s)" would be better, but the exact terminology isn't important at this stage. (See also Note 2 for this chapter.)

5. Since Freud's era there have been three major approaches to mental illness. First, there is "psychological," or talk therapy, which would include psychodynamic (Freudian) as well as more recent "cognitive" accounts. Second, there are the anatomical approaches, which simply point out correlations between certain mental disorders and physical abnormalities in specific structures. For example, there is a presumed link between the caudate nucleus and obsessive-compulsive disorder, or between right frontal lobe hypometabolism and schizophrenia. Third there are neuropharmacological interpretations: think Prozac, Ritalin, Xanax. Of these three, the last approach has paid rich dividends (at least to the pharmaceutical industry) in terms of treating psychiatric disease; for better or worse, it has revolutionized the field.

What is missing, though, and what I have attempted to broach in this book, is what might be called "functional anatomy"—to explain the cluster of symptoms that are unique to a given disorder in terms of functions that are equally unique to certain specialized circuits in the brain. (Here one must distinguish between a vague correlation and an actual explanation.) Given the inherent complexity of the human brain, it is unlikely that there will be a single climactic solution like DNA (although I don't rule it out). But there may well be many instances where such a synthesis is possible on a smaller scale, leading to testable predictions and novel therapies. These examples may even pave the way for a grand unified theory of the mind—of the kind physicists have been dreaming about for the material universe.

6. The idea of a hardwired genetic scaffolding for one's body image was also brought home to me vividly when Paul McGeoch and I recently saw a fifty-five-year-old woman with a phantom hand. She had been born with a birth defect called phocomelia; most of her right arm had been missing since birth except for a hand dangling from her shoulder with only two fingers and a tiny thumb. When she was twenty-one, she was in a car crash that entailed amputation of the crushed hand, but much to her surprise she experienced a phantom hand with four fingers instead of two! It was as if her entire hand was hardwired and lying dormant in her brain, being suppressed and refashioned by the abnormal proprioception (joint and muscle sense) and visual image of her deformed hand. Until the age

of twenty-one, when removal of the deformed hand allowed her dormant hardwired hand to reemerge into consciousness as a phantom. The thumb did not come back initially, but when she used the mirror box (at age fifty-five) her thumb was resurrected as well.

In 1998, in a paper published in *Brain*, I reported that by using visual feedback with mirrors positioned in the right manner, one could make the phantom hand adopt anatomically impossible positions (such as fingers bending backward)—despite the fact that the brain had never previously computed or experienced that before. The observation has since then been confirmed by others.

Findings such as these emphasize the complexity of interactions between nature and nurture in constructing body image.

7. We don't know where the discrepancy between S2 and the SPL is picked up, but my intuition is that the right insula is involved, given the GSR increase. (The insula is partly involved in generating the GSR signal.) Consistent with this, the insula is also involved in nausea and vomiting due to discrepancies between the vestibular and visual senses (which familiarly produces seasickness, for example).

8. Intriguingly, even some otherwise normal men report having mainly phantom erections rather than real ones, as my colleague Stuart Anstis pointed out to me.

9. This "adopting an objective view" toward oneself is also an essential requirement for discovering and correcting one's own Freudian defenses, which is partially achieved through psychoanalysis. The defenses are ordinarily unconscious; the concept of "conscious defenses" is an oxymoron. The therapist's goal, then, is to bring the defenses to the surface of your consciousness so you can deal with them (just as an obese person needs to analyze the source of his obesity to take corrective measures). One wonders whether adopting a *conceptual* allocentric stance (in plain English: encouraging the patient to adopt a realistic detached view of herself and her follies) for psychoanalysis could be aided by encouraging the patient to adopt a *perceptual* allocentric stance (such as pretending she is someone else watching her own lecture). This in turn could, in theory, be facilitated by ketamine anesthesia. Ketamine generates out-of-body experiences, making you see yourself from outside.

Or perhaps we could mimic the effects of ketamine by using mirrors and video cameras, which can also produce out-of-body experiences. It seems ludicrous to suggest the use of optical tricks for psychoanalysis, but believe me, I have seen stranger things in my career in neurology. (For example, Elizabeth Seckel and I used a combination of multiple reflections, delayed video feedback, and makeup to create a temporary out-of-body experience in a patient with fibromyalgia, a mysterious chronic pain disorder that affects the entire body. The patient reported a substantial reduction in pain during the experience. As for all pain disorders, this requires placebo-controlled evaluation.)

Returning to psychoanalysis: surely, removing psychological defenses raises a dilemma for the analyst; it's a double-edged sword. If defenses are normally an adaptive response by the organism (mainly by the left hemisphere) to avoid destabilization of behavior, wouldn't laying bare these defenses be maladaptive, disturbing one's sense of an internally consistent self along with your inner peace? The way out of this dilemma is to realize that mental illness and neuroses arise from a *mis*application of defenses— no biological system is perfect. Such a misapplication would, if anything, lead to additional chaos rather than restoring coherence.

And there are two reasons for this. First, chaos may result from "leakage" of improperly suppressed emotions from the right hemisphere, leading to anxiety—a poorly articulated internal feeling of lacking harmony in one's life. Second, there may be instances in which defenses might be maladaptive for the person in his real life; a little overconfidence is adaptive but too much isn't; it leads to hubris and to unrealistic delusions about one's abilities; you start buying Ferraris you can't afford. There is a fine line between what's maladaptive and what's not, but an experienced therapist knows how to correct only the former (by bringing them out) while preserving the latter, so that she avoids causing what Freudians call a catastrophic reaction (a euphemism for "The patient breaks down and starts crying").

10. Our sense of coherence and unity as a single person may—or may not— require a single brain region, but if it does, reasonable candidates would include the insula and the inferior parietal lobule—each of which receives a convergence of multiple sensory inputs. I mentioned this idea to my colleague Francis Crick just before his death. With a sly conspiratorial wink he told me that a mysterious structure called the claustrum—a sheet of cells buried in the sides of the brain—also receives inputs from many brain regions, and may therefore mediate the unity of conscious experience. (Perhaps we are both right!) He added that he and his colleague Christof Koch had just finished writing a paper on this very topic.

11. This speculation is based on a model proposed by German Berrios and Mauricio Sierra of Cambridge University.

12. The distinction between the "how" and "what" pathways was first made by Leslie Ungerleider and Mortimer Mishkin of the National Institutes of Health; it is based on meticulous anatomy and physiology. The further subdivision of the "what" pathway into pathways 2 (semantics and meaning) and 3 (emotions) is more speculative and based on functional criteria; a combination of neurology and physiology. (For example, cells in the

STS respond to changing facial expressions and biological motion, and the STS has connections with the amygdala and the insula—both involved in emotions.) Postulating a functional distinction between pathways 2 and 3 also helps explain Capgras syndrome and prosopagnosia, which are mirror images of each other, in terms of both symptoms and GSR responses. This cannot occur if messages were processed entirely in a sequence from meaning to emotion and there was no parallel output from the fusiform area to the amygdala (either directly or via the STS).

13. Here and elsewhere, although I invoke the mirror-neuron system as a candidate neural system, the logic of the argument doesn't depend critically on that system. The crux of the argument is that there must be specialized brain circuitry for recursive self-representation and for maintaining a distinction—and reciprocity—between the self and the other in the brain. A dysfunction of this system would contribute to many of the seemingly bizarre syndromes described in this chapter.

14. To complicate matters further, Ali started developing other delusions as well. A psychiatrist diagnosed him as having schizophrenia or "schizoid traits" (in addition to his epilepsy) and prescribed him antipsychotic medication. The last time I saw Ali, in 2009, he was claiming that in addition to being dead he had grown to enormous size, reaching out into the cosmos to touch the moon, becoming one with the Universe—as if nonexistence and union with the cosmos were synonymous. I began to wonder if his seizure activity had spread into his right parietal lobe, where body image is constructed, which might explain why he had lost his sense of scale, but I have not yet had a chance to investigate this hunch.

15. One might expect, therefore, that in Cotard syndrome there would initially be no GSR whatsoever, but it should be partially restored with SSRIs (selective serotonin reuptake inhibitors). This can be tested experimentally.

16. When I make remarks of this nature about God (or use the word "delusion"), I do not wish to imply that God doesn't exist; the fact that some patients develop such delusions doesn't disprove God—certainly not the abstract God of Spinoza or Shankara. Science has to remain silent on such maters. I would argue, like Erwin Schrödinger and Stephen Jay Gould, that science and religion (in the nondoctrinaire philosophical sense) belong to different realms of discourse and one cannot negate the other. My own view, for what it is worth, is best exemplified by the poetry of the bronze *Nataraja* (*The Dancing Shiva*), which I described in Chapter 8.

17. There has long been a tension in biology between those who advocate a purely functional, or black-box approach, and those who champion reductionism, or understanding how component parts interact to generate complex functions. The two groups are often contemptuous of each other.

Psychologists often promote black-box functionalism and attack reductionist neuroscience—a syndrome I have dubbed "neuron envy." The syndrome is partly a legitimate reaction to the fact that most funding from grant-giving agencies tends to be siphoned off, unfairly, by neuroreductionists. Neuroscience also garners the lion's share of attention from the popular press, partly because people (including scientists) like looking at the results of brain imaging; all those pretty colored dots on pictures of brains. At a recent meeting of the Society for Neuroscience, a colleague approached me to describe an elaborate-brain imaging experiment he had done which used a complex cognitive-perceptual task to explore brain mechanisms. "You will never guess which area of the brain lit up, Dr. Ramachandran," he said, brimming with enthusiasm. I responded with a sly wink saying, "Was it the anterior cingulate?" The man was astonished, failing to realize that the anterior cingulate lights up on so many of these tasks that the odds were already stacked in my favor, even though I was just guessing.

But by itself, pure psychology or "black boxology" (which Stuart Sutherland once defined as "the ostentatious display of flow diagrams as a substitute for thought") is unlikely to generate revolutionary advances in biology, where mapping function onto structure has been the most effective strategy. (And I would consider psychology to be a branch of biology.) I will drive home this point using an analogy from the history of genetics and molecular biology.

Mendel's laws of heredity, which established the particulate nature of genes, was an example of the black-box approach. These laws were established by simply studying the patterns of inheritance that resulted from mating different types of pea plants. Mendel derived his laws by simply looking at the surface appearance of hybrids and deducing the existence of genes. But he didn't know what or where genes were. That became known when Thomas Hunt Morgan zapped the chromosomes of fruit flies with X-rays and found that the heritable changes in appearance that occurred in the flies (mutations) correlated with changes in banding patterns of chromosomes. (This would be analogous to lesion studies in neurology.) This discovery allowed biologists to home in on chromosomes—and the DNA within them—as the carriers of heredity. Which in turn paved the way for decoding DNA's double helical structure and the genetic code of life. But once the molecular machinery of life was decoded, it not only explained heredity but a great many other previously mysterious biological phenomena as well.

The key idea came when Crick and Watson saw the analogy between the complementarity of the two strands of DNA and the complementarity between parent and offspring, and recognized that the structural logic of DNA dictates the functional logic of heredity: a high-level phenomenon. That flash of insight gave birth to modern biology. I believe that the same strategy of mapping function onto structure is the key to understanding brain function.

More relevant to this book is the discovery that damage to the hippocampus leads to anterograde amnesia. This allowed biologists to focus on synapses in the hippocampus, leading to the discovery of LTP (long-term potentiation), the physical basis of memory. Such changes were originally discovered by Eric Kandel in a mollusk named *Aplysia*.

In general, the problem with the pure black-box approach (psychology) is that sooner or later you get multiple competing models to explain a small set of phenomena, and the only way to find out which is right is through reductionism—opening the box(es). A second problem is that they very often have an ad hoc "surface level" quality, in that they may partially "explain" a given "high level" or macroscopic phenomenon but don't explain other macroscopic phenomena and their predictive power is limited. Reductionism, on the other hand, often explains not just the phenomenon in question at a deeper level but often also ends up explaining a number of other phenomena as well.

Unfortunately, for many physiologists reductionism becomes an end in itself, a fetish almost. An analogy to illustrate this comes from Horace Barlow. Imagine that an asexual (parthenogenetic) Martian biologist lands on Earth. He has no idea what sex is since he reproduces by dividing into two, like an ameba. He (it) examines a human and finds two round objects (which we call testes) dangling between the legs. Being a reductionist Martian, he dissects them and, looking through microscope, finds them swarming with sperms; but he wouldn't know what they were for. Barlow's point is that no matter how meticulous the Martian is at dissection and how detailed an analysis he performs on them he will never truly understand the function of the testes unless he knew about the "macroscopic" phenomenon of sex; he may even think the sperm are wriggling parasites. Many (fortunately not all!) of our physiologists recording from brain cells are in the same position as the asexual Martian.

The second, related point is that one must have the intuition to focus on the appropriate level of reductionism for explaining a given higher-level function (such as sex). If Watson and Crick had focused on the subatomic level or atomic level of chromosomes instead of the macromolecular level (DNA), or if they had focused on the wrong molecules (the histones in the chromosomes instead of DNA) they would have made no headway in discovering the mechanism of heredity.

18.   Even simple experiments on normal subjects can be instructive in this regard. I will mention an experiment I did (with my student Laura Case) inspired by the "rubber hand illusion" discovered by Botvinick and Cohen (1998) and by the dummy-head illusion (Ramachandran and Hirstein,

1998). You, the reader, stand about a foot behind a bald-headed manikin looking at its head. I stand on the right side of you both and randomly tap and stroke the *back* of your head (especially ears) with my left hand (so you can't see my hand) while simultaneously doing the same thing on the plastic head with my right hand, in perfect synchrony. In about two minutes you will experience that the stroking and tapping on your head is emerging from the dummy you are looking at. Some people develop the illusion of a twin or phantom head in front of them, especially if they get it going by "imagining " their head displaced forward. The brain regards it as highly improbable that the plastic head is *seen* to be tapped in the same precise sequence as you *feel* on own head by chance and so is willing to temporarily to project your head on the manikin's shoulder. This has powerful implications since, contrary to recent proposals, it rules out simple associative learning as the basis of the rubber hand illusion. (Every time you saw your hand touched you *felt* it touched as well.) After all, you have never seen the back of your head being touched. It is one thing to regard your hand sensations as being slightly out of register with your real hand but quite another to project them to the back of a dummy head!

The experiment proves that your brain has constructed an internal model of your head—even unseen parts—and used Bayesian inference to experience (incorrectly) your sensations as arising from the dummy's head even though it is logically absurd. Would doing something like this help alleviate your migraine symptoms ("the dummy is experiencing migraine; not me")? I wonder.

Olaf Blanke and Henrik Ehrsson of the Karolinska Institute in Sweden have shown that out-of-body experiences can also be induced by having subjects watch video images of themselves moving or being touched. Laura Case, Elizabeth Seckel, and I found that such illusions are enhanced if you wear a Halloween mask and introduce a tiny time delay together with a left-right reversal in the image. You suddenly start inhabiting and controlling the "alien" in the video image. Remarkably, if you wear a smiling mask you actually feel happy because "you, out there" look happy! I wonder if you could use it to "cure" depression.

EPILOGUE

1. These two Darwin quotes come from the *London Illustrated News,* April 21, 1862 ("I feel most deeply . . ."), and Darwin's letter to Asa Gray, May 22, 1860 ("I own that I cannot see . . .").

BIBLIOGRAPHY

Entries marked with an asterisk are suggestions for further reading.

Aglioti, S., Bonazzi, A., & Cortese, F. (1994). Phantom lower limb as a perceptual marker of neural plasticity in the mature human brain. *Proceedings of the Royal Society of London, Series B: Biological Sciences, 255,* 273–278.

Aglioti, S., Smania, N., Atzei, A., & Berlucchi, G. (1997). Spatio-temporal properties of the pattern of evoked phantom sensations in a left index amputee patient. *Behavioral Neuroscience, 111,* 867–872.

Altschuler, E. L., & Hu, J. (2008). Mirror therapy in a patient with a fractured wrist and no active wrist extension. *Scandinavian Journal of Plastic and Reconstructive Surgery and Hand Surgery, 42*(2), 110–111.

Altschuler, E. L., Vankov, A., Hubbard, E. M., Roberts, E., Ramachandran, V. S., & Pineda, J. A. (2000, November). *Mu wave blocking by observer of movement and its possible use as a tool to study theory of other minds.* Poster session presented at the 30th annual meeting of the Society for Neuroscience, New Orleans, LA.

Altschuler, E. L., Vankov, A., Wang, V., Ramachandran, V. S., & Pineda, J. A. (1997). *Person see, person do: Human cortical electrophysiological correlates of monkey see monkey do cells.* Poster session presented at the 27th Annual Meeting of the Society for Neuroscience, New Orleans, LA.

Altschuler, E. L., Wisdom, S. B., Stone, L., Foster, C., Galasko, D., Llewellyn, D. M. E., et al. (1999). Rehabilitation of hemiparesis after stroke with a mirror. *The Lancet, 353,* 2035–2036.

Arbib, M. A. (2005). From monkey-like action recognition to human language: An evolutionary framework for neurolinguistics. *The Behavioral and Brain Sciences, 28*(2), 105–124.

Armel, K. C., & Ramachandran, V. S. (1999). Acquired synesthesia in retinitis pigmentosa. *Neurocase, 5*(4), 293–296.

Armel, K. C., & Ramachandran, V. S. (2003). Projecting sensations to external objects: Evidence from skin conductance response. *Proceedings of the Royal Society of London, Series B: Biological Sciences, 270*(1523), 1499–1506.

Armstrong, A. C., Stokoe, W. C., & Wilcox, S. E. (1995). *Gesture and the nature of language.* Cambridge, UK: Cambridge University Press.

Azoulai, S., Hubbard, E. M., & Ramachandran, V. S. (2005). Does synesthesia contribute to mathematical savant skills? *Journal of Cognitive Neuroscience, 69*(Suppl).

Babinski, J. (1914). Contribution a l'étude des troubles mentaux dans l'hémiplégie organique cérébrale (anosognosie). *Revue Neurologique, 12,* 845–847.

Bach-y-Rita, P., Collins, C. C., Saunders, F. A., White, B., & Scadden, L. (1969). Vision substitution by tactile image projection. *Nature, 221,* 963–964.

Baddeley, A. D. (1986). *Working memory.* Oxford, UK: Churchill Livingstone.

*Barlow, H. B. (1987). The biological role of consciousness. In C. Blakemore & S. Greenfield (Eds.), *Mindwaves* (pp. 361–374). Oxford, UK: Basil Blackwell.

Barnett, K. J., Finucane, C., Asher, J. E., Bargary, G., Corvin, A. P., Newell, F. N., et al. (2008). Familial patterns and the origins of individual differences in synaesthesia. *Cognition, 106*(2), 871–893.

Baron-Cohen, S. (1995). *Mindblindness.* Cambridge, MA: MIT Press.

Baron-Cohen, S., Burt, L., Smith-Laittan, F., Harrison, J., & Bolton, P. (1996). Synaesthesia: Prevalence and familiality. *Perception, 9,* 1073–1079.

Baron-Cohen, S., & Harrison, J. (1996). *Synaesthesia: Classic and contemporary readings.* Oxford, UK: Blackwell Publishers.

Bauer, R. M. (1986). The cognitive psychophysiology of prosopagnosia. In H. D. Ellis, M. A. Jeeves, F. Newcombe, & A. W. Young (Eds.), *Aspects of face processing* (pp. 253–278). Dordrecht, Netherlands: Martinus Nijhoff.

Berlucchi, G., & Aglioti, S. (1997). The body in the brain: Neural bases of corporeal awareness. *Trends in Neurosciences, 20*(12), 560–564.

Bernier, R., Dawson, G., Webb, S., & Murias, M. (2007). EEG mu rhythm and imitation impairments in individuals with autism spectrum disorder. *Brain and Cognition, 64*(3), 228–237.

Berrios, G. E., & Luque, R. (1995). Cotard's syndrome. *Acta Psychiatrica Scandinavica, 91*(3), 185–188.

*Bickerton, D. (1994). *Language and human behavior.* Seattle: University of Washington Press.

Bisiach, E., & Geminiani, G. (1991). Anosognosia related to hemiplegia and hemianopia. In G. P. Prigatano and D. L. Schacter (Eds.), *Awareness of deficit after brain injury: Clinical and theoretical issues.* Oxford: Oxford University Press.

Blake, R., Palmeri, T. J., Marois, R., & Kim, C. Y. (2005). On the perceptual reality of synesthetic color. In L. Robertson and N. Sagiv (Eds.), *Synesthesia: Perspectives from cognitive neuroscience* (pp. 47–73). New York: Oxford University Press.

*Blackmore, S. (1999). *The meme machine.* Oxford: Oxford University Press.

Blakemore, S.-J., Bristow, D., Bird, G., Frith, C., & Ward, J. (2005). Somatosensory activations during the observation of touch and a case of vision-touch synaesthesia. *Brain, 128,* 1571–1583.

*Blakemore, S.-J., & Frith, U. (2005). *The learning brain.* Oxford, UK: Blackwell Publishing.

Botvinick, M., & Cohen, J. (1998). Rubber hands "feel" touch that eyes see. *Nature, 391*(6669), 756.

Brang, D., Edwards, L., Ramachandran, V. S., & Coulson, S. (2008). Is the sky 2? Contextual priming in grapheme-color synaesthesia. *Psychological Science, 19*(5), 421–428.

Brang, D., McGeoch, P., & Ramachandran, V. S. (2008). Apotemnophilia: A neurological disorder. *Neuroreport, 19*(13), 1305–1306.

Brang, D., & Ramachandran, V. S. (2007a). Psychopharmacology of synesthesia: The role of serotonin S2a receptor activation. *Medical Hypotheses, 70*(4), 903–904.

Brang, D., & Ramachandran, V. S. (2007b). Tactile textures evoke specific emotions: A new form of synesthesia. Poster session presented at the 48th annual meeting of the Psychonomic Society, Long Beach, CA.

Brang, D., & Ramachandran, V. S. (2008). Tactile emotion synesthesia. *Neurocase, 15*(4), *390–399.*

Brang, D., & Ramachandran, V.S. (2010). Visual field heterogeneity, laterality, and eidetic imagery in synesthesia. *Neurocase, 16*(2), 169–174.

Buccino, G., Vogt, S., Ritzl, A., Fink, G. R., Zilles, K., Freund, H. J., et al. (2004). Neural circuits underlying imitation of hand actions: An event related fMRI study. *Neuron, 42,* 323–334.

Bufalari, I., Aprile, T., Avenanti, A., Di Russo, F., & Aglioti, S. M. (2007). Empathy for pain and touch in the human somatosensory cortex. *Cerebral Cortex, 17,* 2553–2561.

Bujarski, K., & Sperling, M. R. (2008). Post-ictal hyperfamiliarity syndrome in focal epilepsy. *Epilepsy and Behavior, 13*(3), 567–569

Caccio, A., De Blasis, E., Necozione, S., & Santilla, V. (2009). Mirror feedback therapy for complex regional pain syndrome. *The New England Journal of Medicine, 361*(6), 634–636.

Campbell, A. (1837, October). Opinionism [Remarks on "New School Divinity," in *The Cross and Baptist Journal*]. *The Millennial Harbinger* [New Series], *1,* 439. Retrieved August 2010 from http://books.google.com.

Capgras, J., & Reboul-Lachaux, J. (1923). L'illusion des "sosies" dans un délire systématisé chronique. *Bulletin de la Société Clinique de Médecine Mentale*, 11, 6–16.

Carr, L., Iacoboni, M., Dubeau, M. C., Mazziotta, J. C., & Lenzi, G. L. (2003). Neural mechanisms of empathy in humans: A relay from neural systems for imitation to limbic areas. *Proceedings of the National Academy of Sciences of the USA, 100*, 5497–5502.

*Carter, R. (2003). *Exploring consciousness*. Berkeley: University of California Press.

*Chalmers, D. (1996). *The conscious mind*. New York: Oxford University Press.

Chan, B. L., Witt, R., Charrow, A. P., Magee, A., Howard, R., Pasquina, P. F., et al. (2007). Mirror therapy for phantom limb pain. *The New England Journal of Medicine, 357*, 2206–2207.

*Churchland P. S. (1986). *Neurophilosophy: Toward a Unified science of the mind/brain.* Cambridge, MA: MIT Press.

*Churchland, P., Ramachandran, V. S., & Sejnowski, T. (1994). A critique of pure vision. In C. Koch & J. Davis (Eds.), *Large-scale neuronal theories of the brain* (pp. 23–47). Cambridge, MA: MIT Press.

Clarke, S., Regli, L., Janzer, R. C., Assal, G., & de Tribolet, N. (1996). Phantom face: Conscious correlate of neural reorganization after removal of primary sensory neurons. *Neuroreport, 7,* 2853–2857.

*Corballis, M. C. (2002). *From hand to mouth: The origins of language*. Princeton, NJ: Princeton University Press.

Corballis, M. C. (2009). The evolution of language. *Annals of the New York Academy of Sciences, 1156,* 19–43.

*Craig, A. D. (2009). How do you feel—now? The anterior insula and human awareness. *Nature Reviews Neuroscience*, 10, 59–70.

*Crick, F. (1994). *The astonishing hypothesis: The scientific search for the soul*. New York: Charles Scribner's Sons.

*Critchley, M. (1953). *The parietal lobes*. London: Edward Arnold.

*Cytowic, R. E. (1989). *Synesthesia: A union of the senses*. New York; Springer.

*Cytowic, R. E. (2003). *The man who tasted shapes*. Cambridge, MA: MIT Press. (Original work published 1993 by G. P. Putnam's Sons)

*Damasio, A. (1994). *Descartes' error*. New York: G. P. Putnam.

*Damasio, A. (1999). *The feeling of what happens: Body and emotion in the making of Consciousness*. New York: Harcourt.

*Damasio, A. (2003). *Looking for Spinoza: Joy, sorrow and the feeling brain*. New York: Harcourt.

Dapretto, M., Davies, M. S., Pfeifer, J. H., Scott, A. A., Sigman, M., Bookheimer, S. Y., et al. (2006). Understanding emotions in others: Mirror neuron dysfunction in children with autism spectrum disorders. *Nature Neuroscience, 9,* 28–30.

*Dehaene, S. (1997). *The number sense: How the mind creates mathematics*. New York: Oxford University Press.

*Dennett, D. C. (1991). *Consciousness explained*. Boston: Little, Brown.

Devinsky, O. (2000). Right hemisphere dominance for a sense of corporeal and emotional self. *Epilepsy and Behavior, 1*(1), 60–73

*Devinsky, O. (2009). Delusional misidentifications and duplications: Right brain lesions, left brain delusions. *Neurology, 72*(80–87).

Di Pellegrino, G., Fadiga, L., Fogassi, L., Gallese, V., & Rizzolatti, G. (1992). Understanding motor events: A neurophysiological study. *Experimental Brain Research, 91,* 176–180.

Domino, G. (1989). Synesthesia and creativity in fine arts students: An empirical look. *Creativity Research Journal, 2,* 17–29.

*Edelman, G. M. (1989). *The remembered present: A biological theory of consciousness*. New York: Basic Books.

*Ehrlich, P. (2000). *Human natures: Genes, cultures, and human prospect*. Harmondsworth, UK: Penguin Books.

Eng, K., Siekierka, E., Pyk, P., Chevrier, E., Hauser, Y., Cameirao, M., et al. (2007). Interactive visuo-motor therapy system for stroke rehabilitation. *Medical and Biological Engineering and Computing, 45,* 901–907.

*Enoch, M. D., & Trethowan, W. H. (1991). *Uncommon psychiatric syndromes* (3rd ed.). Oxford: Butterworth-Heinemann.

*Feinberg, T. E. (2001). *Altered egos: How the brain creates the self*. Oxford University Press.

Fink, G. R., Marshall, J. C., Halligan, P. W., Frith, C. D., Driver, J., Frackowiak, R. S., et al. (1999). The neural consequences of conflict between intention and the senses. *Brain, 122,* 497–512.

First, M. (2005). Desire for an amputation of a limb: Paraphilia, psychosis, or a new type of identity disorder. *Psychological Medicine, 35,* 919–928.

Flor, H., Elbert, T., Knecht, S., Wienbruch, C., Pantev, C., Birbaumer, N., et al.

(1995). Phantom-limb pain as a perceptual correlate of cortical reorganization following arm amputation. *Nature, 375,* 482–484.

Fogassi, L., Ferrari, P. F., Gesierich B., Rozzi, S., Chersi, F., & Rizzolatti, G. (2005, April 29). Parietal lobe: From action organization to intention understanding. *Science, 308,* 662–667.

Friedmann, C. T. H., & Faguet, R. A. (1982). *Extraordinary disorders of human behavior.* New York: Plenum Press.

Frith, C. & Frith, U. (1999, November 26). Interacting minds—A biological basis. *Science, 286,* 1692–1695.

Frith, U., & Happé, F. (1999). Theory of mind and self consciousness: What is it like to be autistic? *Mind and Language, 14,* 1–22.

Gallese, V., Fadiga, L., Fogassi, L., & Rizzolatti, G. (1996). Action recognition in the premotor cortex. *Brain, 119,* 593–609.

Gallese, V., & Goldman, A. (1998). Mirror neurons and the simulation theory of mind-reading. *Trends in Cognitive Sciences, 12,* 493–501.

Garry, M. I., Loftus, A., & Summers, J. J. (2005). Mirror, mirror on the wall: Viewing a mirror reflection of unilateral hand movements facilitates ipsilateral M1 excitability. *Experimental Brain Research*, 163, 118–122.

*Gawande, A. (2008, June, 30). Annals of medicine: The itch. *New Yorker*, pp. 58–64.

*Gazzaniga, M. (1992). *Nature's mind.* New York: Basic Books.

*Glynn, I. (1999). *An anatomy of thought.* London: Weidenfeld & Nicolson.

*Greenfield, S. (2000). *The human brain: A guided tour.* London: Weidenfeld & Nicolson.

*Gregory, R. L. (1966). *Eye and brain.* London: Weidenfeld & Nicolson.

Gregory, R. L. (1993). *Odd perceptions.* New York: Routledge.

Grossenbacher, P. G., & Lovelace, C. T. (2001). Mechanisms of synesthesia: Cognitive and physiological constraints. *Trends in Cognitive Sciences, 5*(1), 36–41.

Happé, F., & Frith, U. (2006). The weak coherence account: Detail-focused cognitive style in autism spectrum disorders. *Journal of Autism and Developmental Disorders, 36*(1), 5–25.

Happé, F., & Ronald, A. (2008). The "fractionable autism triad": A review of evidence from behavioural, genetic, cognitive and neural research. *Neuropsychology Review, 18*(4), 287–304.

Harris, A. J. (2000). Cortical origin of pathological pain. *The Lancet, 355,* 318–319.

Havas, H., Schiffman, G., & Bushnell, M. (1990). The effect of bacterial vaccine on tumors and immune response of ICR/Ha mice. *Journal of Biological Response Modifiers, 9,* 194–204.

Hirstein, W., Iversen, P., Ramachandran, V. S. (2001). Autonomic responses of autistic children to people and objects. *Proceedings of the Royal Society of London, Series B: Biological Sciences, 268*(1479), 1883–1888.

Hirstein, W., & Ramachandran, V. S. (1997). Capgras syndrome: A novel probe for understanding the neural representation and familiarity of persons. *Proceedings of the Royal Society of London, Series B: Biological Sciences, 264*(1380), 437–444.

Holmes, N. P., & Spence, C. (2005). Visual bias of unseen hand position with a mirror: Spatial and temporal factors. *Experimental Brain Research, 166,* 489–497.

Hubbard, E. M., Arman, A. C., Ramachandran, V. S., & Boynton, G. (2005). Individual differences among grapheme-color synesthetes: Brain-behavior correlations. *Neuron, 45*(6), 975–985.

Hubbard, E. M., Manohar, S., & Ramachandran, V. S. (2006). Contrast affects the strength of synesthetic colors. *Cortex, 42*(2), 184–194.

Hubbard, E. M., & Ramachandran, V. S.(2005). Neurocognitive mechanisms of synesthesia. *Neuron, 48*(3), 509–520.

*Hubel, D. (1988). *Eye, brain, and vision*. Scientific American Library Series. New York: W. H. Freeman.

Humphrey, N. (1992). *A history of the mind*. New York: Simon & Schuster.

Humphrey, N. K. (1980). Nature's psychologists. In B. D. Josephson & V. S. Ramachandran (Eds.), *Consciousness and the physical world: Edited proceedings of an interdisciplinary symposium on consciousness held at the University of Cambridge in January 1978*. Oxford, UK/New York: Pergamon Press.

*Humphreys, G. W., & Riddoch, M. J. (1998). *To see but not to see: A case study of visual agnosia*. Hove, East Sussex, UK: Psychology Press.

*Iacoboni, M. (2008). *Mirroring people: The new science of how we connect with others*. New York: Farrar, Straus.

Iacoboni, M., & Dapretto, M. (2006, December). The mirror neuron system and the consequences of its dysfunction. *Nature Reviews Neuroscience, 7*(12), 942–951.

Iacoboni, M., Molnar-Szakacs, I., Gallese, V., Buccino, G., Mazziotta, J. C., & Rizzolatti, G. (2005). Grasping the intentions of others with one's own mirror neuron system. *PLoS Biology, 3*(3), e79.

Iacoboni, M., Woods, R. P., Brass, M., Bekkering, H., Mazziotta, J. C., & Rizzolatti, G. (1999, December 24). Cortical mechanisms of human imitation. *Science, 286,* 2526–2528.

Jellema, T., Oram, M. W., Baker, C. I., & Perrett, D. I. (2002). Cell populations in the banks of the superior temporal sulcus of the macaque monkey and imitation. In A. N. Melzoff & W. Prinz (Eds.), *The imitative mind: Development, evolution, and brain bases* (pp. 267–290). Cambridge, UK: Cambridge University Press.

Johansson, G. (1975). Visual motion perception. *Scientific American, 236*(6), 76–88.

*Kandel, E. (2005). *Psychiatry, psychoanalysis, and the new biology of the mind*. Washington, DC: American Psychiatric Publishing.

*Kandel, E. R., Schwartz, J. H., & Jessell, T. M. (Eds.). (1991). *Principles of neural science* (3rd ed.). Norwalk, CT: Appleton & Lange.

Kanwisher, N., & Yovel, G. (2006). The fusiform face area: A cortical region specialized for the perception of faces. *Philosophical Transactions of the Royal Society of London, Series B: Biological Sciences, 361,* 2109–2128.

Karmarkar, A., & Lieberman, I. (2006). Mirror box therapy for complex regional pain syndrome. *Anaesthesia, 61,* 412–413.

Keysers, C., & Gazzola, V. (2009). Expanding the mirror: Vicarious activity for actions, emotions, and sensations. *Current Opinion in Neurobiology, 19,* 666–671.

Keysers, C., Wicker, B., Gazzola, V., Anton, J. L., Fogassi, L., & Gallese, V. (2004). A touching sight: SII/PV activation during the observation and experience of touch. *Neuron, 42,* 335–346.

Kim, C.-Y., Blake, R., & Palmeri, T. J. (2006). Perceptual interaction between real and synesthetic colors. *Cortex, 42,* 195–203.

*Kinsbourne, M. (1982). Hemispheric specialization. *American Psychologist, 37,* 222–231.

Kolmel, K. F., Vehmeyer, K., & Gohring, E., et al. (1991). Treatment of advanced malignant melanoma by a pyrogenic bacterial lysate: A pilot study. *Onkologie, 14,* 411–417.

Kosslyn, S. M., Reiser, B. J., Farah, M. J., & Fliegel, S. L. (1983). Generating visual images: Units and relations. *Journal of Experimental Psychology, General, 112,* 278–303.

Lakoff, G., & Johnson, M. (2003). *Metaphors we live by.* Chicago: University of Chicago Press.

Landis, T., & Thut, G. (2005). Linking out-of-body experience and self processing to mental own-body imagery at the temporoparietal junction. *The Journal of Neuroscience, 25,* 550–557.

*LeDoux, J. (2002). *Synaptic self. How our brains become who we are.* New York: Viking Press.

*Luria, A. (1968). *The mind of a mnemonist.* Cambridge, MA: Harvard University Press.

MacLachlan, M., McDonald, D., & Waloch, J. (2004). Mirror treatment of lower limb phantom pain: A case study. *Disability and Rehabilitation, 26,* 901–904.

Matsuo, A., Tezuka, Y., Morioka, S., Hiyamiza, M., & Seki, M. (2008). *Mirror therapy accelerates recovery of upper limb movement after stroke: A randomized cross-over trial* [Abstract]. Paper presented at the 6th World Stroke Conference, Vienna, Austria.

Mattingley, J. B., Rich, A. N., Yelland, G., & Bradshaw, J. L. (2001). Unconscious priming eliminates automatic binding of colour and alphanumeric form in synaesthesia. *Nature, 401*(6828), 580–582.

McCabe, C. S., Haigh, R. C., Halligan, P. W., & Blake, D. R. (2005). Simulating sensory-motor incongruence in healthy volunteers: Implications for a cortical model of pain. *Rheumatology* (Oxford), *44,* 509–516.

McCabe, C. S., Haigh, R. C., Ring, E. F., Halligan, P. W., Wall, P. D., & Blake, D. R. (2003). A controlled pilot study of the utility of mirror visual feedback in the treatment of complex regional pain syndrome (type 1). *Rheumatology* (Oxford), *42,* 97–101.

McGeoch, P., Brang, D., & Ramachandran, V. S. (2007). Apraxia, metaphor and mirror neurons. *Medical Hypotheses, 69*(6), 1165–1168.

*Melzack, R. A., & Wall, P. D. (1965, November 19). Pain mechanisms: A new theory. *Science, 150*(3699), 971–979.

Merzenich, M. M., Kaas, J. H., Wall, J., Nelson, R. J., Sur, M., & Felleman, D. (1983). Topographic reorganization of somatosensory cortical areas 3b and 1 in adult monkeys following restricted deafferentation. *Neuroscience, 8,* 33–55.

*Milner, D., & Goodale, M. (1995). *The visual brain in action.* New York: Oxford University Press.

Mitchell, J. K. (1831). On a new practice in acute and chronic rheumatism. *The American Journal of the Medical Sciences, 8*(15), 55–64.

Mitchell, S. W. (1872). *Injuries of nerves and their consequences.* Philadelphia: J. B. Lippincott.

Mitchell, S. W., Morehouse, G. R., & Keen, W. W. (1864). *Gunshot wounds and other injuries of nerves.* Philadelphia: J. B. Lippincott.

*Mithen, S. (1999). *The prehistory of the mind.* London: Thames & Hudson.

Money, J., Jobaris, R., & Furth, G. (1977). Apotemnophilia: Two cases of self-demand amputation as a paraphilia. *Journal of Sex Research, 13,* 115–125.

Moseley, G. L., Olthof, N., Venema, A., Don, S., Wijers, M., Gallace, A., et al. (2008). Psychologically induced cooling of a specific body part caused by the illusory ownership of an artificial counterpart. *Proceedings of the National Academy of Sciences of the USA, 105*(35), 13169–13173

Moyer, R. S., & Landauer, T. K. (1967). Time required for judgements of numerical inequality. *Nature, 215*(5109), 1519–1520.

Nabokov, V. (1966). *Speak, memory: An autobiography revisited.* New York: G. P. Putnam's Sons.

Naeser, M. A., Martin, P. I., Nicholas, M., Baker, E. H., Seekins, H., Kobayashi M., et al. (2005). Improved picture naming in chronic aphasia after TMS to part of right Broca's area: An open-protocol study. *Brain and Language, 93*(1), 95–105.

Nuckolls, J. B. (1999). The case for sound symbolism. *Annual Review of Anthropology, 28,* 225–252.

Oberman, L. M., Hubbard, E. M., McCleery, J. P., Altschuler, E. L., & Ramachandran, V. S. (2005). EEG evidence for mirror neuron dysfunction in autism spectrum disorders. *Cognitive Brain Research, 24*(2), 190–198.

Oberman, L. M., McCleery, J. P., Ramachandran, V. S., & Pineda, J. A. (2007). EEG evidence for mirror neuron activity during the observation of human and robot actions: Toward an analysis of the human qualities of interactive robots. *Neurocomputing, 70,* 2194–2203.

Oberman, L. M., Pineda, J. A., & Ramachandran, V. S. (2007). The human mirror neuron system: A link between action observation and social skills. *Social Cognitive and Affective Neuroscience, 2,* 62–66.

Oberman, L. M., & Ramachandran, V. S. (2007a). Evidence for deficits in mirror neuron functioning, multisensory integration, and sound-form symbolism in autism spectrum disorders. *Psychological Bulletin, 133*(2), 310–327.

Oberman, L. M., & Ramachandran, V. S. (2007b). The simulating social mind: The role of the mirror neuron system and simulation in the social and communicative deficits of autism spectrum disorders. *Psychological Bulletin, 133*(2), 310–327.

Oberman, L. M., & Ramachandran, V. S. (2008). How do shared circuits develop? *Behavioral and Brain Sciences, 31,* 1–58.

Oberman, L. M., Ramachandran, V. S., & Pineda, J. A. (2008). Modulation of mu suppression in children with autism spectrum disorders in response to familiar or unfamiliar stimuli: the mirror neuron hypothesis. *Neuropsychologia, 46,* 1558–1565.

Oberman, L. M., Winkielman, P., & Ramachandran, V. S. (2007). Face to face: Blocking facial mimicry can selectively impair recognition of emotional faces. *Social Neuroscience, 2*(3), 167–178.

Palmeri, T. J., Blake, R., Marois, R., Flanery, M. A., & Whetsell, W., Jr. (2002). The perceptual reality of synesthetic colors. *Proceedings of the National Academy of Sciences of the USA, 99,* 4127–4131.

Penfield, W., & Boldrey, E. (1937). Somatic motor and sensory representation in the cerebral cortex of man as studied by electrical stimulation. *Brain, 60,* 389–443.

*Pettigrew, J. D., & Miller, S. M. (1998). A "sticky" interhemispheric switch in bipolar disorder? *Proceedings of the Royal Society of London, Series B: Biological Sciences, 265*(1411), 2141–2148.

Pinker, S. (1997). *How the mind works.* New York: W. W. Norton.

*Posner, M., & Raichle, M. (1997). *Images of the mind.* New York: W. H. Freeman.

*Premack, D., & Premack, A. (2003). *Original intelligence.* New York: McGraw-Hill.

*Quartz, S., & Sejnowski, T. (2002). *Liars, lovers and heroes.* New York: William Morrow.

Ramachandran, V. S. (1993). Behavioral and magnetoencephalographic correlates of plasticity in the adult human brain. *Proceedings of the National Academy of Sciences of the USA, 90,* 10413–10420.

Ramachandran, V. S. (1994). Phantom limbs, neglect syndromes, repressed memories, and Freudian psychology. *International Review of Neurobiology, 37,* 291–333.

Ramachandran, V. S. (1996, October). *Decade of the brain.* Symposium organized by the School of Social Sciences, University of California, San Diego, La Jolla.

Ramachandran, V. S. (1998). Consciousness and body image: Lessons from phantom limbs, Capgras syndrome and pain asymbolia. *Philosophical Transactions of the Royal Society of London, Series B: Biological Sciences, 353*(1377), 1851–1859.

Ramachandran, V. S. (2000, June 29). Mirror neurons and imitation as the driving force behind "the great leap forward" in human evolution. *Edge: The Third Culture,* Retrieved from http://www.edge.org/3rd_culture/ramachandran/rama chandran_pl.html., pp. 1–6.

Ramachandran, V. S. (2003). The phenomenology of synaesthesia. *Journal of Consciousness Studies, 10*(8), 49–57.

Ramachandran, V. S. (2004). The astonishing Francis Crick. *Perception, 33*(10), 1151–1154.

Ramachandran, V. S. (2005). Plasticity and functional recovery in neurology. *Clinical Medicine, 5*(4), 368–373.

Ramachandran, V. S., & Altschuler, E. L. (2009). The use of visual feedback, in particular mirror visual feedback, in restoring brain function. *Brain, 132*(7), 16.

Ramachandran, V. S., Altschuler, E. L., & Hillyer, S. (1997). Mirror agnosia. *Proceedings of the Royal Society of London, Series B: Biological Sciences, 264,* 645–647.

Ramachandran, V. S., & Azoulai, S. (2006). Synesthetically induced colors evoke apparent-motion perception. *Perception, 35*(11), 1557–1560.

Ramachandran, V. S., & Blakeslee, S. (1998). *Phantoms in the brain.* New York: William Morrow.

Ramachandran, V. S., & Brang, D. (2008). Tactile-emotion synesthesia. *Neurocase, 14*(5), 390–399.

Ramachandran, V. S., & Brang, D. (2009). Sensations evoked in patients with amputation from watching an individual whose corresponding intact limb is being touched. *Archives of Neurology, 66*(10), 1281–1284.

Ramachandran, V. S., Brang, D., & McGeoch, P. D. (2009). Size reduction using Mirror Visual Feedback (MVF) reduces phantom pain. *Neurocase, 15*(5), 357–360.

Ramachandran, V. S., & Hirstein, W. (1998). The perception of phantom limbs. The D. O. Hebb lecture. *Brain, 121*(9), 1603–1630.

Ramachandran, V. S., Hirstein, W., Armel, K. C., Tecoma, E., & Iragul, V. (1997, October 25–30). *The neural basis of religious experience.* Paper presented at the 27th annual meeting of the Society for Neuroscience, New Orleans, LA.

Ramachandran, V. S., & Hubbard, E. M. (2001a). Psychophysical investigations into the neural basis of synaesthesia. *Proceedings of the Royal Society of London, Series B: Biological Sciences, 268*(1470), 979–983.

Ramachandran, V. S., & Hubbard, E. M. (2001b). Synaesthesia: A window into perception, thought and language. *Journal of Consciousness Studies, 8*(12), 3–34.

Ramachandran, V. S., & Hubbard, E. M. (2002a). Synesthetic colors support symmetry perception and apparent motion. *Abstracts of the Psychonomic Society's 43rd Annual Meeting, 7,* 79.

Ramachandran, V. S., & Hubbard, E. M. (2002b, November). Synesthetic colors support symmetry perception and apparent motion. Poster session presented at the 43rd annual meeting of the Psychonomic Society, Kansas City, MO.

Ramachandran, V. S., & Hubbard, E. M. (2003). Hearing colors, tasting shapes. *Scientific American, 288*(5), 42–49.

Ramachandran, V. S., & Hubbard, E. M. (2005a). The emergence of the human mind: Some clues from synesthesia. In L. C. Robertson & N. Sagiv (Eds.), *Synesthesia: Perspectives from cognitive neuroscience* (pp. 147–190). New York: Oxford University Press.

Ramachandran, V. S., & Hubbard, E. M. (2005b). Synesthesia: What does it tell us about the emergence of qualia, metaphor, abstract thought, and language? In J. L. van Hemmen & T. J. Sejnowski (Eds.), *23 problems in systems neuroscience.* Oxford, UK: Oxford University Press.

Ramachandran, V. S., & Hubbard, E. M. (2006, October). Hearing colors, tasting shapes. Secrets of the senses [Special issue]. *Scientific American,* 76–83.

Ramachandran, V. S., & McGeoch, P. D. (2007). Occurrence of phantom genitalia after gender reassignment surgery. *Medical Hypotheses, 69*(5), 1001–1003.

Ramachandran, V. S., McGeoch, P. D., & Brang, D. (2008). *Apotemnophilia: A neurological disorder with somatotopic alterations in SCR and MEG activation.* Paper presented at the annual meeting of the Society for Neuroscience, Washington, DC.

Ramachandran, V. S., & Oberman, L. M. (2006a, May 13). Autism: The search for Steven. *New Scientist,* pp. 48–50.

Ramachandran, V. S., & Oberman, L. M. (2006b, November). Broken mirrors: A theory of autism. *Scientific American, 295*(5), 62–69.

Ramachandran, V. S., & Rogers-Ramachandran, D. (2008). Sensations referred to a patient's phantom arm from another subject's intact arm: Perceptual correlates of mirror neurons. *Medical Hypotheses, 70*(6), 1233–1234.

Ramachandran, V. S., Rogers-Ramachandran, D., & Cobb, S. (1995). Touching the phantom limb. *Nature, 377,* 489–490.

*Restak, R. (2000). *Mysteries of the mind.* Washington, DC: National Geographic Society.

Rizzolatti, G., & Arbib, M. A. (1998). Language within our grasp. *Trends in Neurosciences, 21,* 188–194.

Rizzolatti, G., & Destro, M. F. (2008). Mirror neurons. *Scholarpedia, 3*(1), 2055.

Rizzolatti, G., Fadiga, L., Fogassi, L., & Gallese, V. (1996). Premotor cortex and the recognition of motor actions. *Cognitive Brain Research, 3,* 131–141.

Rizzolatti, G., Fogassi, L., & Gallese, V. (2001). Neurophysiological mechanisms underlying the understanding and imitation of action. *Nature Reviews Neuroscience, 2,* 661–670.

Ro, T., Farne, A., Johnson, R. M., Wedeen, V., Chu, Z., Wang, Z. J., et al. (2007). Feeling sounds after a thalamic lesion. *Annals of Neurology, 62*(5), 433–441.

*Robertson, I. (2001). *Mind sculpture.* New York: Bantam Books.

Robertson, L. C., & Sagiv, N. (2005). *Synesthesia: Perspectives from cognitive neuroscience.* New York: Oxford University Press.

*Rock, I., & Victor, J. (1964). Vision and touch: An experimentally created conflict between the two senses. *Science, 143,* 594–596.

Rosén, B., & Lundborg, G. (2005). Training with a mirror in rehabilitation of the hand. *Scandinavian Journal of Plastic and Reconstructive Surgery and Hand Surgery, 39*(104–108).

Rouw, R., & Scholte, H. S. (2007). Increased structural connectivity in grapheme-color synesthesia. *Nature Neuroscience, 10*(6), 792–797.

Saarela, M. V., Hlushchuk, Y., Williams, A. C., Schurmann, M., Kalso, E., & Hari, R. (2007). The compassionate brain: Humans detect intensity of pain from another's face. *Cerebral Cortex, 17*(1), 230–237.

Sagiv, N., Simner, J., Collins, J., Butterworth, B., & Ward, J. (2006). What is the relationship between synaesthesia and visuo-spatial number forms? *Cognition, 101*(1), 114–128.

*Sacks, O. (1985). *The man who mistook his wife for a hat.* New York: HarperCollins.

*Sacks, O. (1995). *An anthropologist on Mars.* New York: Alfred A. Knopf.

*Sacks, O. (2007). *Musicophilia: Tales of music and the brain.* New York: Alfred A. Knopf.

Sathian, K., Greenspan, A. I., & Wolf, S. L. (2000). Doing it with mirrors: A case study of a novel approach to neurorehabilitation. *Neurorehabilitation and Neural Repair, 14,* 73–76.

Saxe, R., & Wexler, A. (2005). Making sense of another mind: The role of the right temporo-parietal junction. *Neuropsychologia, 43,* 1391–1399.

*Schacter, D. L. (1996). *Searching for memory.* New York: Basic Books.

Schiff, N. D., Giacino, J. T., Kalmar, K., Victor, J. D., Baker, K., Gerber, M., et al. (2007). Behavioural improvements with thalamic stimulation after severe traumatic brain injury. *Nature, 448,* 600–603.

Selles, R. W., Schreuders, T. A., & Stam, H. J. (2008). Mirror therapy in patients with causalgia (complex regional pain syndrome type II) following peripheral nerve injury: Two cases. *Journal of Rehabilitation Medicine, 40,* 312–314.

*Sierra, M., & Berrios, G. E. (2001). The phenomenological stability of depersonalization: Comparing the old with the new. *The Journal of Nervous and Mental Disease, 189*(9), 629–636.

Simner, J., & Ward, J. (2006). Synaesthesia: The taste of words on the tip of the tongue. *Nature, 444*(7118), 438.

Singer, T, (2006). The neuronal basis and ontogeny of empathy and mind reading: Review of literature and implications for future research. *Neuroscience and Biobehavioral Reviews, 6,* 855–863.

Singer, W., & Gray, C. M. (1995). Visual feature integration and the temporal correlation hypothesis. *Annual Review of Neuroscience, 18,* 555–586.

Smilek, D., Callejas, A., Dixon, M. J., & Merikle, P. M. (2007). Ovals of time: Time-space associations in synaesthesia. *Consciousness and Cognition, 16*(2), 507–519.

Snyder, A. W., Mulcahy, E., Taylor, J. L., Mitchell, D. J., Sachdev, P., & Gandevia, S. C. (2003). Savant-like skills exposed in normal people by suppressing the left fronto-temporal lobe. *Journal of Integrative Neuroscience, 2*(2), 149–158.

*Snyder, A., & Thomas, M. (1997). Autistic savants give clues to cognition. *Perception, 26*(1), 93–96.

*Solms, M., & Turnbull, O. (2002). *The brain and the inner world: An introduction to the neuroscience of subjective experience.* New York: Other Press.

Stevens, J. A., & Stoykov, M. E. (2003). Using motor imagery in the rehabilitation of hemiparesis. *Archives of Physical Medicine and Rehabilitation, 84,* 1090–1092.

Stevens, J. A., & Stoykov, M. E. (2004). Simulation of bilateral movement training through mirror reflection: A case report demonstrating an occupational therapy technique for hemiparesis. *Topics in Stroke Rehabilitation, 11,* 59–66.

Sumitani, M., Miyauchi, S., McCabe, C. S., Shibata, M., Maeda, L., Saitoh, Y., et al. (2008). Mirror visual feedback alleviates deafferentation pain, depending on qualitative aspects of the pain: A preliminary report. *Rheumatology* (Oxford), *47,* 1038–1043.

Sütbeyaz, S., Yavuzer, G., Sezer, N., & Koseoglu, B. F. (2007). Mirror therapy enhances lower-extremity motor recovery and motor functioning after stroke: A randomized controlled trial. *Archives of Physical Medicine and Rehabilitation, 88,* 555–559.

Tang, Z. Y., Zhou, H. Y., Zhao, G., Chai, L. M., Zhou, M., Lu, J., et al. (1991). Preliminary result of mixed bacterial vaccine as adjuvant treatment of hepatocellular carcinoma. *Medical Oncology & Tumor Pharmacotherapy, 8,* 23–28.

Thioux, M., Gazzola, V., & Keysers, C. (2008). Action understanding: How, what and why. *Current Biology, 18*(10), 431–434.

*Tinbergen, N. (1954). *Curious naturalists.* New York: Basic Books.

Tranel, D., & Damasio, A. R. (1985). Knowledge without awareness: An autonomic index of facial recognition by prosopagnosics. *Science, 228*(4706), 1453–1454.

Tranel, D., & Damasio, A. R. (1988). Non-conscious face recognition in patients with face agnosia. *Behavioural Brain Research, 30*(3), 239–249.

*Ungerleider, L. G., & Mishkin, M. (1982). Two visual streams. In D. J. Ingle, M. A. Goodale, & R. J. W. Mansfield (Eds.), *Analysis of visual behavior.* Cambridge, MA: MIT Press.

Vallar, G., & Ronchi, R. (2008). Somatoparaphrenia: A body delusion. A review of the neuropsychological literature. *Experimental Brain Research, 192*(3), 533–551.

Van Essen, D. C., & Maunsell, J. H. (1980). Two-dimensional maps of the cerebral cortex. *Journal of Comparative Neurology, 191*(2), 255–281.

Vladimir Tichelaar, Y. I., Geertzen, J. H., Keizer, D., & Van Wilgen, P. C. (2007). Mirror box therapy added to cognitive behavioural therapy in three chronic complex regional pain syndrome type I patients: A pilot study. *International Journal of Rehabilitation Research, 30,* 181–188.

*Walsh, C. A., Morrow, E. M. , & Rubenstein, J. L. (2008). Autism and brain development. *Cell, 135*(3), 396–400.

Ward, J., Yaro, C., Thompson-Lake, D., & Sagiv, N. (2007). Is synaesthesia associated with particular strengths and weaknesses? UK Synaesthesia association meeting.

*Weiskrantz, L. (1986). *Blindsight: A case study and implications*. New York: Oxford University Press.

Wicker, B., Keysers, C., Plailly, J., Royet, J. P., Gallese, V., & Rizzolatti, G. (2003). Both of us disgusted in my insula: The common neural basis of seeing and feeling disgust. *Neuron, 40,* 655–664.

Winkielman, P., Niedenthal, P. M., & Oberman, L. M. (2008). The Embodied Emotional Mind. In G. R. Smith & E. R. Smith (Eds.), *Embodied grounding: Social, cognitive, affective, and neuroscientific approaches*. New York: Cambridge University Press.

Wolf, S. L., Winstein, C. J., Miller, J. P., Taub, E., Uswatte, G., Morris, D., et al. (2006). Effect of constraint-induced movement therapy on upper extremity function 3 to 9 months after stroke: The EXCITE randomized clinical trial. *Journal of the American Medical Association, 296,* 2095–2104.

Wolpert, L. (2001). *Malignant sadness: The anatomy of depression*. New York: Faber and Faber.

Yang, T. T., Gallen, C., Schwartz, B., Bloom, F. E., Ramachandran, V. S., & Cobb, S. (1994). Sensory maps in the human brain. *Nature, 368,* 592–593.

Yavuzer, G., Selles, R. W., Sezer, N., Sütbeyaz, S., Bussmann, J. B., Köseoğlu, F., et al. (2008). Mirror therapy improves hand function in subacute stroke: A randomized controlled trial. *Archives of Physical Medicine and Rehabilitation, , 89*(3), 393–398.

Young, A. W., Leafhead, K. M., & Szulecka, T. K. (1994). Capgras and Cotard delusions. *Psychopathology, 27,* 226–231.

*Zeki, S. (1993). *A Vision of the Brain*. Oxford: Oxford University Press.

Zeki, S. (1998). Art and the brain. *Proceedings of the American Academy of Arts and Sciences, 127*(2), 71–104.

ILLUSTRATION CREDITS

———

FIGURE 3.8   From Francis Galton, "Visualised Numerals," *Journal of the Anthropological Institute, 10*(1881) 85–102.

CHAPTER 4

FIGURE 4.1   By permission of Giuseppe di Pellegrino
FIGURE 4.2   V. S. Ramachandran

CHAPTER 6

FIGURE 6.1   V. S. Ramachandran
FIGURE 6.2   V. S. Ramachandran

CHAPTER 7

FIGURE 7.1   Illustration from *Animal Architecture* by Karl von Frisch and Otto von Frisch, illustrations copyright © 1974 by Turid Holldober, reprinted by permission of Harcourt, Inc.
FIGURE 7.2   By permission of Amita Chatterjee
FIGURE 7.3   Réunion des Musées Nationaux/Art Resource, NY
FIGURE 7.4   V. S. Ramachandran
FIGURE 7.5   V. S. Ramachandran
FIGURE 7.6   V. S. Ramachandran
FIGURE 7.7   Photograph by Rosemania for Wikicommons
FIGURE 7.8   V. S. Ramachandran

CHAPTER 8

FIGURE 8.1   From *Nadia: A Case of Extraordinary Drawing Ability in an Autistic Child* (1978) by Lorna Selfe
FIGURE 8.2   V. S. Ramachandran
FIGURE 8.3   V. S. Ramachandran
FIGURE 8.4   V. S. Ramachandran
FIGURE 8.5   © The Metropolitan Museum of Art/Art Resource, NY

CHAPTER 9

FIGURE 9.1   V. S. Ramachandran
FIGURE 9.2   V. S. Ramachandran

INDEX

———

Page numbers in *italics* refer to illustrations.
Page numbers beginning with 306 refer to endnotes.

de Clérembault syndrome, 253–54
dementia, creativity and, 7, 224, 227
dendrites, 14, *15*
denial, 269
Dennett, Daniel, 55, 271
depression, 21, 85, 280, 281, 282
Descent of Man (Darwin), 244
design:
    asymmetry in, 235–36
    metaphor in, 237–38, 311
    orderliness in, 233
Devinsky, Orrin, 319
diabetes, 140, 314
Diamond, Jared, 119
Di Pellegrino, Giuseppe, 121
Disraeli, Benjamin, 3, 290
dissociative states, 273–74
Dobzhansky, Theodosius, xiv, 201
Dohle, Christian, 34–35
dolphins, cerebral cortex in, 17
Donald, Merlin, 133
Donovan, Tara, 220
dopamine, 19
dorsolateral prefrontal cortex (DLF), *16,*
    264, 265
dorsomedial prefrontal cortex (DMF), 264,
    265
drugs:
    antidepressant, 282
    autism treatment with, 146–47
    out-of-body experience caused by, 272,
        321
    synesthesia and, 97
Duchamp, Marcel, 193
dummy-head illusion, 325–26

echopraxia, 125
Eddington, Arthur, 292
Ehrsson, Henrik, 326
Einstein, Albert, xii, 6, 224
electroencephalography (EEG), 140–41
Ellis, Hadyn, 307
embodied cognition, 143
emotions, *276, 277*
    in art, 243
    auditory activation of, 71
    color and, 83, 101–2

external input contradicting, 283, 290
generation of, 98
improperly suppressed, 322
perception and, 19, 69–71, 149
scientific measurement of, 70–71
severance of, 281–82
social, 252
and synesthesia, 75, 83, 85–86, 97, 98,
    101–2
temporal lobe epilepsy affecting, 151, 279
theories of, xiii
visual activation of, 65, 68–70, 205–6,
    221, 228, 229–31, 275, 278
see also limbic system
empathy:
    in autistics, 137, 146–47, 149
    body image and, 22, 151
    drug enhancement of, 146–47
    heightening of, 282
    mirror neurons and, xv, 6, 22, 124, 251,
        260, 261, 265, 281
    in primates, 265, 291
epilepsy, 278
    temporal lobe, 151, 224, 279–80, 282
estrus, 42
eugenics, 77
euphoria, 21
evolution, human, 3–4, 5, 290–91, 292–93
    abstract thought and, 106, 128, 133, 183,
        187
    art and, 241–42, 290
    brain development and, xiv–xv, 13, 17,
        19–20, 22–23, 37–38, 118–20, 121,
        130, 131, 133, 145, 178, 179–80, 182,
        189, 204, 246, 264, *276,* 284, 287,
        291, 319
    culture and, xv, xvi, 13, 23, 38, 132, 133,
        134–35, 313–14
    extinct species in, 8–10
    genetics and, 107, 180, 182, 306, 313–14
    humor and, 40
    language and, xv, 13, 43, 120, 122, 130,
        161, 163–76, *177,* 178–80, 181–83,
        189–91, 291
    mathematics and, 113, 134
    mirror neurons and, 134–35, 145, 260,
        313–14

types of, 284
vision and, 60, 65, 67, 244
working, 265
Mendel, Gregor, 54, 324
mental illness:
    approaches to, 320
    see also brain damage
Merzenich, Mike, 37
metaphor:
    aesthetic, 236–41, 244, 311
    autistics' difficulty with, 142–43
    brain areas for, 20, 105, 106, 143, 178,
        180, 237, 244, 291
    cross-sensory, 131–32
    origins of, 168
    synesthesia and, 76, 79, 87, 104–7, 108,
        310, 311
    use of, 104, 105, 163
metaphor blindness, 7, 105–6, 131, 180
Metropolitan Museum of Art, *239*
midbrain, 17
middle temporal (MT) area, movement
        perception in, 60–61, 316–17
Miller, Bruce, 224, 227
Miller, Geoffrey, 241, 242
Milner, Brenda, 286, 312
Minsky, Marvin, 319
mirror neurons, xv–xvi, 117–35, *276,*
        312–13, 314
    anatomical data for, 145
    and autism, xvi, 137, 139–40, 141–47,
        149, 151–52, 179, 262, 314–15
    embodied cognition and, 143
    and empathy, xv, 251, 260, 261, 265, 281
    evidence for, 123–24, 312
    evolution and, 134–35, 145, 260,
        313–14
    exotic syndromes and, 261–62
    function of, 22–23, 37, 43, 127, 128–32,
        145, 251
    in humans, 121, 130, 143, 145, 291
    imitation and, 117–18, 121, 122, 127–28,
        131, 132, 144
    inhibition of, 124–25, 260, 261, 272, 282,
        290
    and language evolution, xvi, 123, 172,
        174, 175–76, 182

learning and, 126–27, 130, 133, 145, 313,
        314
limbic system connections with, 152
in monkeys, 118, 120–23, *122,* 135, 143,
        179, 291, 312
motor system and, 124–25, 131, 143, 179,
        313
network of, 22–23, 37, 145
for pain, 124, 125
posture activation of, 208–9
in primates, xvi, 291
self-awareness and, 128, 253, 260–61,
        277–78, 281, 290, 323
theories about, 126–28, 145
for touch, 124, 125–26, 251, 260
mirror visual feedback (MVF):
    in chronic pain treatment, 36
    in phantom limb treatment, 30, 32–34,
        33, 36, 321
    in stroke paralysis treatment, *33,* 34–35
Mishkin, Mortimer, 307, 322
monkeys:
    communication by, 162
    face recognition in, 59–60
    mirror neurons in, 118, 120–23, *122,* 135,
        143, 175, 179, 291, 312
    visual grouping by, 205
Morgan, Thomas Hunt, 324
motor cortex, 18, 20, 31, 154, *177,* 264
    in autistics, 142
motor system, 17, 18
    in autistics, 137, 141, 142
    cross-activation in, 173–74, 175
    cross-modal abstraction in, 129–31, 173,
        175, *177, 276*
    disorders in, 18–19, 20
    following stroke, 35
    mirror neurons and, 124–25, 131, 143,
        179, 313
    neurons controlling, 121
    perception linked to, 142, 143, 178–79
    pyramidal tracts in, 154
    sensory system interaction with, 30,
        31–32, 143, 144
    speech controlled by, 172, 173, 175
    in speech imitation, 144
    syntactic structure in, 181–82

ABOUT THE AUTHOR

V. S. Ramachandran is director of the Center for Brain and Cognition, Distinguished Professor with the Psychology Department and Neurosciences Program at the University of California, San Diego, and adjunct professor of biology at the Salk Institute.

Ramachandran trained as a physician and subsequently went on to obtain a PhD on a scholarship from Trinity College at the University of Cambridge. His early work was on visual perception, but he is best known for his experiments in behavioral neurology, which, despite their extreme simplicity, have had a profound impact on the way we think about the brain. He has been called "a latter-day Marco Polo" by Richard Dawkins and "the modern Paul Broca" by the Nobel laureate Eric R. Kandel.

In 2005 Ramachandran was awarded the Henry Dale Medal and elected to an honorary life membership by the Royal Institution of Great Britain. His other honors and awards include a fellowship from All Souls College, two honorary doctorates, the annual Ramón y Cajal Award from the International Neuropsychiatric Association, and the Ariens Kappers Medal from the Royal Nederlands Academy of Sciences. In 2003 he gave the annual BBC Reith lectures, the first of which was given by Bertrand Russell in 1949. In 1995 he gave the Decade of the Brain Lecture at the 25th annual (Silver Jubilee) meeting of the Society for Neuroscience. Most recently, the president of India conferred upon him the third-highest civilian award and honorific title in India: the Padma Bhushan.

Ramachandran's much-acclaimed book *Phantoms in the Brain* formed the basis for a two-hour PBS special. He has appeared on the *Charlie Rose* show, and *Newsweek* named him a member of the "Century Club"—one of the one hundred most important people to watch this century.